T0291586

Reservoir Geomechanics

This interdisciplinary book encompasses the fields of rock mechanics, structural geology, and petroleum engineering to address a wide range of geomechanical problems that arise during the exploitation of oil and gas reservoirs.

Covering the exploration, assessment, and production phases of petroleum reservoir development, the book considers key practical issues such as prediction of pore pressure; estimation of hydrocarbon column heights and fault seal potential; determination of optimally stable well trajectories; casing set points and mud weights; changes in reservoir performance during depletion; and production-induced faulting and subsidence. The first part of the book establishes the basic principles of geomechanics in a way that allows readers from different disciplinary backgrounds to understand the key concepts. It then goes on to introduce practical measurement and experimental techniques before illustrating their successful application, through case studies taken from oil and gas fields around the world, to improve recovery and reduce exploitation costs.

Reservoir Geomechanics is a practical reference for geoscientists and engineers in the petroleum and geothermal industries, and for research scientists interested in stress measurements and their application to problems of faulting and fluid flow in the crust.

Mark Zoback is the Benjamin M. Page Professor of Earth Sciences and Professor of Geophysics in the Department of Geophysics at Stanford University. The author/co-author of approximately 250 published research papers, he is a Fellow of the Geological Society of America, the American Geophysical Union, and the American Association for the Advancement of Science. In 2006 he was awarded the Emil Wiechert Medal of the German Geophysical Society and, in 2008, the Walter S. Bucher Medal of the American Geophysical Union.

Reservoir
Geomechanics

Mark D. Zoback

Department of Geophysics, Stanford University

CAMBRIDGE
UNIVERSITY PRESS

University Printing House, Cambridge CB2 8BS, United Kingdom

One Liberty Plaza, 20th Floor, New York, NY 10006, USA

477 Williamstown Road, Port Melbourne, VIC 3207, Australia

314-321, 3rd Floor, Plot 3, Splendor Forum, Jasola District Centre, New Delhi - 110025, India

79 Anson Road, #06-04/06, Singapore 079906

Cambridge University Press is part of the University of Cambridge.

It furthers the University's mission by disseminating knowledge in the pursuit of education, learning and research at the highest international levels of excellence.

www.cambridge.org
Information on this title: www.cambridge.org/9780521146197

© M. Zoback 2007, 2010

First published 2007
Paperback edition 2010
15th printing 2018

A catalogue record for this publication is available from the British Library

ISBN 978-0-521-77069-9 Hardback
ISBN 978-0-521-14619-7 Paperback

This book is dedicated to the Ph.D. students and post-docs at Stanford University whom I've had the privilege to teach, and learn from, over the past 22 years. This book would not have been possible without many contributions from these talented scientists. I thank them for their efforts, enthusiasm and friendship.

Contents

PART III: APPLICATIONS

The plates are to be found between pages 178 and 179.

Preface

This book has its origin in an interdisciplinary graduate class that I've taught at Stanford University for a number of years and a corresponding short course given in the petroleum industry. As befitting the subject matter, the students in the courses represent a variety of disciplines – reservoir engineers and geologists, drilling engineers and geophysicists. In this book, as in the courses, I strive to communicate key concepts from diverse disciplines that, when used in a coordinated way, make it possible to develop a comprehensive geomechanical model of a reservoir and the formations above it. I then go on to illustrate how to put such a model to practical use. To accomplish this, the book is divided into three major sections: The first part of the book (Chapters 1–5) addresses basic principles related to the state of stress and pore pressure at depth, the various constitutive laws commonly used to describe rock deformation and rock failure in compression, tension and shear. The second part of the book (Chapters 6–9) addresses the principles of wellbore failure and techniques for measuring stress orientation and magnitude in deep wells of any orientation. The techniques presented in these chapters have proven to be reliable in a diversity of geological environments. The third part of the book considers applications of the principles presented in the first part and techniques presented in the second. Hence, Chapters 10–12 address problems of wellbore stability, fluid flow associated with fractures and faults and the effects of depletion on both a reservoir and the surrounding formations.

Throughout the book, I present concepts, techniques and investigations developed over the past 30 years with a number of talented colleagues. Mary Lou Zoback (formerly with the U.S. Geological Survey) and I developed the methodologies for synthesis of various types of data that indicate current stress orientations and relative magnitudes in the earth's crust. As summarized in Chapter 1, Mary Lou and I demonstrated that it was possible to develop comprehensive maps of stress orientation and relative magnitude and interpret the current state of crustal stress in terms of geologic processes that are active today. The quality ranking system we developed for application to the state of stress in the conterminous U.S. (and later North America) is presented in Chapter 6. It has been used as the basis for almost all stress mapping endeavors carried out over the past 20 years and provided the basis for the compilation of stress at a global scale (the World Stress Map project), led by Mary Lou.

The examples of regional stress fields from various regions around the world presented in Chapters 1 and 6 are taken from collaborative research done with former Ph.D. students David Castillo, Lourdes Colmenares, Balz Grollimund and Martin Brudy. This work, and much of the other work done with students and post-docs at Stanford, was supported by the companies participating in the Stanford Rock and Borehole Geophysics Consortium (SRB). Chapter 2, on pore pressure, refers to work done with former Ph.D. students Thomas Finkbeiner, David Wiprut, Balz Grollimund and Alvin Chan. The concepts in this chapter benefited from discussions with Peter Flemings (Penn State) and Chris Ward. Chapter 3, on elasticity and constitutive laws, includes a section on viscoelastic and viscoplastic constitutive laws for uncemented reservoir sands that is based on research done in collaboration with former Ph.D. students Carl Chang, Dan Moos and Paul Hagin. Chapter 4, on rock failure, was done in part in collaboration with Lourdes Colmenares, former Ph.D. student John Townend, former post-docs Chandong Chang and Lev Vernik and Dan Moos. James Byerlee (USGS retired) was an inspirational Ph.D. advisor and teacher in rock mechanics. His work on rock friction, discussed in Chapter 4, is of critical importance on establishing bounds on stress magnitudes at depth in the crust. Chapter 5 is on fractures and faults at depth, and is based largely on wellbore imaging studies initiated with former Ph.D. student Colleen Barton and includes applications done with Thomas Finkbeiner and Sneha Chanchani.

At the beginning of the second part of the book on *Measuring Stress Orientation and Magnitude*, Chapter 6 discusses stress concentrations around vertical wells and compressional and tensional wellbore failures. This work was done in part in collaboration with Dan Moos, Martin Brudy, David Wiprut and David Castillo, as well as former post-docs Pavel Peska and Marek Jarosinski. John Healy and Steve Hickman of the USGS were early collaborators on the use of hydraulic fracturing for stress measurements. The stress measurement methods based on wellbore failures in vertical (Chapters 7) and deviated wellbores (Chapter 8) were developed in collaboration with Pavel Peska, Martin Brudy and Dan Moos. Former Ph.D. student Naomi Boness and I developed the methodologies presented in Chapter 8 for utilizing cross-dipole shear velocity logs for mapping stress orientation in deviated wells. The techniques described in these chapters are not intended to be a comprehensive review of the numerous techniques proposed over the years for stress measurement (or stress estimation) at depth. Rather, I emphasize stress measurement techniques that have proven to work reliably in deep wells under conditions commonly found in oil and gas reservoirs. Chapter 9 reviews stress magnitude measurements made in various sedimentary basins around the world in the context of global patterns of *in situ* stress and some of the mechanisms responsible for intraplate stress. Chapter 9 also includes a case study related to deriving stress magnitude information from geophysical logs carried out with former Ph.D. student Amie Lucier.

The final part of the book, *Applications*, starts with a discussion of wellbore stability in Chapter 10. Many of the examples considered in the section are taken from studies done with Pavel Peska and Dan Moos and several of the topics considered (such as the degree of *acceptable* wellbore failure with well deviation, the influence of weak bedding planes on wellbore stability and the circumstances under which it might be possible to drill with mud weights greater than the least principal stress) were undertaken at the suggestions of Steve Willson and Eric van Oort. The theory presented on drilling with mud weights in excess of the least principal stress was developed with Takatoshi Ito of Tohoku University. The wellbore stability study of the SAFOD research borehole was done in collaboration with former Ph.D. student Pijush Paul. At the time of this writing, the principles discussed in the sections dealing with wellbore stability have been successfully applied in over 500 studies carried out over the past several years by colleagues at GeoMechanics International (GMI) and other companies. This success validates the practical utility of both the techniques outlined in Chapters 7 and 8 for estimating *in situ* stress magnitude and orientation and the effectiveness of using a relatively straightforward *strength of materials* approach in assessing wellbore stability in many situations. I thank GMI for use of its software in many of the applications presented in this book.

The brief discussion of formation stability during production (referred to as sand, or solids, production) is based on the work of Martin Brudy and Wouter van der Zee, principally using finite element techniques. The work on flow through fractured reservoirs in Chapter 11 and the importance of critically stressed faults on controlling fluid flow is based on research initially carried out with Colleen Barton and Dan Moos and extended with John Townend. Work on localized fluctuations of stress orientation due to slip on faults was done originally with Gadi Shamir and subsequently extended with Colleen Barton. Extension of this work to the fault seal problem was initially done with David Wiprut. Studies related to dynamic constraints on hydrocarbon migration were done with Thomas Finkbeiner. Roger Anderson (Columbia University) and Peter Flemings played instrumental roles in this research. The work done on the state of stress and hydrocarbon leakage in the northern North Sea was motivated by Bjorn Larsen. Chapter 12 considers a number of topics related to reservoir depletion, including subsidence and production-induced faulting. The majority of this work was done in collaboration with Alvin Chan, with contributions from former post-doc Jens Zinke and former Ph.D. student Ellen Mallman. The work on depletion-induced stress orientation changes was done principally with former Ph.D. student Amy Day-Lewis based on work done originally with Sangmin Kim.

Finally, I'd like to thank Steve Willson, Chris Ward, Dan Moos, John Townend and Mary Lou Zoback for their comments on the first draft of this book.

Mark Zoback
Stanford University
2006

Comments for the paperback edition

Since the first printing of this book in 2007, there have been several areas where application of some of the principles of geomechanics that are discussed have seen such rapid increases of utilization that they deserve mention. One is the widespread utilization of three- (and four)-dimensional numerical geomechanical models. As explained at the end of Chapter 1, utilization of three-dimensional numerical models is essential in the vicinity of salt domes, where both stress magnitudes and orientations are perturbed by the presence of the salt. Examples of cases where four-dimensional reservoir models are important are in cases where the coupling between stress and deformation in (and around) reservoirs is affected by depletion. One obvious example is the coupling between depletion and compaction drive as discussed briefly in Chapter 12 for an idealized hypothetical example.

A second area of recent activities that involves application of the principles discussed in this book is the triggering of slip on critically stressed faults during so-called *slickwater frac'ing* operations. These are being employed in thousands of wells that are drilled to enhance production from shale gas and tight gas (very low permeability) reservoirs, somewhat analogously to what has been done in geothermal reservoirs for several decades. As discussed in Chapters 5 and 11, the induced slip on these faults creates a network of faults with sufficiently enhanced permeability that gas can be exploited economically from otherwise impermeable reservoirs. At the time of this writing, a great deal of research is being done to better understand how to optimize the creation of such networks and to accurately model fluid flow through them.

Finally, many projects involving injection of CO_2 into saline aquifers and depleted oil and gas reservoirs are getting under way as part of efforts to limit greenhouse gas emissions. One critically important issue concerning such activities is whether the increase in formation pressure caused by injection is likely to induce slip on pre-existing faults. Geomechanically constrained reservoir simulations of CO_2 injection in different kinds of reservoirs have been carried out with former Ph.D. students Amie Lucier, Laura Chiaramonte, and Hannah Ross. The techniques described in Chapter 11 are being increasingly used to assess limits on safe injection pressures and maximum rates of CO_2 injection.

Mark Zoback, 2010

Part I Basic principles

1 The tectonic stress field

My goals in writing this book are to establish basic principles, introduce practical experimental techniques and present illustrative examples of how the development of a comprehensive geomechanical model of a reservoir (and overlaying formations) provides a basis for addressing a wide range of problems that are encountered during the *life-cycle* of a hydrocarbon reservoir. These include questions that arise (i) during the exploration and assessment phase of reservoir development such as the prediction of pore pressure, hydrocarbon column heights and fault seal (or leakage) potential; (ii) during the development phase where engineers seek to optimize wellbore stability through determination of optimal well trajectories, casing set points and mud weights and geologists attempt to predict permeability anisotropy in fractured reservoirs; (iii) throughout the production phase of the reservoir that requires selection of optimal completion methodologies, the prediction of changes in reservoir performance during depletion and assessment of techniques, such as repeated hydraulic fracturing, to optimize total recovery; and (iv) during the secondary and tertiary recovery phases of reservoir development by optimizing processes such as water flooding and steam injection. Chapters 1–5 address basic principles related to the components of a comprehensive geomechanical model: the state of stress and pore pressure at depth, the constitutive laws that commonly describe rock deformation and fractures and faults in the formations of interest. Chapters 6–9 address wellbore failure and techniques for using observations of failure to constrain stress orientation and magnitude in wells of any orientation. Chapters 10–12 address case studies that apply the principles of the previous chapters to problems of wellbore stability, flow associated with fractures and faults and the effects of depletion on a reservoir and the surrounding formations.

Why stress is important

The key component of a comprehensive geomechanical model is knowledge of the current state of stress. Wellbore failure occurs because the stress concentrated around the circumference of a well exceeds the strength of a rock (Chapters 6 and 10). A fault will slip when the ratio of shear to effective normal stress resolved on the fault exceeds its frictional strength (Chapters 4, 11 and 12). Depletion causes changes in the stress state of the reservoir that can be beneficial, or detrimental, to production in a number of ways (Chapter 12). As emphasized throughout this book, determination of the state of stress at depth in oil and gas fields is a tractable problem that can be addressed with data that are routinely obtained (or are straightforwardly obtainable) when wells are drilled.

In this chapter, I start with the basic definition of a stress tensor and the physical meaning of principal stresses. These concepts are important to establish a common vocabulary among readers with diverse backgrounds and are essential for understanding how stress fields change around wellbores (Chapters 6 and 8) and in the vicinity of complex structures such as salt domes (as discussed at the end of the chapter). I also introduce a number of fundamental principles about the tectonic stress field at a regional scale in this chapter. These principles are revisited at scales ranging from individual wellbores to lithospheric plates in Chapter 9. While many of these principles were established with data from regions not associated with oil and gas development, they have proven to have broad relevance to problems encountered in the petroleum industry. For example, issues related to global and regional stress patterns are quite useful when working in areas with little pre-existing well control or when attempting to extrapolate knowledge of stress orientation and relative stress magnitudes from one area to another.

Stress in the earth's crust

Compressive stress exists everywhere at depth in the earth. Stress magnitudes depend on depth, pore pressure and active geologic processes that act at a variety of different spatial and temporal scales. There are three fundamental characteristics about the stress field that are of first-order importance throughout this book:

- Knowledge of stress at depth is of fundamental importance for addressing a wide range of practical problems in geomechanics within oil, gas and geothermal reservoirs and in the overlaying formations.
- The *in situ* stress field at depth is remarkably coherent over a variety of scales. These scales become self-evident as data from various sources are analyzed and synthesized.
- It is relatively straightforward to measure, estimate or constrain stress magnitudes at depth using techniques that are practical to implement in oil, gas and geothermal reservoirs. Hence, the state of stress is directly determinable using techniques that will be discussed in the chapters that follow.

In short, the *in situ* stress field in practice is determinable, comprehensible and needed to address a wide range of problems in reservoir geomechanics.

In this chapter I review a number of key points about the state of stress in the upper part of the earth's crust. First, we establish the mathematical terminology that will be used throughout this book and some of the fundamental physical concepts and definitions that make it possible to address many practical problems in subsequent chapters. While there are many excellent texts on elasticity and continuum mechanics that discuss stress at great length, it is useful to set forth a few basics and establish a consistent nomenclature for use throughout this book. Next, the relative magnitudes of *in situ* stresses are discussed in terms of E. M. Anderson's simple, but powerful, classification scheme (Anderson 1951) based on the style of faulting that would be induced by a given stress state. This scheme leads naturally to some general constraints on stress magnitudes as a function of depth and pore pressure. These constraints will be revisited and refined, first in Chapter 4 where we will discuss constraints on stress magnitudes in terms of the strength of the crust and further refined when we incorporate information about the presence (or absence) of wellbore failures (Chapters 7 and 8).

In the next section of this chapter I briefly review some of the stress indicators that will be discussed at length in subsequent chapters. I do so in order to review synoptically some general principles about the state of stress in the crust that can be derived from compilations of stress information at a variety of scales. The overall coherence of the stress field, even in areas of active tectonic deformation and geologic complexity is now a demonstrable fact, based on thousands of observations from sites around the world (in a wide range of geologic settings). We next briefly review several mechanisms that control crustal stress at regional scale. Finally, we consider the localized rotation of stress in the presence of near frictionless interfaces, such as salt bodies in sedimentary basins such as the Gulf of Mexico.

Basic definitions

In simplest terms, stress is defined as a force acting over a given area. To conform with common practice in the oil and gas industry around the world I utilize throughout the book calculations and field examples using both English units (psi) and SI units (megapascals (MPa), where 1 MPa = 145 psi).

To be more precise, stress is a tensor which describes the density of forces acting on all surfaces passing through a given point. In terms of continuum mechanics, the stresses acting on a homogeneous, isotropic body at depth are describable as a second-rank tensor, with nine components (Figure 1.1, left).

$$\mathbf{S} = \begin{bmatrix} S_{11} & S_{12} & S_{13} \\ S_{21} & S_{22} & S_{23} \\ S_{31} & S_{32} & S_{33} \end{bmatrix} \tag{1.1}$$

The subscripts of the individual stress components refer to the direction that a given force is acting and the face of the unit cube upon which the stress component acts. Thus, any given stress component represents a force acting in a specific direction on a unit area of given orientation. As illustrated in the left side of Figure 1.1, a stress tensor can

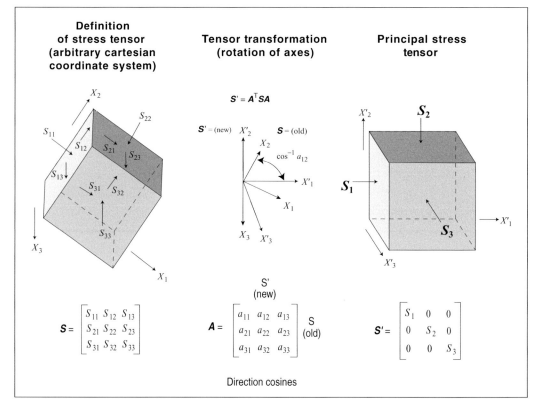

Figure 1.1. Definition of stress tensor in an arbitrary cartesian coordinate system (Engelder and Leftwich 1997), rotation of stress coordinate systems through tensor transformation (center) and principal stresses as defined in a coordinate system in which shear stresses vanish (right).

be defined in terms of any reference system. An arbitrarily oriented cartesian coordinate system is shown. Because of equilibrium conditions

$$s_{12} = s_{21}$$
$$s_{13} = s_{31} \tag{1.2}$$
$$s_{23} = s_{32}$$

so that the order of the subscripts is unimportant. In general, to fully describe the state of stress at depth, one must define six stress magnitudes or three stress magnitudes and the three angles that define the orientation of the stress coordinate system with respect to a reference coordinate system (such as geographic coordinates, wellbore coordinates, etc.).

In keeping with the majority of workers in rock mechanics, tectonophysics and structural geology, I utilize the convention that compressive stress is positive because *in situ* stresses at depths greater than a few tens of meters in the earth are *always* compressive. Tensile stresses do not exist at depth in the earth for two fundamental reasons. First, because the tensile strength of rock is generally quite low (see Chapter 4), significant tensile stress cannot be supported in the earth. Second, because there is always a fluid phase saturating the pore space of rock at depth (except at depths shallower than the water table), the pore pressure resulting from this fluid phase would cause the rock to hydraulically fracture should the least compressive stress reach a value close to the value of the pore pressure (Chapter 4).

Once a stress tensor is known in one coordinate system, it is possible to evaluate stresses in any other coordinate system via tensor transformation. To accomplish this transformation, we need to specify the direction cosines (a_{ij}, as illustrated in Figure 1.1) that describe the rotation of the coordinate axes between the old and new coordinate systems. Mathematically, the equation which accomplishes this is

$$S' = A^{\mathrm{T}}SA \tag{1.3}$$

where,

$$A = \begin{bmatrix} a_{11} & a_{12} & a_{13} \\ a_{21} & a_{22} & a_{23} \\ a_{31} & a_{32} & a_{33} \end{bmatrix}$$

There are two reasons why the ability to transform coordinate systems is of fundamental importance here. First, once we know an *in situ* stress field in some coordinate system, we can compute stresses in any other. For example, if we know the stress state in a geographic coordinate system, we will show how it is possible to derive the stress field surrounding a wellbore of arbitrary orientation (Chapter 8) to address problems of stability (Chapter 10), or along a fault plane (Chapter 5) to gauge its proximity to frictional failure and slip (Chapter 11). Another reason why tensor transformation is important is because we can choose to describe the state of stress at depth in terms of the principal stresses (i.e. those acting in the principal coordinate system), making the issue of describing the stress state *in situ* appreciably easier. The principal coordinate system is the one in which shear stresses vanish and three principal stresses, $S_1 \geq S_2 \geq S_3$ fully describe the stress field (as illustrated in the right side of Figure 1.1). In the principal coordinate system we have diagonalized the stress tensor such that the principal stresses correspond to the eigenvalues of the stress tensor and the principal stress directions correspond to its eigenvectors:

$$S' = \begin{bmatrix} S_1 & 0 & 0 \\ 0 & S_2 & 0 \\ 0 & 0 & S_3 \end{bmatrix} \tag{1.4}$$

The reason this concept is so important is that because the earth's surface is in contact with a fluid (either air or water) which cannot support shear tractions, it is a principal stress plane. Thus, one principal stress is generally normal to the earth's surface with the other two principal stresses acting in an approximately horizontal plane. While it is clear that this must be true close to the earth's surface, compilation of earthquake focal mechanism data and other stress indicators (described below) suggest that it is also generally true to the depth of the brittle–ductile transition in the upper crust at about 15–20 km depth (Zoback and Zoback 1980, 1989; Zoback 1992). Assuming this is the case, we must define only four parameters to fully describe the state of stress at depth: three principal stress magnitudes, S_v, the vertical stress, corresponding to the weight of the overburden; S_{Hmax}, the maximum principal horizontal stress; and S_{hmin}, the minimum principal horizontal stress and one stress orientation, usually taken to be the azimuth of the maximum horizontal compression, S_{Hmax}. This obviously helps make stress determination in the crust (as well as description of the *in situ* stress tensor) a much more tractable problem than it might first appear.

Relative stress magnitudes and E. M. Anderson's classification scheme

In applying these concepts to the earth's crust, it is helpful to consider the magnitudes of the greatest, intermediate, and least principal stress at depth (S_1, S_2, and S_3) in terms of S_v, S_{Hmax} and S_{hmin} in the manner originally proposed by E. M. Anderson and alluded to above. As illustrated in Figure 1.2 and Table 1.1, the Anderson scheme classifies an area as being characterized by normal, strike-slip or reverse faulting depending on whether (i) the crust is extending and steeply dipping normal faults accommodate movement of the *hanging wall* (the block of rock above the fault) downward with respect to the *footwall* (the block below the fault), (ii) blocks of crust are sliding horizontally past one another along nearly vertical strike-slip faults or (iii) the crust is in compression and relatively shallow-dipping reverse faults are associated with the hanging wall block moving upward with respect to the footwall block. The Anderson classification scheme also defines the horizontal principal stress magnitudes with respect to the vertical stress. The vertical stress, S_v, is the maximum principal stress (S_1) in normal faulting regimes, the intermediate principal stress (S_2) in strike-slip regimes and the least principal stress (S_3) in reverse faulting regimes. The dip and strike of expected normal, strike-slip and reverse faults with respect to the principal stress are discussed in Chapter 4.

Table 1.1. *Relative stress magnitudes and faulting regimes*

Regime	Stress		
	S_1	S_2	S_3
Normal	S_v	S_{Hmax}	S_{hmin}
Strike-slip	S_{Hmax}	S_v	S_{hmin}
Reverse	S_{Hmax}	S_{hmin}	S_v

The magnitude of S_v is equivalent to integration of rock densities from the surface to the depth of interest, z. In other words,

$$S_v = \int_0^z \rho(z)g\mathrm{d}z \approx \overline{\rho}gz \tag{1.5}$$

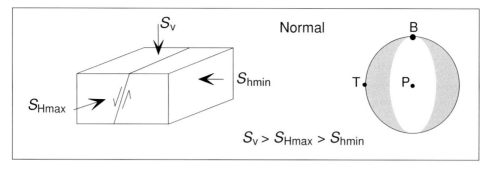

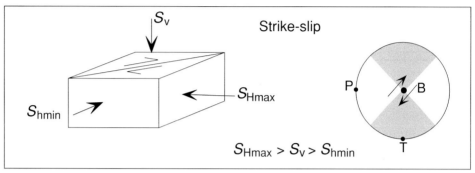

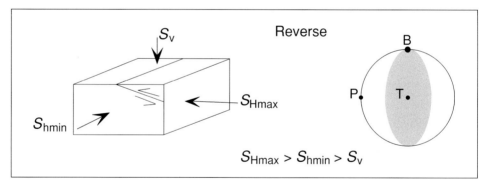

Figure 1.2. E. M. Anderson's classification scheme for relative stress magnitudes in normal, strike-slip and reverse faulting regions. Earthquake focal mechanisms, the *beach balls* on the right, are explained in Chapter 5.

where $\rho(z)$ is the density as a function of depth, g is gravitational acceleration and $\overline{\rho}$ is the mean overburden density (Jaeger and Cook 1971). In offshore areas, we correct for water depth

$$S_v = \rho_w g z_w + \int_{z_w}^{z} \rho(z) g \mathrm{d}z \approx \rho_w g z_w + \overline{\rho} g(z - z_w) \tag{1.6}$$

where ρ_w is the density of water and z_w is the water depth. As $\rho_w \sim 1\,\mathrm{g/cm^3}$ (1.0 SG), water pressure (hydrostatic pressure) increases at a rate of $10\,\mathrm{MPa/km}$ ($0.44\,\mathrm{psi/ft}$). Most clastic sedimentary rock has an average density of about 2.3 g/cm^3 which corresponds to a porosity of about 15%. This results in a vertical principal stress that increases with depth at a rate of 23 MPa/km (or conveniently, $\sim$1 psi/ft). Correspondingly, the magnitudes of the two horizontal principal stresses increase with depth. Some of the practical problems associated with the computation of S_v using equations (1.5) and (1.6) relate to the facts that density logs frequently measure anomalously low density when the well is rugose and density is often not measured all the way up to the seafloor when drilling offshore. This is illustrated by the density log in Figure 1.3. The density log (top figure) is somewhat noisy and no data are available between the seafloor (1000 ft below the platform) and 3600 ft. This makes it necessary to extrapolate densities to the seafloor where the density is quite low. Integration of the density log using equation (1.6) yields the overburden stress as a function of depth (middle figure). The rate at which the overburden stress gradient increases with depth is shown in the lower figure. Note that because of the water depth and low densities immediately below the seafloor (or mud line), the overburden stress gradient is only 0.9 psi/ft at a depth of 14,000 ft, even though density exceeds 2.3 g/cm^3 below 8000 ft.

According to the Anderson classification scheme, the horizontal principal stresses may be less than, or greater than, the vertical stress, depending on the geological setting. The relative magnitudes of the principal stresses are simply related to the faulting style currently active in a region. As illustrated in Figure 1.2, the vertical stress dominates in normal faulting regions ($S_1 = S_v$), and fault slip occurs when the least horizontal principal stress (S_{hmin}) reaches a sufficiently low value at any given depth depending on S_v and pore pressure (Chapter 4). Conversely, when both horizontal stresses exceed the vertical stress ($S_3 = S_v$) crustal shortening is accommodated through reverse faulting when the maximum horizontal principal stress (S_{Hmax}) is sufficiently larger than the vertical stress. Strike-slip faulting represents an intermediate stress state ($S_2 = S_v$), where the maximum horizontal stress is greater than the vertical stress and the minimum horizontal stress is less ($S_{Hmax} \geq S_v \geq S_{hmin}$). In this case, faulting occurs when the difference between S_{Hmax} and S_{hmin} is sufficiently large. The angle between the principal stress directions and the strike and dip of active faults is discussed in Chapter 5.

Third, an implicit aspect of Andersonian faulting theory is that the magnitudes of the three principal stresses at any depth are limited by the strength of the crust at depth. An obvious upper limit for stress magnitudes might be the compressive strength of rock. In fact, a more realistic upper limit for the magnitudes of principal stresses *in situ* is the frictional strength of previously faulted rock, as essentially all rocks at depth contain pre-existing fractures and faults (Chapter 4).

Of critical interest in this book is the current state of stress (or perhaps that which existed at the onset of reservoir exploitation) because that is the stress state applicable in the problems of reservoir geomechanics considered in this book. Hence, a point about Figure 1.2 worth emphasizing is that the figure shows the relationship between states of stress and the style of faulting consistent with that stress state. In some parts of the world

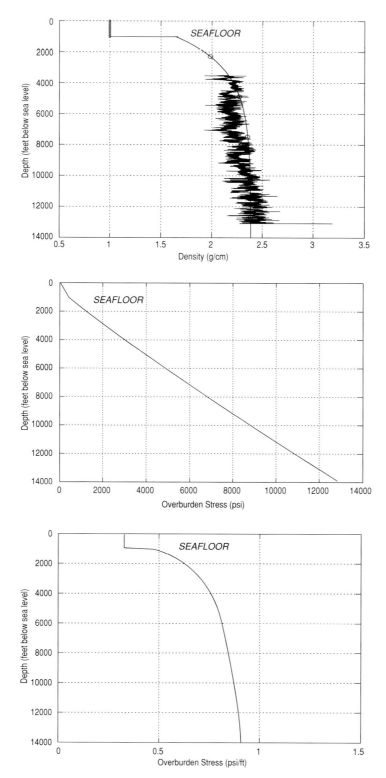

Figure 1.3. Illustration of how integration of density logs (upper figure) can be used to estimate overburden stress at depth (center figure) or overburden stress gradient (lower figure). Variability of the density logs, as well as the fact that they are often not obtained over the total depth of interest, leads to uncertainty in the calculated overburden (see text).

there is a close correspondence between the current stress field and large-scale active faults in the region. Western California (discussed below) is such a region. However, in other regions, the current stress state is not consistent with large-scale geologic structures because those structures evolved during previous tectonic regimes, in some cases, regimes that have not been active for tens, or even hundreds, of millions of years. In fact, in some parts of the world there is a marked disagreement between currently active tectonic stresses and the large-scale geologic structures defining oil and gas. One example of this is the Tampen Spur area of the northern North Sea (mentioned below and discussed in detail in Chapter 9) where earthquake focal mechanisms and direct stress measurements indicate that there is currently a compressional (strike-slip and reverse faulting) state of stress in much of the area, but the principal geologic structures are those associated with extension and basin formation (normal faulting and subsidence) at the time of opening of the North Atlantic in Cretaceous time, more than 70 million years ago. As discussed in Chapter 9, the compressional stresses in this area appear to arise from lithospheric flexure associated with deglaciation and uplift of Fennoscandia in only the past 20,000 years. In some places in the northern North Sea, after tens of millions of years of fault dormancy, some of the normal faults in the region are being reactivated today as strike-slip and reverse faults in the highly compressional stress field (Wiprut and Zoback 2000). The opposite is true of the eastern foothills of the Andes in Colombia and the Monagas basin of eastern Venezuela. Although extremely high horizontal compression and reverse faulting were responsible for formation of the large-scale reverse faults of the region, the current stress regime is much less compressive (strike-slip to normal faulting) (Colmenares and Zoback 2003).

Stress magnitudes at depth

To consider the ranges of stress magnitudes at depth in the different tectonic environments illustrated in Figure 1.2, it is necessary to evaluate them in the context of the vertical stress and pore pressure, P_p. Figure 1.4 schematically illustrates possible stress magnitudes for normal, strike-slip and reverse faulting environments when pore pressure is hydrostatic (a–c) and when pore pressure approaches lithostatic (overburden) values at depth (d–f). At each depth, the range of possible values of S_{hmin} and S_{Hmax} are established by (i) Anderson faulting theory (which defines the relative stress magnitude), (ii) the fact that the least principal stress must always exceed the pore pressure (to avoid hydraulic fracturing) and (iii) the difference between the minimum and maximum principal stress which cannot exceed the strength of the crust (which depends on depth and pore pressure as discussed in Chapter 4). Note in Figure 1.4a, for an extensional (or normal faulting) regime, that if pore pressure is close to hydrostatic, the least principal stress can be significantly below the vertical stress (it will be shown in Chapter 4 that the lower bound on S_{hmin} is approximately $0.6S_v$). In this case, the maximum horizontal stress, S_{Hmax}, must be between S_{hmin} and S_v. Alternatively, for the same pore pressure

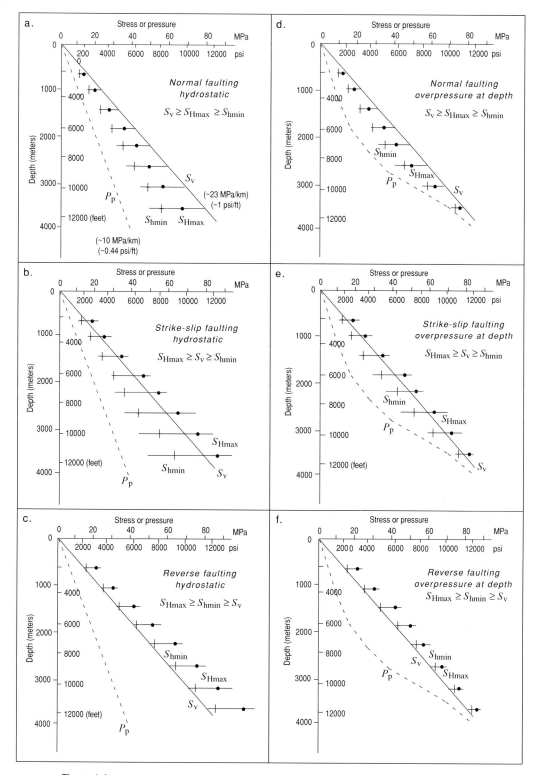

Figure 1.4. Variation of stress magnitudes with depth in normal, strike-slip and reverse faulting stress regimes for hydrostatic (a–c) and overpressure conditions (d–f). Note that the difference between principal stresses increases with depth (due to the increase of the frictional crustal strength of the crust with depth – see Chapter 4) but decreases as severe overpressure develops due to the decrease of frictional strength with elevated pore pressure (also discussed in Chapter 4).

conditions, if S_{hmin} increases more rapidly than $0.6S_v$ (as shown in Figure 1.4b), a more compressional stress state is indicated and S_{Hmax} may exceed S_v, which would define a strike-slip faulting regime. If the least principal stress is equal to the overburden, a reverse faulting regime is indicated as both horizontal stresses would be greater than the vertical stress (Figure 1.4c). As seen in Figure 1.4a–c, the differences between the three principal stresses can be large and grow rapidly with depth when pore pressure is close to hydrostatic. This will be especially important when we consider wellbore failure in Chapter 10. Again, in all cases shown in Figure 1.4, the maximum differential stress $(S_1 - S_3)$ is constrained by the frictional strength of the crust, as described in Chapter 4.

When there are severely overpressured formations at depth (Figures 1.4d–f) there are consequently small differences among the three principal stresses. In normal and strike-slip faulting domains S_{hmin}, the least principal stress $(S_{hmin} = S_3)$ must increase as P_p increases because, with the exception of transients, the least principal stress can never be less than the pore pressure. In strike-slip and reverse faulting regimes $(S_{Hmax} = S_1)$, the upper bound value of S_{Hmax} is severely reduced by high pore pressure (see Chapter 4). Thus, when pore pressure approaches the vertical stress, both horizontal stresses must also be close to the vertical stress, regardless of whether it is a normal, strike-slip or reverse faulting environment.

Measuring *in situ* stress

Over the past ~25 years, stress measurements have been made in many areas around the world using a variety of techniques. The techniques that will be described in this book have proven to be most reliable for measuring stress at depth and are most applicable for addressing the types of geomechanical problems considered here. Stress measurement techniques such as overcoring and strain relief measurements (Amadei and Stephansson 1997; Engelder 1993) are not discussed here because, in general, they are useful only when one can make measurements close to a free surface. Such *strain recovery* techniques require azimuthally oriented core samples from wells (which are difficult to obtain) and analysis of the data requires numerous environmental corrections (such as temperature and pore pressure) as well as detailed knowledge of a sample's elastic properties. If the rock is anisotropic (due, for example, to the existence of bedding) interpreting strain recovery measurements can be quite difficult.

A general overview of the strategy that we will use for characterizing the stress field is as follows:

- Assuming that the overburden is a principal stress (which is usually the case), S_v can be determined from integration of density logs as discussed previously. In Chapter 8 we discuss how observations of drilling-induced tensile fractures are an effective way to test whether the vertical stress is a principal stress.

- The orientation of the principal stresses is determined from wellbore observations (Chapter 6), recent geologic indicators and earthquake focal mechanisms (Chapter 5).
- S_3 (which corresponds to S_{hmin}, except in reverse faulting regimes) is obtained from mini-fracs and leak-off tests (Chapter 6).
- Pore pressure, P_p, is either measured directly or estimated from geophysical logs or seismic data (Chapter 2).
- With these parameters constrained, it is only necessary to constrain S_{Hmax} in order to have a reliable estimate of the complete stress tensor as part of a comprehensive geomechanical model of the subsurface. Constraints on the frictional strength of the crust (discussed in Chapter 4) provide general bounds on S_{Hmax} (as a function of depth and pore pressure). Having observations of wellbore failures (breakouts and drilling-induced tensile fractures) allows for much more precise estimates of S_{Hmax}. This is discussed for vertical wells in Chapter 7 and for deviated and horizontal wells in Chapter 8.

This strategy for *in situ* stress measurement at depth was first employed to estimate the magnitude of the three principal stresses in the Cajon Pass and KTB (Kontinentale Tiefbohrprogramm der Bundesrepublik Deutschland) scientific drilling projects (Zoback and Healy 1992; Zoback, Apel *et al.* 1993; Brudy, Zoback *et al.* 1997) and is referred to as an *integrated stress measurement strategy* as it utilizes a wide variety of observations (Zoback, Barton *et al.* 2003). Geomechanical models determined with these techniques appear in the case histories discussed in Chapters 9–12. Table 1.2 provides an overview of horizontal principal stress determination methods discussed in the chapters that follow.

Table 1.2. *Summary of horizontal principal stress measurement methods*

Stress orientation
  Stress-induced wellbore breakouts (Chapter 6)
  Stress-induced tensile wall fractures (Chapter 6)
  Hydraulic fracture orientations (Chapter 6)
  Earthquake focal plane mechanisms (Chapter 5)
  Shear velocity anisotropy (Chapter 8)

Relative stress magnitude
  Earthquake focal plane mechanisms (Chapter 5)

Absolute stress magnitude
  Hydraulic fracturing/leak-off tests (Chapter 7)
  Modeling stress-induced wellbore breakouts (Chapter 7, 8)
  Modeling stress-induced tensile wall fractures (Chapter 7, 8)
  Modeling breakout rotations due to slip on faults (Chapter 7)

Indicators of contemporary stress orientation and relative magnitude

Zoback and Zoback (1980) showed that a variety of different types of stress-related data could be used to produce comprehensive maps of stress orientation and relative magnitude at regional scales. A stress measurement quality criterion for different types of stress indicators was later proposed by Zoback and Zoback (1989, 1991) which is discussed in detail in Chapter 6. A key decision that Mary Lou Zoback and I made in these initial compilations was to consider only stress data from depths greater than several hundred meters. This was to avoid a myriad of non-tectonic, surface-related sources of stress (due, for example, to topography, thermal effects and weathering) from having a large effect where tectonic stresses are small (see Zoback and Zoback 1991). The success of our initial stress mapping efforts demonstrated that with careful attention to data quality, coherent stress patterns over large regions of the earth can be mapped with reliability and interpreted with respect to large scale geological processes. The Zoback and Zoback criterion was subsequently utilized in the International Lithosphere Program's World Stress Map Project, a large collaborative effort of data compilation and analyses by scientists from 30 different countries led by Mary Lou Zoback (Zoback 1992). Today, the World Stress Map (WSM) database has almost 10,000 entries and is maintained at the Heidelberg Academy of Sciences and the Geophysical Institute of Karlsruhe University, Germany (http://www-wsm.physik.uni-karlsruhe.de/).

The following provides a brief description of stress indicators described in the stress compilations presented throughout this book. As indicated in Table 1.2, these techniques are discussed in detail in subsequent chapters.

Wellbore stress measurements

The most classic stress measurement technique used in wellbores at depth is the hydraulic fracturing technique (Haimson and Fairhurst 1970). When a well or borehole is drilled, the stresses that were previously supported by the exhumed material are transferred to the region surrounding the well. The resultant stress concentration is well understood in terms of elastic theory, and amplifies the stress difference between far-field principal stresses by a factor of 4 (see Chapter 6). Under ideal circumstances, recording the trace of a hydraulic fracture on a wellbore wall can be used to determine stress orientation. However, such measurements are usually limited to *hard rock* sites and relatively shallow depths (<3 km) where *open-hole* hydraulic fracturing is possible. In most oil and gas wells, hydraulic fracturing cannot be used to determine stress orientation because the wells must be cased in order to carry out hydraulic fracturing without endangering the downhole equipment and wellbore. As discussed in Chapter 6, hydraulic fracturing enables the least principal stress magnitude to be determined with some accuracy (Zoback and Haimson 1982).

Observations of stress-induced wellbore breakouts are a very effective technique for determining stress orientation in wells and boreholes (Chapter 6). Breakouts are related to a natural compressive failure process that occurs when the maximum hoop stress around the hole is large enough to exceed the strength of the rock. This causes the rock around a portion of the wellbore to fail in compression (Bell and Gough 1983; Zoback, Moos et al. 1985; Bell 1989). For the simple case of a vertical well when S_v is a principal stress, this leads to the occurrence of stress-induced borehole breakouts that form at the azimuth of the minimum horizontal compressive stress. Breakouts are an important source of stress information because they are ubiquitous in oil and gas wells drilled around the world and because they also permit stress orientations to be obtained over a range of depths in an individual well. Detailed studies have shown that these orientations are quite uniform with depth, and independent of lithology and age (e.g. Plumb and Cox 1987; Castillo and Zoback 1994). Breakouts occurring in deviated wells are somewhat more complicated to analyze (Peska and Zoback 1995), but as discussed in Chapter 8, have the potential for providing information about stress orientation and stress magnitude.

Drilling-induced tensile fractures are another type of wellbore failure yielding useful information about stress orientations (Moos and Zoback 1990; Brudy and Zoback 1999). These fractures form in the wall of the borehole at the azimuth of the maximum horizontal compressive stress when the circumferential stress acting around the well locally goes into tension. As shown by Wiprut, Zoback et al. (2000), drilling-induced tensile fractures can define stress orientations with great detail and precision. As with breakouts, drilling-induced tensile fractures observed in deviated wells (Brudy and Zoback 1993; Peska and Zoback 1995) have the potential for providing information about stress orientation and stress magnitude (Chapter 8).

Earthquake focal mechanisms

Because they are so widespread, earthquake focal plane mechanisms would seem to be a ubiquitous indicator of stress in the crust. While there is indeed important information about stress magnitudes and relative orientations inherent in focal mechanism observations, these data must be interpreted with caution. Focal mechanisms are discussed at greater length in Chapter 5. The pattern of seismic radiation from the focus of an earthquake permits construction of earthquake focal mechanisms as illustrated by the figures (beach ball diagrams) in the right column of Figure 1.2. At this point, it is only necessary to recognize that there are two types of information about stress that are obtainable from well-constrained focal mechanisms of crustal earthquakes. (By well-constrained we mean that the earthquake is recorded at a sufficient number of seismographs that the orientation of the focal planes can be reliably determined.) First, the style of faulting that occurred in the earthquake can be determined (i.e. normal, strike-slip, or reverse faulting) which, in turn defines the relative magnitudes of S_{Hmax},

S_{hmin} and S_v. Second, the orientation of the P (compressional), B (intermediate), and T (extensional) axes (which are defined with respect to the orientation of the fault plane and auxiliary plane) give an approximate sense of stress directions. Unfortunately, these axes are sometimes incorrectly assumed to be the same as the orientation of S_1, S_2 and S_3 but the P, B and T axes are only approximate indicators of stress orientation as discussed in Chapter 5; nevertheless a collection of diverse focal mechanisms in a given area can be inverted to determine a best-fitting stress field. Focal mechanisms from earthquakes along plate-bounding faults, such as the San Andreas fault in California, cannot be used to determine stress orientation because of their low frictional strength. In such cases, the focal plane mechanisms are indicators of the kinematics of fault slip (and relative plate motion) and not closely related to principal stress orientations (MacKenzie 1969).

Geologic stress indicators

There are two general types of relatively recent geologic data that can be used for *in situ* stress determinations: (1) the orientations of igneous dikes or cinder cone alignments, both of which form in a plane normal to the least principal stress (Nakamura, Jacob *et al.* 1977) in the manner of a magma-filled hydraulic fracture (see Chapter 7); and (2) fault slip data, particularly the inversion of sets of striae (i.e. slickensides) on faults as for earthquake focal mechanisms as mentioned above. Of course, the term *relatively young* is often quite subjective but essentially means that the features in question are characteristic of the tectonic processes currently active in the region of question. In most cases, data that are Quaternary in age are used to represent the youngest episode of deformation in an area. Like focal mechanisms, a collection of observations of recent fault slip can be inverted to find a best-fitting stress tensor.

Regional stress patterns

Zoback and Zoback (1980) showed that it was possible to define specific *stress provinces*, regions of relatively uniform stress orientation and magnitude that correlate with physiographic provinces defined by the topography, tectonics and crustal structure. Figure 1.5 shows maximum horizontal stress orientations for North America taken from the WSM database. The legend identifies the different types of stress indicators. Because of the density of data, only highest quality data are plotted (A and B quality, as defined in Chapter 6, are distinguished by lines of different length). Where known, the tectonic regime (i.e. normal faulting, strike-slip faulting or reverse faulting) is given by the symbol color. The data principally come from wellbore breakouts, earthquake focal mechanisms, *in situ* stress measurements greater than 100 m depth, and young (<2 Ma old) geologic indicators. These data, originally presented and described by Zoback and Zoback (1991) building upon work in the conterminous United States

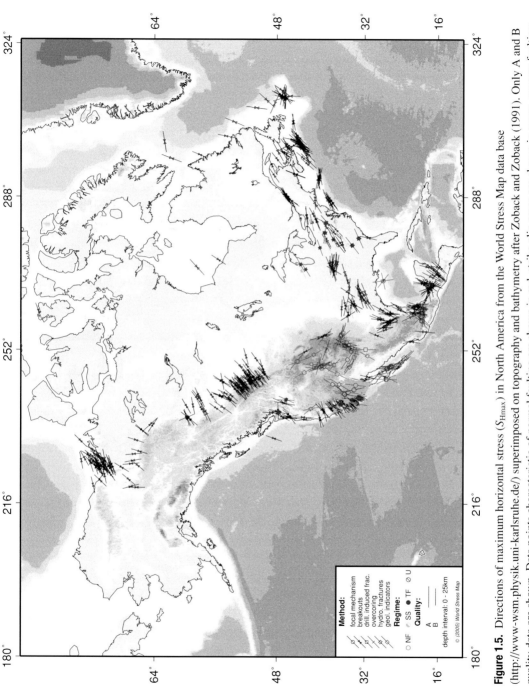

Figure 1.5. Directions of maximum horizontal stress (S_{Hmax}) in North America from the World Stress Map data base (http://www-wsm.physik.uni-karlsruhe.de/) superimposed on topography and bathymetry after Zoback and Zoback (1991). Only A and B quality data are shown. Data points characteristic of normal faulting are shown in red, strike-slip areas are shown in green, reverse faulting areas are shown in blue and indicators with unknown relative stress magnitudes are shown in black. (*For colour version see plate section.*)

(Zoback and Zoback 1980, 1989), demonstrate that large regions of the North American continent (most of the region east of the Rocky Mountains) are characterized by relatively uniform horizontal stress orientations. Furthermore, where different types of stress orientation data are available, see, for example, the eastern U.S., the correlation between the different types of stress indicators is quite good. The distribution of data is quite uneven throughout North America as the absence of data from wells, earthquakes or young geologic phenomenon in the much of the intraplate region leave large regions where the state of stress is unknown. In contrast, well-constrained earthquake focal plane mechanisms are ubiquitous in southern California such that the data are so dense that individual data points cannot be identified at the scale of this map.

Two straightforward observations about crustal stress can be made by comparison of different types of stress indicators. First, no major changes in the orientation of the crustal stress field occur between the upper 2–5 km, where essentially all of the wellbore breakout and stress measurement data come from, and 5–20 km where the majority of crustal earthquakes occur. Second, a consistent picture of the regional stress field is observed despite the fact that the measurements are made in different rock types and geologic provinces. Finally, the criterion used to define reliable stress indicators discussed in subsequent chapters appears to be approximately correct. Data badly contaminated by non-tectonic sources of stress or other sources of noise appear to have been effectively eliminated from the compilations. The state of stress in the crust at very shallow depth (*i.e.* within ∼100 m of the surface) is not discussed here for two reasons. First, this topic is outside the scope of this book (see, for example, Amadei and Stephansson 1997). Second, *in situ* stress measurements at shallow depth cannot be used in tectonic stress compilations because tectonic stresses are very small at shallow depth (because of the low frictional strength and tensile strength of near-surface rock) and a number of non-tectonic processes, including thermal effects, strongly affect *in situ* stresses near the earth's surface (Engelder and Sbar 1984). In general, only *in situ* stress measurements made at depths greater than ∼100 m seem to be independent of rock type, are spatially uniform and consistent with earthquake focal plane mechanism data coming from much greater depths. This means that techniques applied in wells and boreholes, and earthquake data can be used together (with sufficient care) to characterize the crustal stress field.

It is important to point out that the relative uniformity of stress orientations and relative magnitudes observed in Figure 1.5 is also seen at a variety of smaller scales. For example, the stress field in central California near the San Andreas fault (an actively deforming fold and thrust belt in a transpressional plate tectonic setting) is generally quite uniform (Figure 1.6, after Castillo and Zoback 1994). With the exception of the southernmost San Joaquin valley (which is discussed below), an overall NE–SW maximum horizontal stress direction is implied by both wellbore breakouts (inward pointing arrows) and earthquake focal mechanisms (lines with open circles) correlate extremely well. Both sets of data are consistently perpendicular to the trend of currently

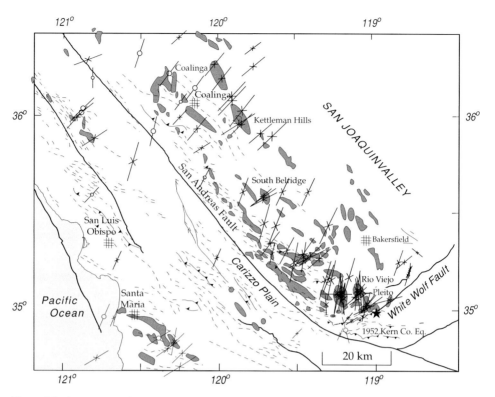

Figure 1.6. Stress map of central California (after Castillo and Zoback 1994) showing S_{Hmax} directions obtained from wellbore breakouts (inward pointed arrows) and earthquake focal plane mechanisms (symbols with open circle). *AAPG©1994 reprinted by permission of the AAPG whose permission is required for futher use.*

active fold axes (dashed lines) and thrust faults (see also Mount and Suppe 1987; Zoback, Zoback *et al.* 1987). Note that as the strike of the San Andreas fault and subparallel folds and thrust bends to a more easterly trend in the southern part of area shown, the direction of maximum horizontal stress also rotates at a scale of ∼100 km and becomes more northerly.

While there appears to be a great deal of scatter in the data from the southernmost San Joaquin valley shown in Figure 1.7, there are, in fact, relatively uniform stresses acting within the individual oil and gas fields in this region. Stress orientations in the southernmost San Joaquin valley appear to be affected by the M7.8 1952 Kern county earthquake (Figure 1.7) that occurred prior to drilling the wells used in the Castillo and Zoback (1995) stress study. Careful study of the stress field in this area illustrates that while the changes in the stress field in this area are quite pronounced, they are also systematic. The state of stress in the fields closest to the faults involved in the 1952 earthquake (San Emidio, Los Lobos, Pleito, Wheeler Ridge and North Tejon) are strongly affected by the change of stress in the crust caused by the earthquake. Fields

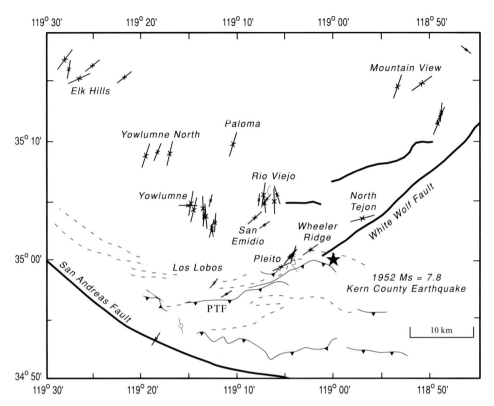

Figure 1.7. Stress map of the southernmost San Joaquin Valley from wellbore breakouts (after Castillo and Zoback 1994). The state of stress in this area was severely affected by the 1952 Kern county earthquake. *AAPG© 1994 reprinted by permission of the AAPG whose permission is required for futher use.*

further to the north (Yowlumne, Yowlumne North, Paloma, and Rio Viejo) seem not to be appreciably influenced by the 1952 earthquake as by the regional change in strike of the San Andreas fault and associated folds and faults mentioned above. So even in this geologically complex area, the observed pattern of stresses (which could not have been predicted a priori), can be measured with routinely collected data and utilized to address problems of hydraulic fracture propagation and wellbore stability. Localized faulting-induced stress perturbations are also observed on the scale of observations in single wells and boreholes as discussed in Chapter 11.

Drilling-induced tensile wall fractures (discussed in Chapters 6 and 7) also reveal a consistent picture of stress orientation in the oil fields in the northern North Sea (Figure 1.8, data after Grollimund and Zoback 2000). Once again, generally uniform stresses are observed with minor rotations occurring over spatial scales of 40–100 km. This area is a passive continental margin where stress magnitudes are currently affected by glacial unloading and lithospheric flexure. It should be noted that while the state of stress in western Europe is generally NNW–SSE compression and a strike-slip/normal

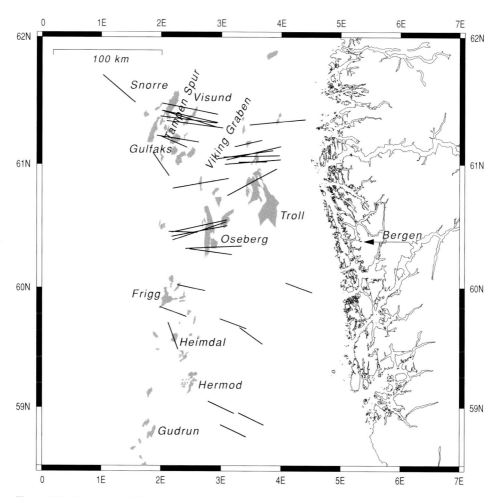

Figure 1.8. Stress map of the northern North Sea as determined principally from drilling induced tensile fractures and wellbore breakouts in wells (modified from Grollimund and Zoback 2000; Grollimund, Zoback *et al.* 2001).

stress field (see Chapter 9), the state of stress in the northern North Sea represents both a counter-clockwise rotation of stress orientation and an increase in stress magnitudes (to a strike-slip/reverse stress field) in areas most affected by the former ice sheet margin. As I discuss in Chapter 9, this modification of the stress field may be the result of deglaciation in just the past ~15,000 years.

Figure 1.9 presents a generalized stress and seismotectonic map of northern South America (Colmenares and Zoback 2003). The east–west oriented strongly compressive stresses observed in the Ecuadorian Andes province reflect the influence of convergence between the Nazca and the South American plates as the direction of maximum compression is the same as the direction of motion of the Nazca plate (single arrow) with respect to the stable interior of South America. To the north, the compression direction

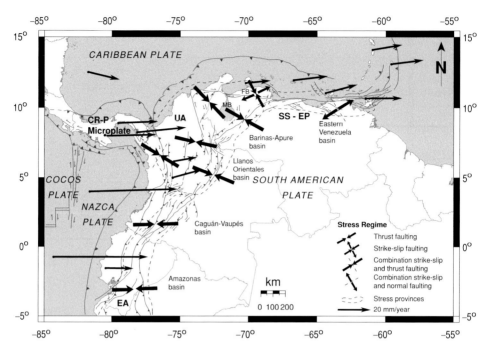

Figure 1.9. Generalized tectonic map of northern South America. The inward-pointed double arrows indicate the direction of either S_{Hmax} whereas the outward pointed double arrows indicate the direction of S_{hmin} (as explained in the inset). The stress provinces shown in the figure are discussed by Colmenares and Zoback (2003) and are abbreviated as follows: Ecuadorian Andes (EA), Upper Andes (UA), San Sebastian – El Pilar (SS-EP). GPS (Global Positioning System) velocity vectors (single arrows) denote velocities with respect to South America.

rotates to northwest–southeast and is slightly less compressive as more strike-slip faulting is observed. Toward the Merida Andes and the Maracaibo basin in Venezuela, the subduction of the Caribbean plate beneath the South American plate may affect the observed direction of maximum compression in the area. Further to the east, the stress orientation continues to rotate and stress magnitudes continue to decrease. Overall, the stress field in northern South America is affected by a diversity of complex geologic processes. Nonetheless, as was the case in the southern San Joaquin valley, careful analysis of the available data reveals uniform stress fields within specific regions and systematic variations of the stress field from region to region.

Frictionless interfaces

Because principal stresses are perpendicular and parallel to any plane without shear stress, the orientation of principal stresses is likely to be affected by the presence of weak salt bodies or severely overpressured formations. In the case of both formations, the

a.

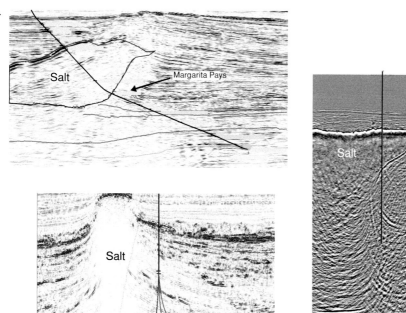

b.

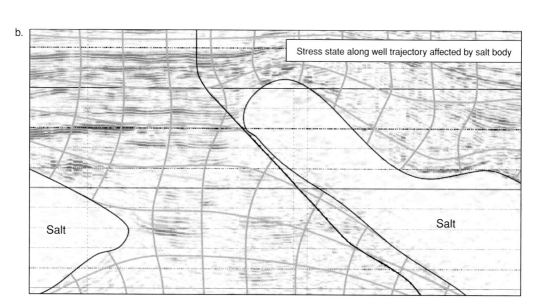

Stress state along well trajectory affected by salt body

Salt

Salt

Figure 1.10. (a) Seismic lines and well paths in the vicinity of salt bodies in deepwater Gulf of Mexico where the stress state around and within the salt bodies may have a significant affect on drilling and wellbore stability (after Fredrich, Coblentz *et al.* 2003). (b) A seismic line in the vicinity of two other salt bodies in the Gulf of Mexico. The presence of the salt is expected to significantly deflect stress trajectories away from horizontal and vertical due to the fact that the salt interface can support no shear stress.

materials are so extremely weak that there can be essentially no shear stress acting on the interface between the salt (or overpressured shale) and the adjacent formations. This means that there will be a tendency for principal stresses to re-orient themselves to be parallel and perpendicular to these weak planes. Yassir and Bell (1994) argue that maximum horizontal stress orientations on the Scotian shelf are controlled by the sloping interface of severely overpressured shale at depth. Yassir and Zerwer (1997) show that stress orientations in the Gulf of Mexico are locally affected by salt bodies.

It is clear that ignoring how stress orientation and magnitude are affected by the presence of salt bodies can lead to costly well failures. In the cases illustrated in Figure 1.10, wells were planned that would be influenced by the modified state of stress around the salt body. Because principal stresses must align parallel and perpendicular to the salt interface, the lines in Figure 1.10b show schematically how the maximum and minimum principal stresses might be deflected by the presence of the salt. In this case the well trajectory tracks beneath the salt body. Needless to say, the principal stresses along the well trajectory deviate markedly from horizontal and vertical and this must be taken into account when considering the stability of such a well. In two of the cases illustrated in Figure 1.10a, the well trajectories involve drilling through the salt such that markedly different stress fields are expected above, within and below the salt bodies.

A theoretical analysis of idealized salt bodies at depth has been considered by Fredrich, Coblentz *et al.* (2003) who used non-linear finite element modeling to illustrate the variation of stress around and within the bodies. Such calculations, when used in accordance with the principles of the stability of deviated wells (as discussed in Chapters 8 and 10) could be quite effective in preventing costly well failures in areas of particularly complicated *in situ* stress. Some of the more interesting findings of the Fredrich, Coblentz *et al.* (2003) study are that stresses within the salt body may not be uniformly equal to lithostatic (as commonly assumed) and that some of the drilling problems encountered when drilling through the bottom of a salt structure that are usually attributed to very weak rock in a hypothesized *rubble-zone*, may actually be associated with concentrated stresses at the bottom of the salt body.

2 Pore pressure at depth in sedimentary basins

Pore pressure at depth is of central importance in reservoir geomechanics. In Chapter 1, I referred to the fact that pore pressure and stress magnitudes are closely coupled (Figure 1.4). The importance of pore fluids and pore fluid pressure on the physical properties of reservoirs is discussed in Chapter 3 in the context of effective stress (the difference between external stresses acting on the rock matrix and pore pressure) and poroelasticity. In Chapter 4, pore pressure is shown to have an effect on the strength of both intact and faulted rock. Elevated pore pressures pose a severe risk during drilling when hydrocarbons are present and place important constraints on the density of drilling mud (*i.e. mud weights*) used during drilling (Chapter 10). Elevated pore pressure also influences maximum hydrocarbon column height in some reservoirs as well as the leakage potential of reservoir-bounding faults (Chapter 11). Reductions in reservoir pore pressure with production (*depletion*) can cause significant deformation in a reservoir including compaction and permeability loss (especially in poorly consolidated and weak formations) and, perhaps counter-intuitively, induce faulting in some reservoirs in normal faulting regimes or the surrounding region (Chapter 12).

I review several fundamental principles about pore pressure in this chapter. First, I define pore pressure and discuss variations of pore pressure with depth. Second, I discuss the way in which a reservoir can be hydrologically subdivided (*compartmentalized*) into distinct pressure and flow units. Third, I briefly discuss some of the mechanisms of overpressure generation that have been proposed. Finally, I discuss the relationship between pore pressure, effective stress and porosity. The ways in which porosity decreases with depth can be used to estimate pore pressure from either seismic data (before drilling) or in relatively impermeable formations (such as shales) in wells already drilled using geophysical well logs. There are a number of compilations of papers on pore pressure in sedimentary basins (Law, Ulmishek *et al.* 1998; Huffman and Bowers 2002; Mitchell and Grauls 1998) that discuss the subjects addressed in this chapter in more detail.

Basic definitions

As illustrated in Figure 2.1, pore pressure is defined as a scalar hydraulic potential acting within an interconnected pore space at depth. The value of pore pressure at depth is

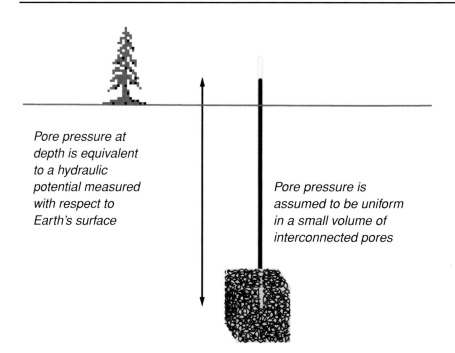

Figure 2.1. Pore pressure at depth can be considered in terms of a hydraulic potential defined with reference to earth's surface. Conceptually, the upper bound for pore pressure is the overburden stress, S_v.

usually described in relation to *hydrostatic* (or *normal*) pressure, the pressure associated with a column of water from the surface to the depth of interest. Hydrostatic pore pressure (P_p^{hydro}) increases with depth at the rate of 10 MPa/km or 0.44 psi/ft (depending on salinity). Hydrostatic pore pressure, P_p^{hydro}, implies an open and interconnected pore and fracture network from the earth's surface to the depth of measurement:

$$P_p^{hydro} \equiv \int_0^z \rho_w(z)gdz \approx \rho_w g z_w \tag{2.1}$$

Pore pressure can exceed hydrostatic values in a confined pore volume at depth. Conceptually, the upper bound for pore pressure is the overburden stress, S_v, and it is sometimes convenient to express pore pressure in terms of λ_p, where $\lambda_p = P_p/S_v$, the ratio of pore pressure to the vertical stress. *Lithostatic* pore pressure means that the pressure in the pores of the rock is equivalent to the weight of the overburden stress S_v. Because of the negligibly small tensile strength of rock (Chapter 4), pore pressure will always be less than the least principal stress, S_3.

In general, I will consider most issues involving pore pressure in quasi-static terms. That is, I will generally disregard pressure gradients that might be associated with fluid flow. With the exception of the problem of how drawdown (the difference between the pressure in the wellbore and that in the reservoir) affects well stability (Chapter 10), in the chapters that follow we will assume that pore pressure is constant at the moment a given calculation is performed.

Figure 2.2 shows the variation of pore pressure with depth from observations in the Monte Cristo field along the Texas Gulf coast (after Engelder and Leftwich 1997). The way in which pore pressure varies with depth in this field is similar to what is seen throughout the Gulf of Mexico oil and gas province and many sedimentary basins where overpressure is encountered at depth. At relatively shallow depths (in this case to about 8000 ft), pore pressures are essentially hydrostatic, implying that a continuous, interconnected column of pore fluid extends from the surface to that depth. Between 8000 ft and 11,000 ft pore pressure increases with depth very rapidly indicating that these formations are hydraulically isolated from shallower depths. By 11,000 ft, pore pressures reach values close to that of the overburden stress, a condition sometimes referred to as *hard* overpressures. Note that the ratio of the pore pressure to the overburden stress (λ_p) reaches a value of 0.91 at depth whereas in the hydrostatic pressure interval λ_p is about half that value.

Figure 2.3 (after Grollimund and Zoback 2001) demonstrates that in addition to variations of pore pressure with depth, lateral variations of pore pressure are quite pronounced in some sedimentary basins. The data shown are color-contoured values of λ_p from wells in the Norwegian sector of the northern North Sea. The color scale ranges from essentially hydrostatic pore pressure to nearly lithostatic values. Note that in some areas (in general, mostly close to the coast) pore pressure remains hydrostatic at 1500, 2000 and 3000 m. In other areas, however, pore pressure is seen to increase from hydrostatic values at 1500 m depth to much higher values at greater depths. Thus, the detailed manner in which pore pressure changes with depth varies from area to area and at any given depth there can be important lateral variations of pore pressure.

Figure 2.3 is a good illustration of why one must use caution when extrapolating average pore pressure trends from one region to another in the manner that Breckels and Van Eekelen (1981), for example, present trends of pore pressure and the least principal stress with depth for a number of oil and gas producing regions around the world. While such trends are representative of regional averages, one can see from Figure 2.3 how variable the change in pore pressure at a given depth can be from one area to another in the same region. Thus, it is always important to consider pore pressure (especially overpressure) in the context of the mechanisms responsible for it (see below) and local geologic conditions.

Reservoir compartmentalization

The observation that a given reservoir can sometimes be *compartmentalized* and hydraulically isolated from surrounding formations has received a lot of attention over

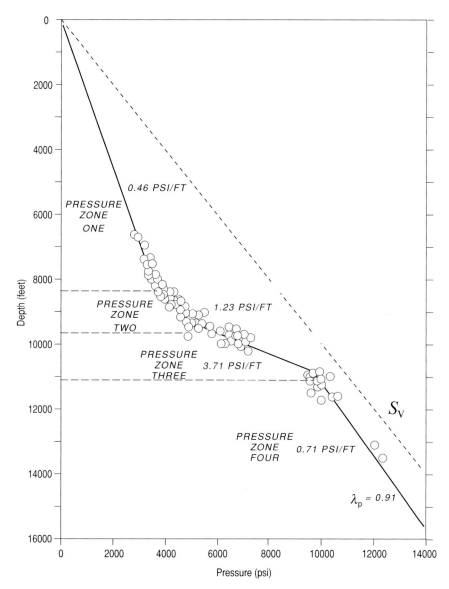

Figure 2.2. Pore pressure measurements in the Monte Cristo field, located onshore near the Gulf of Mexico in south Texas (after Engelder and Leftwich 1997). Such data are typical of many sedimentary basins in which overpressure is encountered at depth. Hydrostatic pore pressures is observed to a certain depth (in this case ~8300 ft), a transition zone is then encountered in which pore pressure increases rapidly with depth (in this case at an apparent gradient of 3.7 psi/ft) and extremely high pore pressures are observed at depths greater than ~11,000 ft. *AAPG© 1997 reprinted by permission of the AAPG whose permission is required for futher use.*

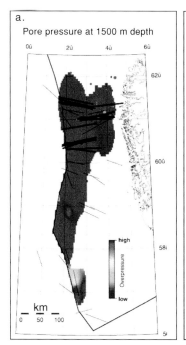
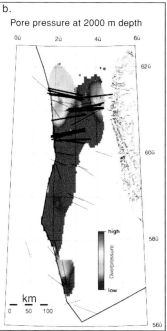
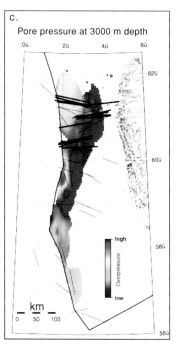

Figure 2.3. Spatial variations of pore pressure at various depths in the Norwegian sector of the northern North Sea (after Grollimund, Zoback *et al.* 2001). Note that at 1500 m depth, near hydrostatic values of pore pressure are observed. At greater depths, regions of elevated pore pressure are observed to develop in several areas. "Hard" overpressure (i.e. values near lithostatic) is observed in only a few restricted areas. Black lines indicate the direction of maximum horizontal compression determined from the orientation of drilling-induced tensile fractures and wellbore breakouts, as described in Chapter 6. (*For colour version see plate section.*)

the past decades. The economic reason for this interest is obvious as production from distinct compartments has a major impact on the drilling program required to achieve reservoir drainage. Ortoleva (1994) presents a compilation of papers related to the subject of reservoir compartmentalization.

    The easiest way to think about separate reservoir compartments is in the context of a series of permeable sands separated by impermeable shales (Figure 2.4) assuming, for the moment, that the lateral extent of each sand is limited. Pressure within each sand layer pore pressure increases with a local hydrostatic gradient because there is an open, interconnected pore space within the layer. The absolute pressure of an isolated layer can either be greater, or less than, normal pressure (Powley 1990). The case shown in Figure 2.4 is from a well in Egypt (Nashaat 1998) and illustrates this principle quite well. Note that while the pressures in compartments IIIC and IIC are appreciably above normal (and differ from each other markedly), pressures within each compartment increase with a localized hydrostatic gradient.

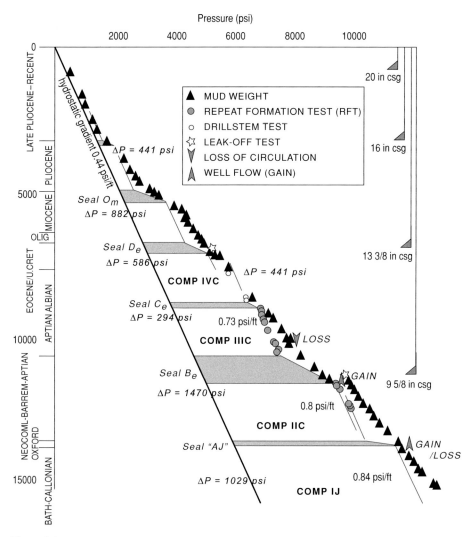

Figure 2.4. Pore pressure, mud weight and related parameters in the Mango-1 well in northern Egypt (after Nashaat 1998). The pore pressure measurements in compartments IIC and IIIC confirm that pore pressure increases with a local hydrostatic gradient within a compartment even though the absolute value of pore pressure is well above normal pressure values. *AAPG© 1998 reprinted by permission of the AAPG whose permission is required for futher use.*

Pore pressures in the reservoirs of the South Eugene Island (SEI) Block 330 field in the Gulf of Mexico provide a good illustration of pressure compartments. The sand reservoirs of the South Eugene Island field are quite young (Plio-Pleistocene in age, <4 million years) and are found mostly in a salt-withdrawal mini-basin bounded by the southwest-dipping normal faults shown in Figure 2.5 (Alexander and Flemings 1995). Localized subsidence and sedimentation (and slip along the normal fault shown) occurred when salt at depth was extruded to the southeast (Hart, Flemings *et al.* 1995).

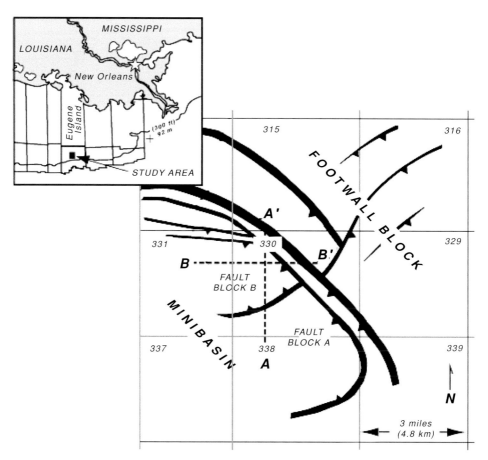

Figure 2.5. Map of the South Eugene Island (SEI) 330 field in the Gulf of Mexico (modified after Finkbeiner, Zoback *et al.* 2001). SEI 330 is one of the world's largest Plio-Pleistocene oil and gas fields. Studies of pore pressure, *in situ* stress and hydrocarbon migration in SEI 330 are referred to in subsequent chapters. *AAPG© 2001 reprinted by permission of the AAPG whose permission is required for futher use.*

A schematic geologic section along section A–A′ of Figure 2.5 is shown in Figure 2.6. Note that the individual sand reservoirs (shaded in the figure) are (i) separated by thick sequences of shale (not shaded), (ii) laterally discontinuous and (iii) frequently truncated by growth faults that provide up-dip closure (Alexander and Flemings 1995).

That many of these sand reservoirs of SEI 330 act as separate compartments is indicated by a variety of data. For example, Figure 2.7 is a map of the OI sand, one of the deeper producing intervals shown in Figure 2.6 (Finkbeiner, Zoback *et al.* 2001), that was significantly overpressured prior to depletion. The reservoirs associated with this sand were subdivided into different fault blocks on the basis of normal faults mapped using 3D seismic data. Note that the distributions of water, oil (shaded) and gas (stippled) are markedly different in adjacent fault blocks. In fault blocks, A, D and E,

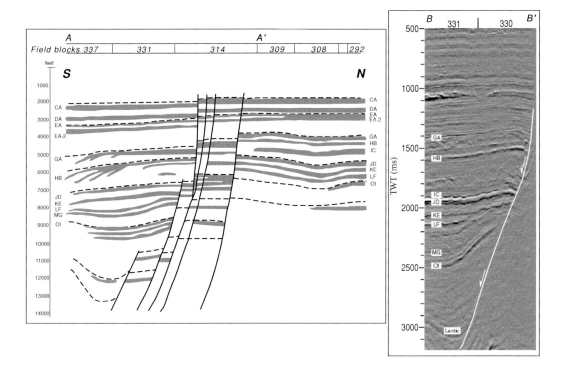

Figure 2.6. Geologic cross-section along line A–A′ in Figure 2.5 and a seismic cross-section along section B–B′ (modified after Alexander and Flemings 1995 and Finkbeiner, Zoback *et al.* 2001). In the geologic cross-section the permeable sands are shown in gray, shales are shown in white. Individual sands are identified by the alphabetic nomenclature shown. Note that slip decreases markedly along the *growth* faults as they extend upward. *AAPG© 1995 and 2001 reprinted by permission of the AAPG whose permission is required for futher use.*

for example, there are relatively small oil columns present whereas in fault blocks B and C there are significant gas columns and relatively small oil columns. Clearly, the faults separating these fault blocks are hydraulically separating the different compartments of the OI sand reservoir. Note the relatively minor offsets (indicated by the contour lines) associated with some of these faults.

It is noteworthy that in the OI sand the water phase pore pressures at the oil/water contact (the base of the oil columns) are quite different. This is shown in Figure 2.8a which presents pressure data for the fault block A (FB-A) and fault block B (FB-B) compartments of the OI reservoir which have different water phase pore pressures. When hydrocarbon columns are added to the water phase pore pressure, very high pressure is seen at the top of the hydrocarbon columns. There is an obvious physical limit to how high pressure in a compartment can become (as discussed in Chapter 11), and high initial water phase pore pressure will be shown to be one reason why

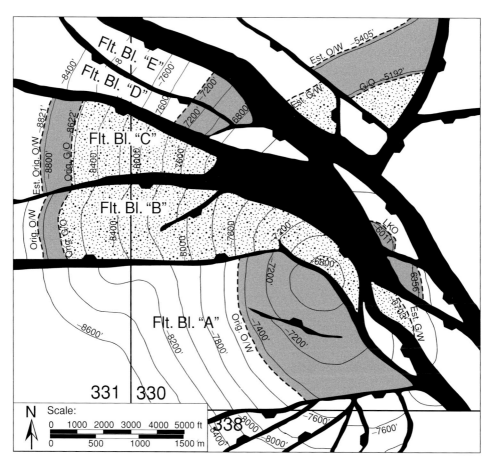

Figure 2.7. Structure contour map of the OI sand in the SEI 330 field (modified after Finkbeiner, Zoback *et al.* 2001). On the down-thrown side (hanging wall) of the major fault striking NW–SE, the sand is divided into structural fault blocs (A, B, C, D and E) based on seismically identified faults. Note the markedly different oil and gas columns in some of the fault blocks shows clear evidence of the compartmentalization of the OI reservoir. Oil is indicated shading, gas by stippling. *AAPG© 2001 reprinted by permission of the AAPG whose permission is required for futher use.*

different reservoir compartments can contain different amounts of hydrocarbons. The pore pressures estimated in the shales (Figure 2.8b) are discussed below.

The most obvious manifestation of reservoir compartmentalization is that as pressure depletion occurs over time, the entire compartment responds in the manner of a single interconnected hydraulic unit. This is clearly the case for Sand 1 (also from SEI 330) shown in Figure 2.9a (Finkbeiner 1998). As above, this reservoir was defined using 3D seismic data which delineated both the stratigraphic horizon associated with the reservoir and its lateral extent. Note that initial pore pressure in this sand was approximately 1000 psi (~7 MPa) above hydrostatic when production started in 1973. During the first five years, pressure dropped by about 1500 psi (~10 MPa) to sub-hydrostatic

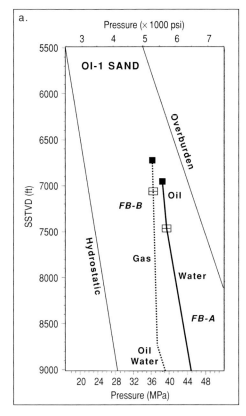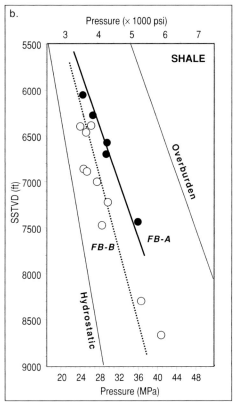

Figure 2.8. (a) Variations of pressure with depth in the OI sand based on direct measurements and extrapolations based on known column heights and fluid densities (Finkbeiner, Zoback *et al.* 2001). Note the markedly different pressures and hydrocarbon columns in these two adjacent fault blocks. Fault block A has a markedly higher water phase pressure and smaller hydrocarbon columns. (b) Shale pore pressures estimated from geophysical logs and laboratory studies of core compaction (after Flemings, Stump *et al.* 2002). Note that shale pressure is also higher in the fault block (FB)-A.

values. As different wells penetrated this compartment in different places and different times, all of the measured pressures fall along the same depletion trend, clearly indicating the compartmentalized nature of this reservoir. Note that after ~1976 pressure remained relatively constant in the reservoir despite continuing production, presumably due to reaching the reservoir's *bubble point* (the pressure where gas comes out of solution and supports reservoir pressure) or aquifer support.

However Sand 2, shown in Figure 2.9b, illustrates completely different behavior, even though it is not far away. While most of the individual wells show pressure depletion associated with production, as different wells were drilled into this sand over time, they do not seem affected by the pore pressure reduction associated with

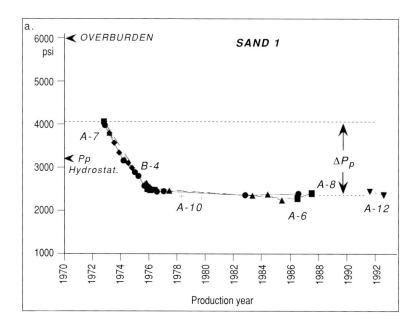

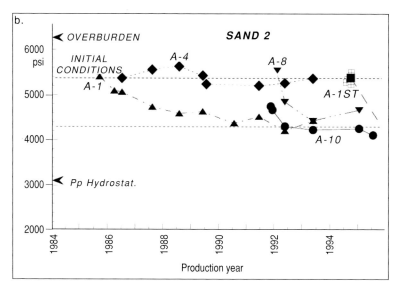

Figure 2.9. Production-induced pore pressure variations within two reservoirs in the South Eugene Island field, Gulf of Mexico (modified after Finkbeiner 1998). (a) Sand 1 appears to behave as a fully interconnected reservoir. Production from the first wells in the field causes pressure to drop from moderately overpressure values to sub-hydrostatic. (b) Production from Sand 2 demonstrates that this reservoir appears to be compartmentalized. The pressure decline associated with production from well A-1 was not sensed in the sections of the reservoir penetrated by the other wells. Instead, A-4, A-8 and A-1ST appear to have penetrated essentially undepleted sections of the reservoir.

production from previously drilled wells in what was mapped as the same reservoir. Thus, this reservoir appears to be compartmentalized at a smaller scale than that mapped seismically, presumably by relatively small, *sub-seismic* sealing faults that subdivide the sand into small compartments.

Figure 2.10 illustrates compartmentalization in a Miocene sand (the Pelican sand) in Southern Louisiana (Chan and Zoback 2007). A structure contour map indicating the presence of faults that compartmentalize the reservoir is superimposed on an air photo of the region. As shown in the inset, the pressure in the wells penetrating this sand in fault blocks I, II and III were initially at a pressure of ∼60 MPa. By 1980 fault blocks I and III had depleted along parallel, but independent depletion paths to ∼5 MPa (all pressures are relative to a common datum at 14,600 ft). Wells B and C are clearly part of the same fluid compartment despite being separated by a fault. Note that in ∼1975, the pressure in the fault blocks I and III differed by about 10 MPa. However, the pressure difference between fault blocks I and III and fault block II after five years of production is quite dramatic. Even though the first two fault blocks were signicantly depleted when production started in fault block II in the early 1980s, the pressure was still about 60 MPa. In other words, the pressure in wells E and F was about 55 MPa higher than that in wells B and C in the same sand. The fault separating these two groups of wells is clearly a sealing fault whereas the fault between wells B and C is not.

An important operational note is that drilling through severely depleted sands (such as illustrated in Figure 2.10a) to reach deeper reservoirs, can often be problematic (Addis, Cauley *et al.* 2001). Because of the reduction of stress with depletion described in Chapter 3, a mud weight sufficient to exceed pore pressure at greater depth (and required to prevent flow into the well from the formation) might inadvertently hydraulically fracture the depleted reservoir (Chapter 6) causing lost circulation. This is addressed in Chapter 12 both in terms of such drilling problems but also from the perspective of the opportunity reservoir depletion offers for refracturing a given formation.

It is worth briefly discussing how pore pressure can appear to increase with depth at gradients greater than hydrostatic. In Figure 2.2, at depths greater than ∼11,000 ft, pore pressure increases with depth at approximately the same rate as the overburden stress increases with depth. This would suggest that a series of compliant, isolated reservoirs is being encountered in which the reservoir pressure is supporting the full overburden stress. However, an extremely high pressure gradient is seen between 9000 ft and 11,000 ft (much greater than the overburden stress gradient). One should keep in mind that data sets that appear to show pressure gradients in excess of hydrostatic are compilations of data from multiple wells which penetrate different reservoirs at different depths, even though a hydrostatic pressure gradient is observed within each individual reservoir (assuming that water is in the pores).

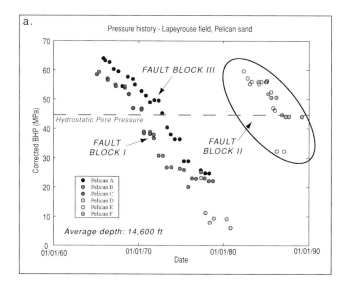

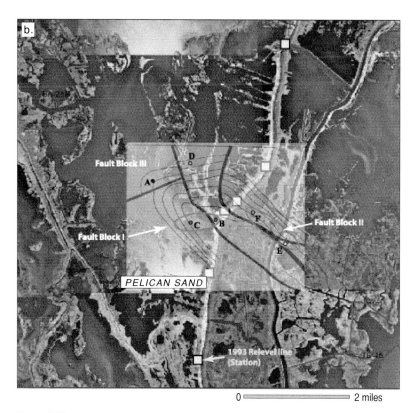

Figure 2.10. The Pelican sand of the Lapeyrouse field in southern Louisiana is highly compartmentalized. Note that in the early 1980s, while fault blocks I and III are highly depleted, fault block II is still at initial reservoir pressure (modified from Chan and Zoback 2006). Hence, the fault separating wells E and F from B and C is a sealing fault, separating compartments at pressures which differ by ~55 MPa whereas the fault between wells B and C is not a sealing fault. (*For colour version see plate section.*)

Mechanisms of overpressure generation

A variety of mechanisms have been proposed to explain the occurrence of overpressure in sedimentary basins. We briefly discuss these mechanisms below. The reader is directed to the papers in Law, Ulmishek *et al.* (1998) and Mitchell and Grauls (1998) and to the review of mechanisms of overpressure generation presented by Swarbrick and Osborne (1998). It should be noted that there are several natural mechanisms leading to underpressure (lower than normal pressure). Because such situations are quite rare in the context of the cases being considered here, underpressure is not discussed (see Ingebritsen, Sanford *et al.* 2006).

Disequilibrium compaction (which is often called *undercompaction*) is perhaps the most easily understood physical mechanism leading to overpressure (and the most important, according to Swarbrick and Osborne 1998). At a given depth, ongoing sedimentation increases the overburden stress which, in turn, will tend to cause compaction and porosity loss. In a hydraulically *open* system, that is, in sufficiently permeable formations to be hydrologically connected to earth's surface, the compaction and porosity loss associated with burial can be accommodated by fluid flow without excess pressure build up. This is apparently the case with the formations at depths less than 8000 ft in Figure 2.2. However, in a low permeability formation (such as a shale), in confined sands isolated from other sands (such as with the formations deeper than 9000 ft in Figure 2.2), or in regions of such rapid sedimentation and compaction fluid expulsion cannot keep pace with the porosity loss. In this case, the increasing overburden stress driving compaction will cause increases in pore pressure as the overburden stress is *carried* by the pore fluid pressure. This state, in which externally applied stresses are supported by pore fluid pressure, is related to the concept of effective stress (Terzaghi 1923), which is discussed at length in Chapter 3.

In the Gulf of Mexico, it is fairly obvious that sedimentation from the Mississippi River has caused large accumulations (many km) of sediment over the past several million years. This sedimentation has caused compaction-induced pore pressures, which can reach extremely high values at great depth where extremely thick sequences of impermeable shales prevent drainage (Gordon and Flemings 1998). It should be noted that under conditions of severe undercompaction, porosity is quite high and the lithostatic gradient can be significantly below 23 MPa/km (1 psi/ft). The transition from hydrostatic pressure to overpressure is highly variable from place to place (depending on local conditions) but is often at depths of between 5000 ft and 10,000 ft. That said, there are some areas in the Gulf of Mexico where overpressure is found at very shallow depth and is responsible for *shallow-water flow* zones. As will be discussed throughout this book, compaction-induced pore pressures have a marked effect on *in situ* stress, reservoir processes, physical properties, etc. Of course, the sedimentation driving compaction at depth need not be a steady-state process. For example, in various parts of the North Sea, there have been appreciable accumulations of sediments in some places since the Pleistocene (the past $\sim$15,000 years) due to locally rapid erosion caused by post-glacial uplift (Riis 1992; Dore and Jensen 1996).

It is relatively straightforward to understand the conditions under which compaction disequilibrium will result in overpressure development. The characteristic time, τ, for linear diffusion is given by

$$\tau = \frac{l^2}{\kappa} = \frac{(\phi\beta_f + \beta_r)\eta l^2}{k} \tag{2.2}$$

where l is a characteristic length-scale of the process, $\kappa \approx k/(\phi\beta_f + \beta_r)$ is the hydraulic diffusivity, β_f and β_r are the fluid and rock compressibilities, respectively, ϕ is the rock porosity, k is the permeability in m^2 (10^{-12} m^2 = 1 Darcy), and η is the fluid viscosity. For relatively compliant sedimentary rocks, equation (2.2) gives

$$\log \tau = 2\log l - \log k - 16 \tag{2.3}$$

where τ and l are in years and kilometers, respectively. In low-permeability sands with a permeability of about 10^{-15} m^2 ($\sim$1 md), the characteristic time for fluid transport over length-scales of 0.1 km is of the order of years, a relatively short amount of time in geological terms. However, in low-permeability shales where $k \sim 10^{-20}$ m^2 ($\sim$10 nd) (Kwon, Kronenberg *et al.* 2001), the diffusion time for a distance of 0.1 km is $\sim 10^5$ years, which is clearly sufficient time for increases in compressive stresses due to sediment loading, or tectonic compression, to enable compaction-driven pressure to build up faster than it can diffuse away.

Tectonic compression is a mechanism for pore pressure generation that is analogous to compaction disequilibrium if large-scale tectonic stress changes occur over geologically short periods of time. Reservoirs located in areas under tectonic compression are the most likely places for this process to be important, such as the coast ranges of California (Berry 1973), or the Cooper basin in central Australia although changes in intraplate stress due to plate tectonic processes can also lead to pore pressure changes (Van Balen and Cloetingh 1993). In the northern North Sea offshore Norway, as well as along the mid-Norwegian margin, Grollimund and Zoback (2001) have shown that compressive stresses associated with lithospheric flexure resulting from deglaciation between 15,000 and 10,000 years ago appear capable of explaining some of the pore pressure variations in Figure 2.3, with higher pore pressures in areas of induced compression and lower pore pressures in the areas of induced extension. Thus, along the Norwegian margin there appear to be three mechanisms for generating excess pore pressure, two mechanisms related to deglaciation – changes in the vertical stress due to recently rapid sedimentation and increases in horizontal compression due to lithospheric flexure – and hydrocarbon generation (discussed below). Interestingly, the two mechanisms associated with deglaciation have resulted in spatially variable pore pressure changes over the past few thousand years.

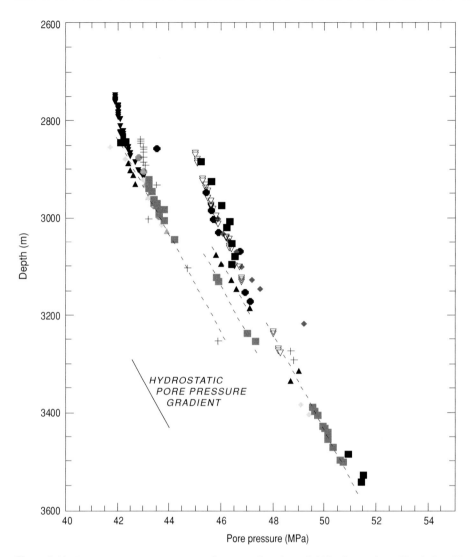

Figure 2.11. Pore pressure measurements from an oil and gas field in the northern North Sea. Note the distinct, low gradient, hydrocarbon *legs* associated with reservoirs encountered in a number of wells.

Hydrocarbon column heights can result in substantial overpressure at the top of reservoir compartments, especially when appreciable amounts of buoyant gas are present. This was seen for the FB-A and FB-B OI sands in South Eugene Island in Figure 2.8a (Finkbeiner, Zoback *et al.* 2001) and as mentioned above can ultimately limit the size of hydrocarbon columns in some reservoirs. Figure 2.11 shows excess pore pressure at the top of hydrocarbon columns from a field in the North Sea.

Centroid effects refer to the fact that relatively high pore pressure occurs at the top of a tilted sand body encased in shale. As shown in Figure 2.12, the pore pressure at the top of

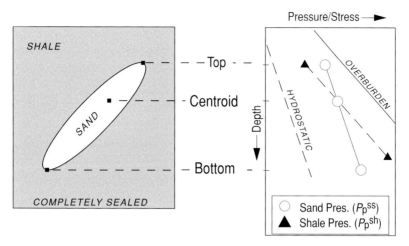

Figure 2.12. Illustration of the centroid effect where the tilting of a sand body encased in a low permeability shale results in a higher pore pressure at the top of the sand than in the shale body at the equivalent depth (from Finkbeiner, Zoback *et al.* 2001). *AAPG© 2001 reprinted by permission of the AAPG whose permission is required for futher use.*

the sand body is higher than that in the adjacent shale at the same elevation. Pressure in very weak shales is presumed to increase with a lithostatic gradient (as below 11,000 ft in Figure 2.2). The depth at which pore pressure is the same in the two bodies is referred to as the centroid. This concept was first described by Dickinson (1953), and expanded upon by England, MacKenzie *et al.* (1987) and Traugott and Heppard (1994). It is clear that drilling into the top of a sand body with pore pressure significantly higher than the adjacent shale poses an appreciable drilling hazard. Finkbeiner, Zoback *et al.* (2001) discussed centroid effects in the context of pressure-limited column heights in some reservoirs (in contrast with column heights controlled by structural closure or sand-to-sand contacts). We revisit this topic in Chapter 11.

Aquathermal pressurization refers to a mechanism of overpressure generation stemming from the fact that as sediments are buried, they are heated. Temperature increases with depth in the earth due to heat produced by radioactive decay of crystalline basement rocks and heat flowing upward through the crust from the mantle. Because heating causes expansion of pore fluid at depth, in a confined and relatively incompressible rock matrix, expanding pore fluid pore would lead in theory to pressure increases. The reason aquathermal pressure increases are not, in general, thought to be a viable mechanism of overpressure generation in most places is that the time-scale for appreciable heating is far longer than the time-scales at which overpressures develop in active sedimentary systems (Daines 1992; Luo and Vasseur 1992) such that a *near-perfect* seal would be required for long periods of geologic time.

Dehydration reactions associated with mineral diagenesis have been proposed as another mechanism that could lead to overpressure development. Smectite dehydration is a complex process (Hall 1993) but can lead to overall volume increases of both the rock matrix and the pore water system. One component of this process is the phase transition from montmorillonite to illite, which involves the expulsion of water from the crystal lattice of montmorillonite. The transition occurs at a temperature of about $100\,°C$ in the Gulf of Mexico, which is often correlative with the depth at which overpressures are observed to develop (Bruce 1984). The transition of anhydrite to gypsum is another dehydration reaction that can lead to overpressure development, but only at relatively shallow depths as the temperature at which this dehydration occurs is only about half that of the smectite–illite transition.

The exact manner in which dehydration reactions may generate overpressure is quite complicated. For example, in the case of the smectite–illite transition, the overall volume change associated with the transition is poorly known and the phase transition may work in conjunction with compaction disequilibrium (due to increased compressibility) and silica deposition (which lowers permeability). Nonetheless, a number of authors (*e.g.* Alnes and Lilburn 1998) argue that dehydration reactions are an important mechanism for generating overpressure in some sedimentary basins around the world.

Hydrocarbon generation from the thermal maturation of kerogen in hydrocarbon source rocks is associated with marked increases in the volume of pore fluid and thus can also lead to overpressure generation. This is true of the generation of both oil and gas from kerogen, although the latter process is obviously more important in terms of changes in the volume of pore fluids. As discussed in detail by Swarbrick and Osborne (1998), this mechanism appears to operate in some sedimentary basins where there is an apparent correlation between the occurrence of overpressure and maturation. In the North Sea, for example, the Kimmeridge clay is deeply buried and presently at appropriate temperatures for the generation of oil or gas (Cayley 1987). However, some younger formations (at depths well above the maturation temperatures) are also overpressured (Gaarenstroom, Tromp *et al.* 1993), so other pore pressure mechanisms are also operative in the area.

Estimating pore pressure at depth

Direct measurement of pore pressure in relatively permeable formations is straightforward using a variety of commercially available technologies conveyed either by wireline (samplers that isolate formation pressure from annular pressure in a small area at the wellbore wall) or pipe (packers and drill-stem testing tools that isolate intervals of a formation). Similarly, mud weights are sometimes used to estimate pore pressure in permeable formations as they tend to *take* drilling mud if the mud pressure is significantly in excess of the pore pressure and *produce* fluids into the well if the converse is true.

There are two circumstances in which estimation of pore pressure from geophysical data is very important. First, is the estimation of pore pressure from seismic reflection data in advance of drilling. This is obviously needed for the safe planning of wells being drilled in areas of possible high pore pressure. Second, is estimation of the pore pressure in shales even after wells are drilled, which tend to be so impermeable that direct measurement is quite difficult. In the sections below, we discuss how geophysical logging data (augmented by laboratory measurements on core, when available) are used to estimate pore pressure in shales. In both cases, techniques which have proven to work well in some areas have failed badly in others. We discuss the reasons why this appears to be the case at the end of the chapter.

Most methods for estimating pore pressure from indirect geophysical measurements are based on the fact that the porosity (ϕ) of shale is expected to decrease monotonically as the vertical effective stress ($S_v - P_p$) increases. The basis for this assumption is laboratory observations such as that shown in Figure 2.13 (Finkbeiner, Zoback et al. 2001) which shows the reduction in porosity with effective stress for a shale sample from SEI 330 field. It should be noted that these techniques are applied to shales (and not sands or carbonates) because diagenetic processes tend to make the reduction of porosity with effective confining pressures in sands and carbonates more complicated than the simple exponential decrease illustrated in Figure 2.13. If one were attempting to estimate pore pressure from seismic data before drilling in places like the Gulf of Mexico, for example, one would first estimate pore pressure in shales using techniques such as described below, and then map sand bodies and correct for the centroid effect, as illustrated in Figure 2.12.

There are basically two types of direct compaction experiments used to obtain the type of data shown in Figure 2.13: *hydrostatic compression tests* in which the applied stress is a uniform confining pressure and an impermeable membrane separates the pore volume of the rock from the confining pressure medium; and *confined compaction tests* in which the sample is subjected to an axial load while enclosed in a rigid steel cylinder that prevents lateral expansion of the sample. The data shown in Figure 2.13 were collected with the latter type of apparatus because it was thought to be more analogous to vertical loading *in situ*. Note that the application of moderate effective stresses results in a marked porosity reduction. If we assume an overburden gradient of ∼23 MPa/km (1 psi/ft), if hydrostatic pore pressure is encountered at depth, the vertical effective stress would be expected to increase at a rate of 13 MPa/km. Correspondingly, shale porosity would be expected to decrease from ∼0.38 near the surface to ∼0.11 at depths of approaching 3 km (∼10,000 ft). As indicated in the figure, in the case of moderate and high overpressure ($\lambda_p = 0.65$ and $\lambda_p = 0.8$), respectively higher porosities would be encountered at the same depth. It should also be noted that one can empirically calibrate porosity as a function of effective stress data in areas with known overburden and pore pressure.

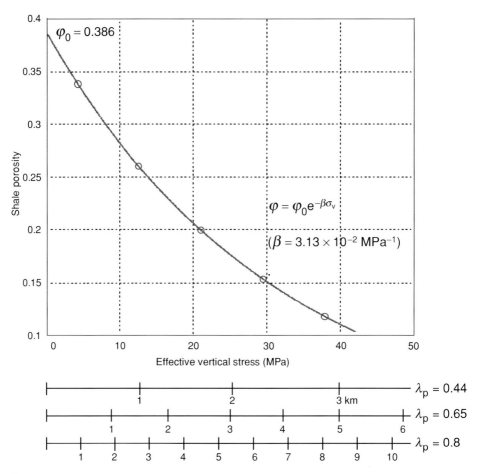

Figure 2.13. The decrease in porosity with effective stress in a SEI-330 shale sample subject to confined uniaxial compression test (Finkbeiner, Zoback *et al.* 2001). The effective stress refers to the difference between the applied uniaxial stress and pore pressure.

Shale compaction data such as that shown in Figure 2.13 can often be described by a relatively simple exponential relation first described by Athy (1930):

$$\phi = \phi_0 e^{-\beta \sigma_v} \tag{2.4}$$

where the porosity, ϕ, is related to an empirically determined initial porosity ϕ_0. σ_v is simply the effective stress, which will be discussed more fully in Chapter 3 ($\sigma_v = S_v - P_p$), and β is a second empirical constant. For the SEI 330 shale studied by Flemings, Stump *et al.* (2002) and illustrated in Figure 2.13, $\phi_0 = 0.386$ and $\beta = 3.13 \times 10^{-2}$ MPa^{-1}.

It is fairly obvious how one could exploit the relation shown in Figure 2.13 to estimate pore pressure at depth because it is relatively straightforward to estimate S_v, the vertical stress, at any given depth and porosity can be measured directly through geophysical logging or estimated from sonic velocity or resistivity data. Thus, if there is a unique relation between porosity and the vertical effective stress, $S_v - P_p$, pore pressure can be estimated because it is the only unknown.

It is helpful to consider a simple example. If a porosity of 0.17 was measured at 2 km depth, one would infer approximately hydrostatic pore pressures in the shale at that depth because the effective stress expected to result in a porosity of 0.17 for this shale at that depth ($\sim$26 MPa) is equivalent to the difference between the vertical stress ($\sim$46 MPa) and hydrostatic pore pressure (20 MPa). If, however, a porosity of 0.26 was measured at the same depth, one would infer that there was anomalously low effective stress ($\sim$10 MPa) implying that the pore pressure was anomalously high ($\sim$36 MPa), or approximately $0.8S_v$.

Near the end of this chapter I review a number of important geologic reasons why the quantitative application of this type of analysis must be used with caution. Qualitatively, however, the simple concept of an expected *compaction trend* has utility for detecting the onset of anomalously high pore pressure at depth. This is illustrated in Figure 2.14 (Burrus 1998). The increase in the vertical effective stress to depth C in Figure 2.14a (the maximum depth at which pore pressure is hydrostatic) is associated with a uniform increase in the vertical effective stress and a corresponding decrease in porosity as predicted using equation (2.4). The onset of overpressure at depths greater than C is associated with a decrease in effective stress (the difference between the overburden and pore pressure) as well as a reversal of the increase in effective stress with depth. This corresponds to anomalously high porosity at depth due to the anomalously high pressure. Note that the porosity at depth E in Figure 2.14b is the same as at the depth A, even though it is buried much more deeply. The existence of anomalously high pore pressure at depth E can be inferred from the marked deviation from the normal compaction trend.

This concept can be formalized in a rather straightforward manner when attempting to evaluate shale pore pressure, P_p^{sh}, from geophysical log data. M. Traugott (written communication, 1999) has proposed the following equation for porosity measurements derived from sonic velocity measurements

$$P_p^{sh} = \left(S_v - P_p^{hydro}\right)\left(\frac{1 - \phi_v}{1 - \phi_n}\right)^x \tag{2.5}$$

where x is an empirical coefficient, ϕ_v is the porosity from shale travel time, ϕ_n is the expected porosity from the *normal* trend, and P_p^{hydro} is the equivalent hydrostatic pore pressure at that depth. Because resistivity measurements can also be used to infer porosity, Traugott has also proposed

$$P_p^{sh} = Z\left[\frac{S_v}{Z} - \left(\frac{S_v}{Z} - \frac{P_p^{hydro}}{Z}\right)\left(\frac{R_o}{R_n}\right)^{1.2}\right] \tag{2.6}$$

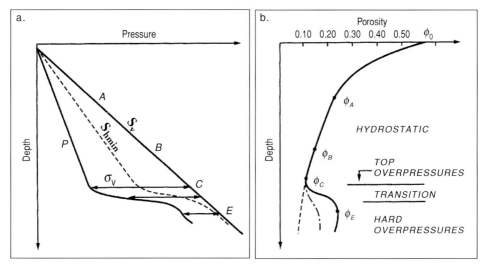

Figure 2.14. At depths A, B and C where pore pressure is hydrostatic (as shown in (a)), there is linearly increasing effective overburden stress with depth causing a monotonic porosity reduction (as shown in (b), after Burrus 1998). If overpressure develops below depth C, the vertical effective stress is less at a given depth than it would be if the pore pressure was hydrostatic. In fact, the vertical effective stress can reach extremely low values in cases of hard overpressure. Geophysical data that indicate abnormally high porosity at a given depth (with respect to that expected from the *normal* compaction trend) can be used to infer the presence of overpressure. *AAPG© 1998 reprinted by permission of the AAPG whose permission is required for futher use.*

where R_o is the observed shale resistivity and R_n is the expected resistivity at a given depth from the normal trend. The utilization of these types of equations is illustrated in Figure 2.15 with data from Trinidad (Heppard, Cander *et al.* 1998). From a qualitative perspective, it is clear that in the center of the basin (Figure 2.15a,b) abnormal pressure is indicated at ~11,500 ft considering both the sonic-derived porosities and resistivity data. In the Galeota ridge area (Figure 2.15c,d), abnormal pressure is indicated at much shallower depth (~5000 ft) by both sets of data.

An alternative way to view the determination of pore pressure in shale from sonic-derived porosity is to simply consider rewriting equation (2.4) as

$$P_p = S_v - \left(\frac{1}{\beta_c} \ln \left(\frac{\phi_0}{\phi} \right) \right) \tag{2.7}$$

where ϕ_0 is the initial porosity (at zero effective pressure), and the porosity ϕ is determined from the sonic travel time, Δt, based on geophysical P-wave velocity (V_p) measurements ($V_p^{-1} = \Delta t$) by

$$\phi = 1 - \left(\frac{\Delta t_{ma}}{\Delta t} \right)^{1/f} \tag{2.8}$$

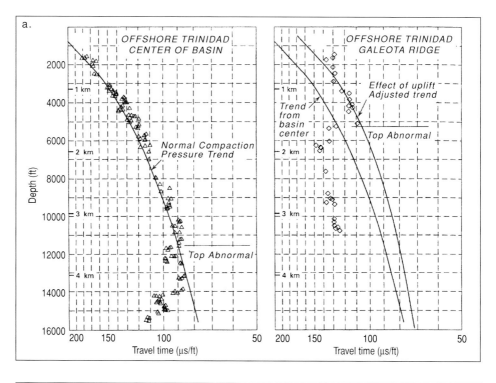

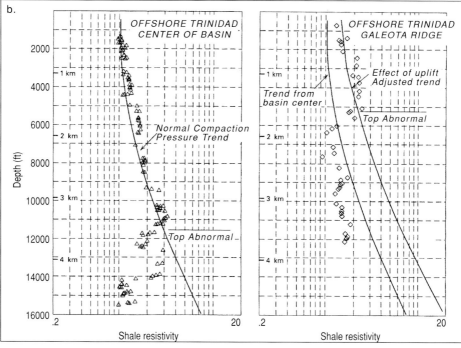

Figure 2.15. Data indicating compaction trends in Trinidad as inferred from (a) travel time data and (b) resistivity data (from Heppard, Cander *et al.* 1998). Assuming that the deviations from the normal compaction trends is evidence of overpressure (abnormally low effective stress), both sets of data indicate an onset of overpressure in the center of the basin at ~11,500 ft and 5,200 ft in the vicinity of the Galeota ridge. *AAPG© 1998 reprinted by permission of the AAPG whose permission is required for futher use.*

where f is the acoustic formation factor and Δt_{ma} is the matrix travel time. Flemings, Stump *et al.* (2002) determined ϕ_0 and β_c from the compaction trend of shales in SEI 330 using data from the hydrostatically pressured section at shallow depth (Figure 2.16a). They went on to determine f and Δt_{ma} from laboratory measurements on the core (Figure 2.16b). These measurements were used to estimate the shale pore pressures shown in Figure 2.8b. Note in that figure that both the direct pore pressure measurements in the sands and the estimate of pore pressure in the shale from the sonic porosity data indicate that fault block A is more overpressured than B, presumably because it did not drain as effectively during burial. Also note the continuity and overall coherence of the shale pressure estimates.

There are many cases in which it is necessary to estimate pore pressure from seismically derived velocity prior to drilling. This is illustrated in the example shown in Figure 2.17 (Dugan and Flemings 1998). Figure 2.17a shows the analysis of RMS (root-mean-square) compressional wave velocities obtained from relatively conventional normal moveout analysis of an east–west seismic line from the South Eugene Island field along the northern edge of the area shown in Figure 2.7. Overall, the RMS velocities increase with depth as expected, although unusually low velocities are seen at depth between shot points 1500 and 1600. Figure 2.17b shows interval velocities (the velocities of individual formations) that were derived from the normal moveout velocities. Again, interval velocities generally increase with depth, as expected for compacting sediments, but two areas of unusual interval velocity are seen – relatively high velocity just west of shot point 1600 at the depth of the JD sand, and relatively low interval velocities in the vicinity of the GA, JD and OI sands just east of the fault near shot point 1700.

To interpret these interval velocities in terms of pore pressure, one can use empirical equations such as

$$V_i = 5000 + A\sigma_v{}^B \tag{2.9}$$

(Bowers 1994) where V_i is the interval velocity (in ft/sec), A and B are empirical constants and $A = 19.8$ and $B = 0.62$ (Stump 1998). Because σ_v increases by about 0.93 psi/ft, this leads to

$$P_p = 0.93z - \left(\frac{V_i - 5000}{19.8}\right)^{\frac{1}{0.62}} \tag{2.10}$$

where the depth, z, is in feet. Utilization of this equation to infer pore pressure at depth is illustrated in Figure 2.17c. Note that the unusually low interval velocity east of the fault at shot point 1700 implies unusually high pore pressure at the depths of the JD and OI sands. The pore pressure is expressed in terms of equivalent mud weight because information such as that shown in Figure 2.17c is especially important for drillers who need to know about excess pore pressure at depth in order to determine the mud weight required for safe drilling (see Chapter 10).

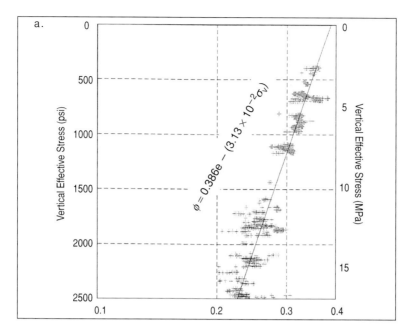

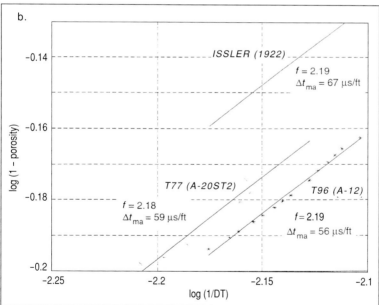

Figure 2.16. (a) Decrease in porosity of shale in the SEI-330 field with effective stress based on geophysical logs at depths characterized by hydrostratic pressures (Finkbeiner, Zoback *et al.* 2001; Flemings, Stump *et al.* 2002). Such measurements are used to determine the compaction parameter, β_c, described in the text. (b) Laboratory measurements of the variation of porosity with P-wave transit time (V_p^{-1}) to determine the acoustic formation factor, *f*, as described in the text. *Reprinted with permission from American Journal Science.*

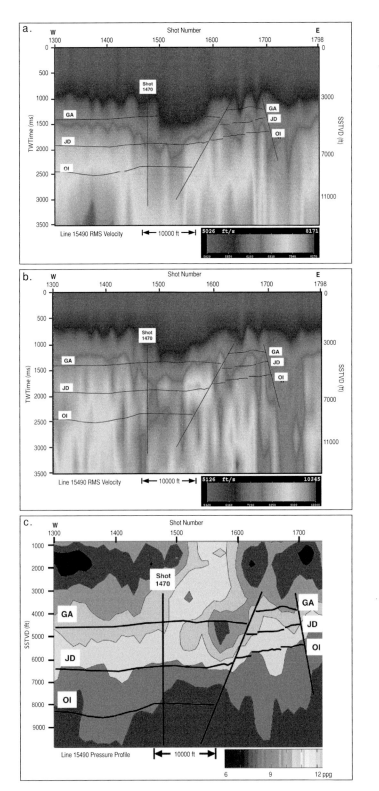

Figure 2.17. Pore pressure estimation from seismic reflection data along an E–W seismic line in the SEI-330 field (after Dugan and Flemings 1998). (a) Stacking, or RMS, velocities determined from normal moveout corrections. (b) Interval velocities determined from the RMS velocities (c) Pore pressures inferred from interval velocity in the manner discussed in the text. (*For colour version see plate section.*)

A somewhat more direct approach to estimation of pore pressure from seismic interval velocity data is based on empirical correlations between P-wave velocity, V_p, S-wave velocity, V_s, both in units of km/s, mean effective stress, σ, in units of kbar (1 kbar = 100 MPa), porosity, ϕ, and clay content, C, based on point counts of thin sections $(0 \le C \le 1)$. The following formulae were derived by Eberhart-Phillips, Han *et al.* (1989) for 64 different sandstones with varying amounts of shale from a comprehensive suite of laboratory measurements of Han, Nur *et al.* (1986)

$$V_p = 5.77 - 6.94\phi - 1.73\sqrt{C} + 0.446(\sigma - e^{-16.7\sigma}) \tag{2.11}$$

$$V_s = 3.70 - 4.94\phi - 1.57\sqrt{C} + 0.361(\sigma - e^{-16.7\sigma}) \tag{2.12}$$

While there are obviously a number of required parameters needed to isolate the relationship between V_p and V_p/V_s and effective stress, approaches to pore pressure prediction using such relations have proven useful in many cases. Because of the non-uniqueness of the relation between V_p/V_s and effective stress, porosity and clay content, an increase of V_p/V_s could indicate a decrease in effective stress (increase in pore pressure), an increase of clay content or some combination of the two.

When using the methodology outlined above, it is important to be aware of a number of complicating factors. First, it is important to note that these methodologies apply best to shales because in sands and carbonates variations in cementation and diagenesis affect how they compact with depth such that relations such as equation (2.4) are not applicable (Burrus 1998). Second, because the method assumes that all shales in a given section follow the same compaction trend, variations of shale lithology with depth represent a decrease in effective stress, whereas they could result from a change of lithology. Third, there are a variety of opinions about how pressure, or stress, in the lab should be related to depth in the earth. Does hydrostatic confining pressure correspond to the overburden stress, S_v; or does a laterally confined uniaxial compression test correspond to S_v; or does hydrostatic confining pressure (or mean stress in a uniaxial compression test) correspond to mean stress in the earth, thus requiring knowledge (or estimates) of S_{hmin} and S_{Hmax}? This is discussed at length by Harrold, Swarbrick *et al.* (1999). The reasons underlying their concerns are discussed in Chapter 3 where we consider porosity losses as a function of pressure and stress.

In addition to the complicating factors just mentioned (and perhaps more significantly), there are reasons why the types of pore pressure prediction methodologies discussed above should not be used in certain geologic environments. Compaction-based methods assume a prograde burial path in which effective stress monotonically increases with burial and time. Hence, in such regions, the deviation from the expected porosity at a given depth is evidence of anomalously high pore pressure. In regions with a complex burial history and/or a history of pore pressure generation, the fundamental assumption of a monotonic increase of effective stress with depth and time is incorrect. As pointed out by Burrus (1998) and schematically illustrated for a laboratory

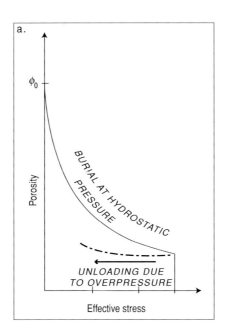

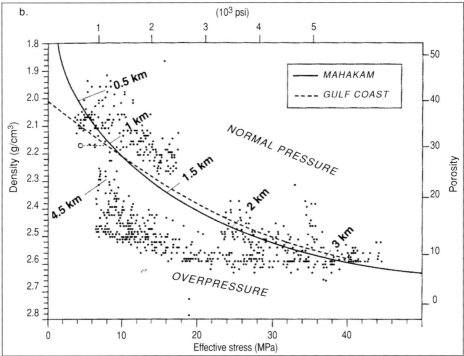

Figure 2.18. (a) Schematic illustration of the small porosity *recovery* that occurs upon unloading in weak sediments. Thus, if the current state of a reservoir represents a formation that has been unloaded, a nearly constant porosity is observed over a wide range of effective stresses (i.e. pore pressures). (b) Field data from the Mahakam Delta (Burrus 1998) which demonstrate that because overpressure at depths > 3 km developed after burial, the relation between porosity and effective stress during loading makes it difficult to infer pore pressure from porosity during unloading (see text). *AAPG© 1998 reprinted by permission of the AAPG whose permission is required for futher use.*

test in Figure 2.18a, should effective stress decrease after loading, there is almost no change in porosity along the *unloading* path. This is because the majority of the porosity loss associated with compaction is associated with unrecoverable deformation. As can be readily seen in Figure 2.18a, essentially constant porosity along the unloading path is associated with a large range ($\sim$16–40 MPa) of effective stresses, making it impossible to estimate pore pressure at a given depth from the porosity–effective-stress relationship. That said, there tends to be a unique geophysical signature to unloading, expecially at low effective stress ($<$ 15 MPa in Figure 2.18b) that helps identify cases when this occurs (Bowers 1994). For example, induced micro-fracturing would tend to affect measurements of sonic velocity or formation resistivity logs but not bulk measurements such as density or neutron porosity.

There are a variety of geologic processes that could be responsible for such an unloading effect. These include an increase in pore pressure after initial burial (associated with the various types of processes described above) or rapid uplift and erosion. Areas with complicated burial and tectonic histories would also be areas where prediction of pore pressure with the technique described above should be considered with great caution. In such places the porosity reduction is more correctly predicted using relationships derived from mean effective stress increases ($S_{Hmax} + S_{hmin} + S_v$)/3 $- P_p$, see Chapter 3. Hence, in areas of significant tectonic compression, pore pressure prediction methods require knowledge of all three principal stresses. A number of the papers in Law, Ulmishek *et al.* (1998) and Huffman and Bowers (2002) discuss pore pressure prediction in areas where compressive stress significantly complicates the use of the relatively simple techniques outlined above.

Field data from the Mahakam delta (Figure 2.18b after Burrus 1998) show such an effect. Note that porosity decreases (and density increases) uniformly with increasing effective stress to $\sim$3 km depth where there is normal compaction and hydrostatic pore pressure. However, when overpressure is encountered at depth of $>$ 3 km, there is a wide range of effective stresses (a wide range of overpressure) associated with almost constant porosity. As the overburden stress at 4 km is expected to be $\sim$90 MPa, an effective stress of $\sim$15 MPa implies a pore pressure of 75 MPa ($\lambda = 0.83$). However, the porosity at this depth ($\sim$12%), if interpreted using the loading curve, would imply an effective stress of about 35 MPa. This would result in an under-prediction of pore pressure by 20 MPa ($\sim$3000 psi), a very appreciable amount. Again, simple compaction curves are not adequate for pore pressure prediction in cases of high compressive stress or when pore pressure increases appreciably after burial and initial compaction.

3 Basic constitutive laws

In this chapter I briefly review a number of the constitutive laws governing rock deformation. Fundamentally, a constitutive law describes the deformation of a rock in response to an applied stress (or vice versa). Because of the breadth of this subject, the material below is restricted to covering key principles that will be referred to later in the text.

One unconventional topic discussed at some length at the end of this chapter is the viscous compaction of uncemented sands. As explained below, the presence of water or oil in the pores of a rock will result in time-dependent deformation of any porous elastic (*poroelastic*) solid. The topic I discuss below is the viscous deformation of *dry* uncemented sands. In other words, in addition to the poroelastic deformation, there is also time-dependent deformation of the dry matrix. There are two main reasons for this. First, many oil and gas reservoirs in the world occur in such formations. Thus, it is important to accurately predict: (i) how they will compact with depletion (especially as related to compaction drive); (ii) what the effects of compaction will be on reservoir properties (such as permeability); and (iii) what the effects will be on the surrounding formations (such as surface subsidence and induced faulting). The basic principles of viscous compaction of uncemented sands are outlined in this chapter, whereas the compaction of reservoirs composed of such materials is addressed in more detail in Chapter 12. Second, while there are several exellent texts on the subject of constitutive laws applicable to rocks (*e.g.* Charlez 1991; Paterson and Wong 2005; Pollard and Fletcher 2005), the subject of viscous deformation in very weak formations, while dealt with extensively in soil mechanics and geotechnical engineering, has not been discussed extensively in the context of hydrocarbon reservoirs.

The schematic diagrams in Figure 3.1 illustrate four generic types of constitutive laws for homogeneous and isotropic materials. Even though each of these constitutive laws is described in greater detail later in this chapter, the following introduction may be useful.

A *linearly elastic* material (Figure 3.1a) is one in which stress and strain are linearly proportional and deformation is reversible. This can be conceptualized in terms of a force applied to a spring where the constant of proportionality is the spring constant, k. An ideal elastic rock strains linearly in response to an applied stress in which the stiffness of the rock is E, Young's modulus. An actual rock mechanics test is presented

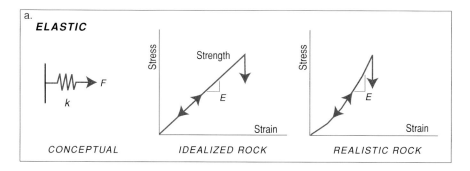

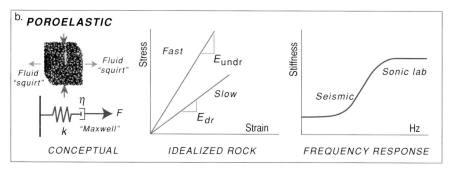

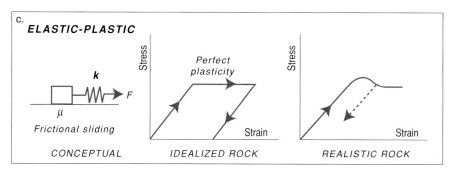

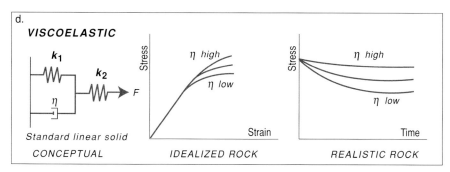

Figure 3.1. Schematic illustration of elastic, poroelastic, elastic–plastic and viscoelastic constitutive laws. In the left panels, analogous physical models are shown, in the center, idealized stress–strain curves, and in right panels, schematic diagrams representing more realistic rock behavior are shown.

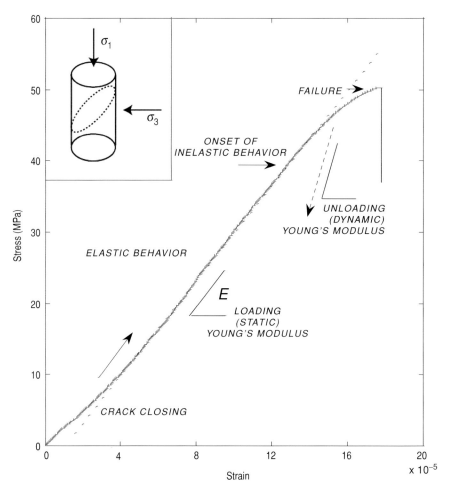

Figure 3.2. Typical laboratory stress–strain data for a well-cemented rock being deformed uniaxially. There is a small degree of crack closure upon initial application of stress followed by linear elastic behavior over a significant range of stresses. Inelastic deformation is seen again just before failure due to damage in the rock.

in Figure 3.2 to illustrate how a relatively well-cemented sandstone exhibits nearly ideal elastic behavior over a considerable range of applied stresses. As axial stress is applied to this rock, there is some curvature in the stress–deformation curve upon initial loading due to the closure of micro-cracks. Once these cracks are closed (at a stress of about 9 MPa), the rock exhibits linear elastic behavior until a stress of about 45 MPa is reached. At this pressure, the stress applied to the rock is so large that it begins to damage the rock such that permanent, or *plastic,* deformation is observed prior to eventual failure of the sample at a stress of about 50 MPa (rock failure is discussed at length in Chapter 4).

Figure 3.2 illustrates why a *strength of materials* approach to rock failure, that will be utilized to consider wellbore failure in the chapters to follow, is frequently a useful simplification of the inherently complex process of rock failure. Using a strength of materials approach assumes that one can consider rock deformation to be elastic until the point of failure. It is evident in Figure 3.2 that a small degree of inelastic deformation precedes failure such that this assumption is not strictly correct. Nonetheless, for well-cemented rocks, a strength of materials approach does a good job of approximating failure. In weak, poorly cemented formations, the applicability of this approach is more questionable.

Figure 3.1b schematically illustrates the profound effect that water (or oil) in the pores of a rock has on its behavior. A porous rock saturated with fluid will exhibit *poroelastic* behavior. One manifestation of poroelasticity is that the stiffness of a fluid-saturated rock will depend on the rate at which external force is applied. When force is applied quickly, the pore pressure in the rock's pores increases because the pore fluid is carrying some of the applied stress and the rock behaves in an undrained manner. In other words, if stress is applied faster than fluid pressure can drain away, the fluid carries some of the applied stress and the rock is relatively stiff. However, when an external force is applied slowly, any increase in fluid pressure associated with compression of the pores has time to drain away such that the rock's stiffness is the same as if no fluid was present. It is obvious that there is a trade-off between the loading rate, permeability of the rock and viscosity of the pore fluid, which is discussed further below. For the present discussion, it is sufficient to note that the deformation of a poroelastic material is time dependent, a property shared with *viscoelastic* materials, as illustrated in Figure 3.1d and discussed further below.

Figure 3.1c illustrates *elastic–plastic* behavior. In this case, the rock behaves elastically to the stress level at which it yields and then deforms plastically without limit. Upon unloading the rock would again behave elastically. Although some highly deformable rocks behave this way in laboratory testing (right panel), we discuss in Chapter 4 how this type of behavior is also characteristic of deformation in the upper crust being taken up by fault slip. In this case, appreciable deformation (*i.e.* fault slip) can occur at a relatively constant stress level (*i.e.* that required to cause optimally oriented faults to slip).

As alluded to previously, a viscoelastic rock (Figure 3.1d) is one in which the deformation in response to an applied stress or strain is rate dependent. The stress required to cause a certain amount of deformation in the rock depends on the apparent viscosity, η, of the rock (center panel). One can also consider the stress resulting from an instantaneously applied deformation (right panel) which will decay at a rate depending on the rock's viscosity. The conceptual model shown in the left panel of Figure 3.1d corresponds to a specific type of viscoelastic material known as a *standard linear solid*. A variety of other types of viscous materials are described below. A viscous material that exhibits permanent deformation after application of a load is described as *viscoplastic*.

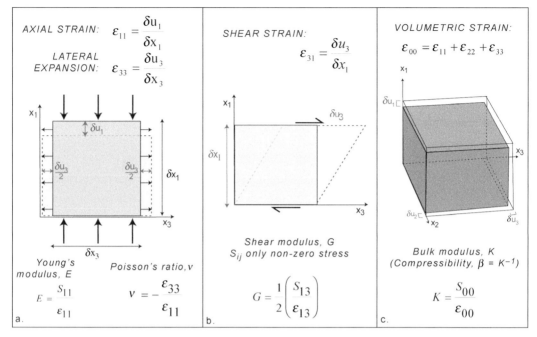

Figure 3.3. Schematic illustration of the relationships between stress, strain and the physical meaning of frequently used elastic moduli in different types of idealized deformation measurements.

To describe constitutive laws more precisely, it is necessary to have a rigorous definition of deformation by considering the components of the second-order strain tensor, ε_{ij} defined as

$$\varepsilon_{ij} = \frac{1}{2} \left(\frac{\delta u_i}{\delta x_j} + \frac{\delta u_j}{\delta x_i} \right) \tag{3.1}$$

In a homogeneous and isotropic material, principal stresses and principal strains act in the same directions.

Several physically meaningful strain components are illustrated in Figure 3.3: axial strain and lateral expansion in a sample compressed uniaxially (Figure 3.3a); shear strain resulting from application of a simple shear stress (Figure 3.3b); and volumetric strain resulting from compressing a body under isostatic mean stress (which corresponds to uniform confining pressure in laboratory experiments), S_{00} (Figure 3.3c) where,

$$\varepsilon_{00} = \varepsilon_{11} + \varepsilon_{22} + \varepsilon_{33} \qquad S_{00} = \frac{1}{3}(S_{11} + S_{22} + S_{33}) \tag{3.2}$$

Linear elasticity

The theory of elasticity typically is discussed in terms of infinitesimally small deformations. In this case, no significant damage or alteration of the rock results from an applied stress and the assumption that stress and strain are linearly proportional and fully reversible is likely to be valid. In such a material, stress can be expressed in terms of strain by the following relation

$$S_{ij} = \lambda \delta_{ij} \varepsilon_{00} + 2G\varepsilon_{ij} \tag{3.3}$$

where the Kronecker delta, δ_{ij}, is given by

$$\delta_{ij} = 1 \quad i = j$$
$$\delta_{ij} = 0 \quad i \neq j$$

Upon expansion, equation (3.3) yields

$$S_1 = (\lambda + 2G)\,\varepsilon_1 + \lambda\varepsilon_2 + \lambda\varepsilon_3 = \lambda\varepsilon_{00} + 2G\varepsilon_1$$
$$S_2 = \lambda\varepsilon_1 + (\lambda + 2G)\,\varepsilon_2 + \lambda\varepsilon_3 = \lambda\varepsilon_{00} + 2G\varepsilon_2$$
$$S_3 = \lambda\varepsilon_1 + \lambda\varepsilon_2 + (\lambda + 2G)\,\varepsilon_3 = \lambda\varepsilon_{00} + 2G\varepsilon_3$$

and λ (Lame's constant), K (bulk modulus) and G (shear modulus) are all elastic moduli.

There are five commonly used elastic moduli for homogeneous isotropic rock. As illustrated in Figure 3.3, the most common is the bulk modulus, K, which is the stiffness of a material in hydrostatic compression and given by

$$K = \frac{S_{00}}{\varepsilon_{00}}$$

The physical representation of K is displayed in Figure 3.3c. The compressibility of a rock, β, is given simply by $\beta = K^{-1}$. Young's modulus, E, is simply the stiffness of a rock in simple (unconfined) uniaxial compression (S_{11} is the only non-zero stress)

$$E = \frac{S_{11}}{\varepsilon_{00}}$$

which is shown in Figure 3.3a (and Figure 3.2). In the same test, Poisson's ratio, ν, is simply the ratio of lateral expansion to axial shortening

$$\nu = \frac{\varepsilon_{33}}{\varepsilon_{11}}$$

In an incompressible fluid, $\nu = 0.5$. The shear modulus, G, is physically shown in Figure 3.3b. It is the ratio of an applied shear stress to a corresponding shear strain

$$G = \frac{1}{2}\left(\frac{S_{13}}{\varepsilon_{13}}\right)$$

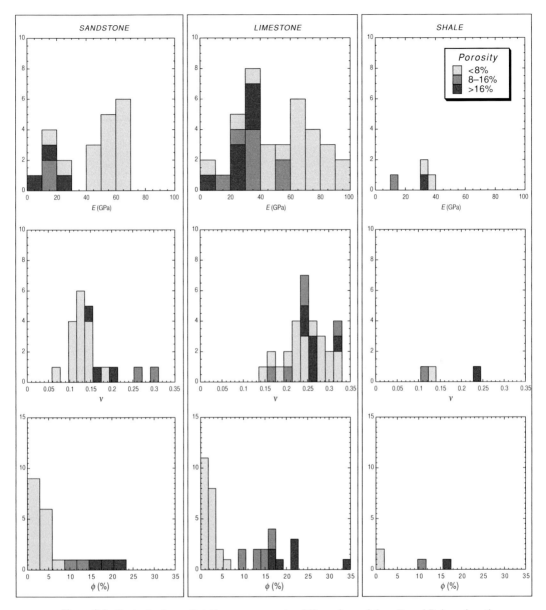

Figure 3.4. Typical values of static measurements of Young's modulus, E, and Poisson's ratio, ν, for sandstone and limestone and porosity, ϕ (from Lama and Vutukuri 1978).

The shear modulus of a fluid is obviously zero. The fifth common elastic modulus, Lame's constant, λ, does not have a straightforward physical representation. In a material in which Poisson's ratio is equal to 0.25 (a Poisson solid), $\lambda = G$.

Figure 3.4 is a compilation of *static* laboratory measurements of E and ν for sandstone, limestone and shale samples for which porosity was also measured (Lama and

Vutukuri 1978). *Static* means that measurements were made in a manner similar to the idealized experiments illustrated in Figure 3.3. Such measurements are distinguished from *dynamic* measurements of elastic moduli utilizing laboratory seismic velocity measurements at ultrasonic frequency ($\sim 10^6$ Hz), which are discussed below. As seen in Figure 3.4, relatively low-porosity sandstones (0–6%) have Young's moduli around 50 MPa and Poisson's ratios of ~ 0.125, whereas relatively high-porosity sandstones (>16%) have appreciably lower Young's moduli and higher Poisson's ratios. Similarly, relatively low porosity limestones have very high Young's moduli (~ 80 MPa), whereas higher porosity limestones have appreciably lower stiffness. The relation between porosity and Poisson's ratio for limestones is less clear than for the sandstone samples because most values are 0.2–0.3, irrespective of porosity. So few shale samples are in this data set that it is difficult to generalize about any relation between porosity and Poisson's ratio.

An important aspect of the theory of elasticity in homogeneous, isotropic material is that only two elastic moduli are needed to describe fully material behavior. Strain can be expressed in terms of applied stress utilizing the following relation

$$\varepsilon_{ij} = \frac{1}{2G}(S_{ij} - \delta_{ij}S_{00}) + \frac{1}{3K}\delta_{ij}S_{00} \tag{3.4}$$

Because only two elastic moduli are needed, it is often convenient to express elastic moduli with respect to each other. For example, if one has knowledge of the bulk modulus and Young's modulus, one can compute the shear modulus from

$$G = \frac{3KE}{9K - E}$$

A fairly complete table of such equivalencies (after Birch 1961) is presented in Table 3.1.

It should be noted that the crack closure associated with initial loading of a sample at very low stress (as illustrated in Figure 3.2) is an example of non-linear elasticity. As the cracks close with increased load the sample becomes stiffer, and as the stress on the sample decreases so does the stiffness of the sample, in a repeatable manner. Hence, stress and strain are proportional and deformation is fully reversible, but stress and strain are not linearly proportional.

Elastic moduli and seismic wave velocity

In an elastic, isotropic, homogeneous solid the elastic moduli also can be determined from the velocity of compressional waves (V_p) and shear waves (V_s) using the following relations

$$V_p = \sqrt{\frac{K + 4G/3}{\rho}} \qquad V_s = \sqrt{\frac{G}{\rho}} \tag{3.5}$$

Table 3.1. *Relationships among elastic moduli in an isotropic material*

K	E	λ	ν	G	M
$\lambda + \dfrac{2G}{3}$	$G\dfrac{3\lambda + 2G}{\lambda + G}$	—	$\dfrac{\lambda}{2(\lambda + G)}$	—	$\lambda + 2G$
—	$9K\dfrac{K - \lambda}{3K - \lambda}$	—	$\dfrac{\lambda}{3K - \lambda}$	$3\dfrac{K - \lambda}{2}$	$3K - 2\lambda$
—	$\dfrac{9KG}{3K + G}$	$K - \dfrac{2G}{3}$	$\dfrac{3K - 2G}{2(3K + G)}$	—	$K + 4\dfrac{G}{3}$
$\dfrac{EG}{3(3G - E)}$	—	$G\dfrac{E - 2G}{3G - E}$	$\dfrac{E}{2G} - 1$	—	$G\dfrac{4G - E}{3G - E}$
—	—	$3K\dfrac{3K - E}{9K - E}$	$\dfrac{3K - E}{6K}$	$\dfrac{3KE}{9K - E}$	$3K\dfrac{3K + E}{9K - E}$
$\lambda\dfrac{1 + \nu}{3\nu}$	$\lambda\dfrac{(1 + \nu)(1 - 2\nu)}{\nu}$	—	—	$\lambda\dfrac{1 - 2\nu}{2\nu}$	$\lambda\dfrac{1 - \nu}{\nu}$
$G\dfrac{2(1 + \nu)}{3(1 - 2\nu)}$	$2G(1 + \nu)$	$G\dfrac{2\nu}{1 - 2\nu}$	—	—	$G\dfrac{2 - 2\nu}{1 - 2\nu}$
—	$3K(1 - 2\nu)$	$3K\dfrac{\nu}{1 + \nu}$	—	$3K\dfrac{1 - 2\nu}{2 + 2\nu}$	$3K\dfrac{1 - \nu}{1 + \nu}$
$\dfrac{E}{3(1 - 2\nu)}$	—	$\dfrac{E\nu}{(1 + \nu)(1 - 2\nu)}$	—	$\dfrac{E}{2 + 2\nu}$	$\dfrac{E(1 - \nu)}{(1 + \nu)(1 - 2\nu)}$

It is obvious from these relations that V_p is always greater than V_s (when $\nu = 0.25$, $V_p/V_s = \sqrt{3} = 1.73$) and that $V_s = 0$ in a fluid.

It is also sometimes useful to consider relative rock stiffnesses directly as determined from seismic wave velocities. For this reason the so-called M modulus has been defined:

$$M = V_p^2 \rho = K + \frac{4G}{3}$$

Poisson's ratio can be determined from V_p and V_s utilizing the following relation

$$\nu = \frac{V_p^2 - 2V_s^2}{2\left(V_p^2 - V_s^2\right)} \tag{3.6}$$

Because we are typically considering porous sedimentary rocks saturated with water, oil or gas in this book, it is important to recall that poroelastic effects result in a frequency dependence of seismic velocities (termed *dispersion*), which means that elastic moduli are frequency dependent. This is discussed below in the context of poroelasticity and viscoelasticity.

Elasticity anisotropy

A number of factors can make a rock mass anisotropic – aligned microcracks (Hudson 1981), high differential stress (Nur and Simmons 1969), aligned minerals (such as mica and clay) (Sayers 1994) along bedding planes (Thomsen 1986), macroscopic fractures and faults (Mueller 1991). Elastic anisotropy can have considerable effects on seismic wave velocities, and is especially important with respect to shear wave propagation. Although a number of investigators have argued that stress orientation can be determined uniquely from shear velocity anisotropy (*e.g.* Crampin 1985), in many cases it is not clear whether shear velocity anisotropy in a volume of rock is correlative with the current stress state or the predominant orientation of fractures in a rock (*e.g.* Zinke and Zoback 2000). Because of this, measurements of shear velocity anisotropy must be used with care in stress mapping endeavors. That said, shear wave velocity anisotropy measured in vertical wellbores often does correlate with modern stress orientations (Yale 2003; Boness and Zoback 2004). We discuss this at greater length in Chapter 8.

With respect to elastic anisotropy, the general formulation that relates stress to strain is

$$S_{ij} = c_{ijkl}\varepsilon_{kl} \tag{3.7}$$

where c_{ijkl}, the elastic stiffness matrix, is a fourth-rank tensor with 81 constants and summation is implied over repeated subscripts k and l. It is obviously not tractable to consider wave propagation through a medium that has to be defined by 81 elastic constants. Fortunately, symmetry of the stiffness matrix (and other considerations) reduces this tensor to 21 constants for the general case of an anistropic medium. Even more fortunately, media that have some degree of symmetry require even fewer elastic constants. As mentioned above, an isotropic material is defined fully by two constants, whereas a material with cubic symmetry is fully described by three constants, and a material characterized by transverse isotropy (such as a finely layered sandstone or shale layer) is characterized by five constants, and so on. Readers interested in wave propagation in rocks exhibiting weak transverse anisotropy are referred to Thomsen (1986) and Tsvankin (2001).

Elastic anisotropy is generally not very important in geomechanics, although, as noted above, shear wave velocity anisotropy can be related to principal stress directions or structural features. On the other hand, anisotropic rock strength, due, for example, to the presence of weak bedding planes, has a major affect on wellbore stability as is discussed both in Chapter 4 and Chapter 10.

Poroelasticity and effective stress

In a porous elastic solid saturated with a fluid, the theory of poroelasticity describes the constitutive behavior of rock. Much of poroelastic theory derives from the work of

Biot (1962). This is a subject dealt with extensively by other authors (*e.g.* Kuempel 1991; Wang 2000) and the following discussion is offered as a brief introduction to this topic.

The three principal assumptions associated with this theory are similar to those used for defining pore pressure in Chapter 2. First, there is an interconnected pore system uniformly saturated with fluid. Second, the total volume of the pore system is small compared to the volume of the rock as a whole. Third, we consider pressure in the pores, the total stress acting on the rock externally and the stresses acting on individual grains in terms of statistically averaged uniform values.

The concept of effective stress is based on the pioneering work in soil mechanics by Terzaghi (1923) who noted that the behavior of a soil (or a saturated rock) will be controlled by the effective stresses, the differences between externally applied stresses and internal pore pressure. The so-called "simple" or Terzaghi definition of effective stress is

$$\sigma_{ij} = S_{ij} - \delta_{ij} P_{\mathrm{p}} \tag{3.8}$$

which means that pore pressure influences the normal components of the stress tensor, $\sigma_{11}, \sigma_{22}, \sigma_{33}$ and not the shear components $\sigma_{12}, \sigma_{23}, \sigma_{13}$. Skempton's coefficient, B, is defined as the change in pore pressure in a rock, ΔP_{p}, resulting from an applied pressure, S_{00} and is given by $B = \Delta P_{\mathrm{p}}/S_{00}$.

In the context of the cartoon shown in Figure 3.5a, it is relatively straightforward to see that the stresses acting on individual grains result from the difference between the externally applied normal stresses and the internal fluid pressure. If one considers the force acting at a single grain contact, for example, all of the force acting on the grain is transmitted to the grain contact. Thus, the force balance is

$$F_{\mathrm{T}} = F_{\mathrm{g}}$$

which, in terms of stress and area, can be expressed as

$$S_{ii} A_{\mathrm{T}} = A_{\mathrm{c}} \sigma_{\mathrm{c}} + (A_{\mathrm{T}} - A_{\mathrm{c}}) P_{\mathrm{P}}$$

where A_{c} is the contact area of the grain and σ_{c} is the effective normal stress acting on the grain contact. Introducing the parameter $a = A_{\mathrm{c}}/A_{\mathrm{T}}$, this is written as

$$S_{ii} = a\sigma_{\mathrm{c}} + (1 - a) P_{\mathrm{P}}$$

The intergranular stress can be obtained by taking the limit where a becomes vanishingly small

$$\lim_{a \to 0} a\sigma_{\mathrm{c}} = \sigma_{\mathrm{g}}$$

such that the "effective" stress acting on individual grains, σ_{g}, is given by

$$\sigma_{\mathrm{g}} = S_{ii} - (1 - a) P_{\mathrm{p}} = S_{ii} - P_{\mathrm{P}} \tag{3.9}$$

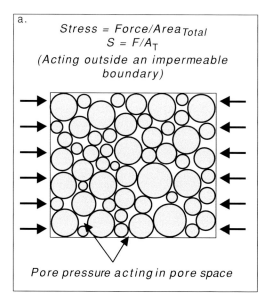

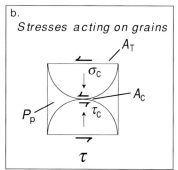

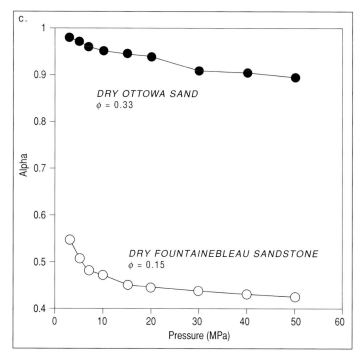

Figure 3.5. (a) Schematic illustration of a porous solid with external stress applied outside an impermeable boundary and pore pressure acting within the pores. (b) Considered at the grain scale, the force acting at the grain contact is a function of the difference between the applied force and the pore pressure. As A_c/A_T goes to zero, the stress acting on the grain contacts is given by the Terzaghi effective stress law (see text). (c) Laboratory measurements of the Biot coefficient, α, for a porous sand and well-cemented sandstone courtesy J. Dvorkin.

for very small contact areas. It is clear in Figure 3.5b that pore fluid pressure does not affect shear stress components, S_{ij}.

Empirical data have shown that the effective stress law is a useful approximation which works well for a number of rock properties (such as intact rock strength and the frictional strength of faults as described in Chapter 4), but for other rock properties, the law needs modification. For example, Nur and Byerlee (1971) proposed an "exact" effective stress law, which works well for volumetric strain. In their formulation

$$\sigma_{ij} = S_{ij} - \delta_{ij}\alpha P_{\mathrm{p}} \qquad (3.10)$$

where α is the Biot parameter

$$\alpha = 1 - K_{\mathrm{b}}/K_{\mathrm{g}}$$

and K_{b} is drained bulk modulus of the rock or aggregate and K_{g} is the bulk modulus of the rock's individual solid grains. It is obvious that $0 \leq \alpha \leq 1$. For a nearly *solid* rock with no interconnected pores (such as quartzite),

$$\lim_{\phi \to 0} \alpha = 0$$

such that pore pressure has no influence on rock behavior. Conversely, for a highly porous, compliant formation (such as uncemented sands)

$$\lim_{\phi \to 1} \alpha = 1$$

and pore pressure has maximum influence. Figure 3.5c shows measured values of the Biot parameter for two materials: a compliant unconsolidated sand in which α is high and a dry, well-cemented sandstone in which α has intermediate values (J. Dvorkin, written communication). In both cases, α decreases moderately with confining pressure. Hofmann (2006) has recently compiled values for α for a wide range of rocks.

Thus, to consider the effect of pore fluids on stress we can re-write equation (3.3) as follows

$$S_{ij} = \lambda \delta_{ij}\varepsilon_{00} + 2G\varepsilon_{ij} - \alpha \delta_{ij} P_0 \qquad (3.11)$$

such that the last term incorporates pore pressure effects.

The relation of compressive and tensile rock strength to effective stress will be discussed briefly in Chapter 4. With respect to fluid transport, both Zoback and Byerlee (1975) and Walls and Nur (1979) have shown that permeabilities of sandstones containing clay minerals are more sensitive to pore pressure than confining pressure. This results in an effective stress law for permeability in which another empirical constant replaces that in equation (3.10). This constant is generally ≥ 1 for sandstones (Zoback and Byerlee 1975) and appears to depend on clay content (Walls and Nur 1979). More recently, Kwon, Kronenberg *et al.* (2001) have shown that this effect breaks down in shales with extremely high clay content. For such situations, permeability seems to depend on the simple form of the effective stress law (equation 3.8) because there is no stiff rock matrix to support externally applied stresses. Other types of effective stress laws describe the dependence of rock permeability on external "confining" pressure and internal pore pressure.

Poroelasticity and dispersion

As mentioned above, the stiffness (elastic moduli) of a poroelastic rock is rate dependent. In regard to seismic wave propagation, this means that P-wave and S-wave velocities will be frequency dependent. Figure 3.6a illustrates the difference between laboratory bulk modulus measurements of an uncemented Gulf of Mexico sand determined statically, and using ultrasonic ($\sim$1 MHz) laboratory velocity measurements. Note that at low confining pressure, there is about a factor of 2 difference between the moduli determined the two different ways. As confining pressure increases the difference increases significantly. Thus, there can be significant differences in velocity (or the elastic modulus) depending on the frequency of seismic waves. Seismic-wave frequencies typical of a reflection seismic measurement ($\sim$10–50 Hz) are slower (yield lower moduli) than sonic logs (typically $\sim$10 kHz), and sonic logs yield slower velocities than ultrasonic laboratory measurements (typically $\sim$1 MHz). As illustrated in Figure 3.6b, this effect is much more significant for P-wave velocity than S-wave velocity.

Figure 3.7a (after Zimmer 2004) clearly demonstrates the difference between static and dynamic bulk modulus in an uncemented Gulf of Mexico sand. As shown by the hydrostatic loading cycles, the static stiffness (corresponding to the slope of the loading line) is much lower than the dynamic stiffness (indicated by the slope of the short lines) determined from ultrasonic velocity measurements (see expanded scale in Figure 3.7b). Upon loading, there is both elastic and inelastic deformation occurring whereas upon unloading, the initial slope corresponds to mostly elastic deformation. Hence, the unloading stiffnesses (as illustrated in Figure 3.2) are quite similar to the dynamically measured stiffnesses during loading.

There are a number of different processes affecting seismic wave propagation that contribute to the effects shown in Figure 3.6. First, the seismic waves associated with seismic reflection profiling, well logging and laboratory studies sample very different volumes of rock. Second, when comparing static measurements with ultrasonic measurements, it is important to remember that the amount of strain to which the samples are subjected is markedly different, which can affect the measurement stiffness (Tutuncu, Podio *et al.* 1998a,b). Finally, pore fluid effects can contribute dramatically to dispersion at high frequencies. SQRT (squirt, or local flow) is a theory used to explain the dependence of wave velocity on frequency in a saturated poroelastic rock at high frequency (see the review by Dvorkin, Mavko *et al.* 1995). Fundamentally, SQRT (and theories like it) calculates the increase in rock stiffness (hence the increase

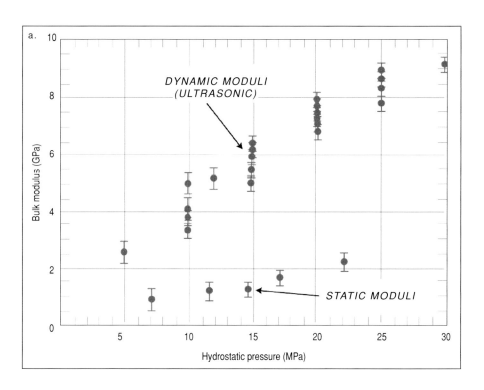

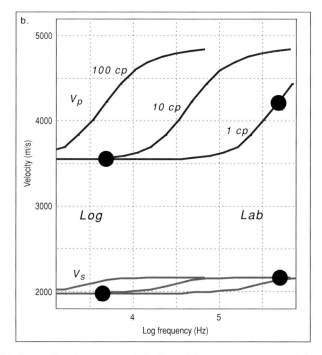

Figure 3.6. (a) Comparison between static bulk modulus measurements and dynamic (ultrasonic) measurements for a dry, uncemented sand as a function of pressure. (b) Comparison between P- and S-wave velocity measurements in a water saturated sample at frequencies corresponding to geophysical logs (several kilohertz) and laboratory measurements (∼ one megahertz). Note the significant difference for P-wave velocity. The curves correspond to SQRT theory (Dvorkin, Mavko *et al.* 1995). Note that the theory explains the increase in velocity with frequency between log and lab measurements for water in the pores of the rock.

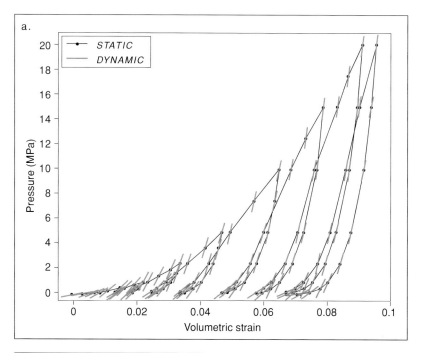

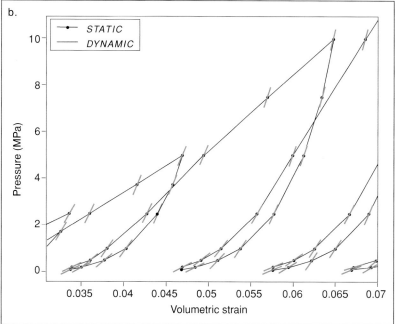

Figure 3.7. (a) Hydrostatic loading and unloading cycles of a saturated, uncemented Gulf of Mexico sand that shows the clear difference between static and dynamic stiffness, especially for loading cycles where both elastic and inelastic deformation is occurring (after Zimmer 2004). (b) An expanded view of several of the cycles shown in (a).

in velocity) associated with the amount of pressure "carried" by pore fluids as seismic waves pass through rock. At very high (ultrasonic) frequencies, there is insufficient time for localized fluid flow to dissipate local pressure increases. Hence, the rock appears quite stiff (corresponding to the undrained modulus and fast ultrasonic P-wave velocities as measured in the lab) because the pore fluid pressure is contributing to the stiffness of the rock. Conversely, at relatively low (seismic or well logging) frequencies, the rock deforms with a "drained" modulus. Hence, the rock is relatively compliant (relatively slow P-wave velocities would be measured *in situ*). It is intuitively clear why the permeability of the rock and the viscosity of the fluid affect the transition frequency from drained to undrained behavior. This is illustrated in Figure 3.6b. SQRT theory predicts the observed dispersion for a viscosity of 1 cp, which is appropriate for the water filling the pores of this rock. Had there been a more viscous fluid in the pores (or if the permeability of the rock was lower), the transition frequency would shift to lower frequencies, potentially affecting velocities measured with sonic logging tools ($\sim 10^4$ Hz). This type of phenomenon, along with related issues of the effect of pore fluid on seismic velocity (the so-called fluid substitution effect), are discussed at length by Mavko, Mukerjii *et al*. (1998), Bourbie, Coussy *et al*. (1987) and other authors.

Viscous deformation in uncemented sands

Although cemented sedimentary rocks tend to behave elastically over a range of applied stresses (depending on their strength), uncemented sands and immature shales tend to behave viscously. In other words, they deform in a time-dependent manner, or creep, as illustrated schematically in Figure 3.1d. Such behavior has been described in several studies (*e.g.* de Waal and Smits 1988; Dudley, Meyers *et al*. 1994; Ostermeier 1995; Tutuncu, Podio *et al*. 1998a; Tutuncu, Podio *et al*. 1998b). In this section we will concentrate on the behavior of the uncemented turbidite sand from a reservoir at a depth of about 1 km in the Wilmington field in Long Beach, California. Chang, Moos *et al*. (1997) discussed viscoelastic deformation of the Wilmington sand utilizing a laboratory test illustrated in Figure 3.8a (from Hagin and Zoback 2004c). Dry samples (traces of residual hydrocarbons were removed prior to testing) were subjected to discrete steps of hydrostatic confining pressure. After several pressure steps were applied, loading was stopped and the sample was allowed to creep. After this creeping period, the sample was partially unloaded, then reloaded to a higher pressure level and allowed to creep again. Note that after each loading step, the sample continued to strain at constant confining pressure and there was appreciable permanent deformation at the conclusion of an experiment. One point of note that will be important when we consider viscoplastic compaction of reservoirs in Chapter 12 is that viscous effects are not seen until the pressure exceeds the highest pressure previously experienced by a sample. In other words, viscous compaction will only be important when depletion results in an effective

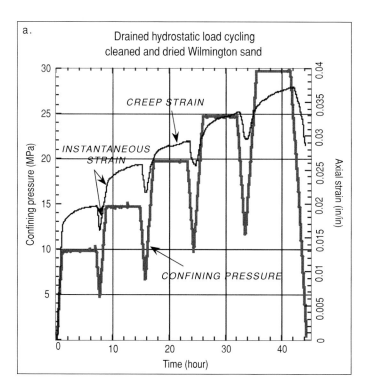

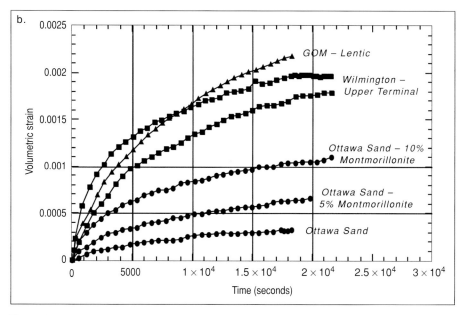

Figure 3.8. (a) Incremental instantaneous and creep strains corresponding to 5 MPa incremental increases in pressure. The data plotted at each pressure reflect the increases in strain that occurred during each increase in pressure. Note that above 15 MPa the incremental creep strain is the same magnitude as the incremental instantaneous strain from Hagin and Zoback (2004b). (b) Creep strain is dependent upon the presence of clay minerals and mica. In the synthetic samples, the amount of creep is seen to increase with clay content. The clay content of the Wilmington sand samples is approximately 15% (modified from Chang, Moos *et al.* 1997). *Reprinted with permission of Elsevier.*

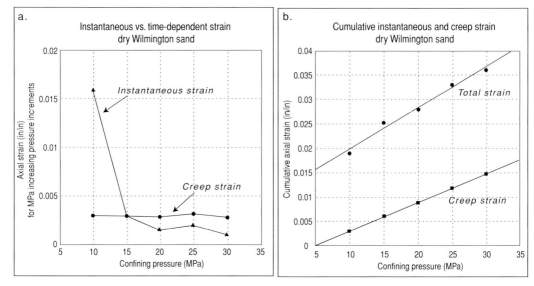

Figure 3.9. (a) Incremental instantaneous and creep strains corresponding to 5 MPa incremental increases in pressure. The data plotted at each pressure reflect the increases in strain that occurred during each increase in pressure. Note that above 15 MPa the incremental creep strain is the same magnitude as the incremental instantaneous strain (from Hagin and Zoback 2004b). The cumulative instantaneous and total (instantaneous plus creep) volumetric strain as a function of pressure. Note that above 10 MPa, both increase by the same amount with each increment of pressure application.

stress that is greater than the sample experienced *in situ*. However, as will be shown below, once the previous highest load experienced has been exceeded, viscoplastic compaction can be quite appreciable. One would dramatically underpredict reservoir compaction from laboratory experiments on uncemented sands if one were to neglect viscoplastic effects.

   In an attempt to understand the physical mechanism responsible for the creep in these samples, Figure 3.8b illustrates an experiment by Chang, Moos *et al.* (1997) that compares the time-dependent strain of Wilmington sand (both dry and saturated) with Ottawa sand, a commercially available laboratory standard that consists of pure, well-rounded quartz grains. Note that in both the dry and saturated Wilmington sand samples, a 5 MPa pressure step at 30 MPa confining pressure results in a creep strain of 2% after 2×10^4 sec (5.5 hours). In the pure Ottawa samples, very little creep strain is observed, but when 5% and 10% montmorillonite was added, respectively, appreciably more creep strain occurred. Thus, the fact that the grains in the sample were uncemented to each other allowed the creep to occur. The presence of montmorillonite clay enhances this behavior. The Wilmington samples are composed of ~20% quartz, ~20% feldspar, 20% crushed metamorphic rocks, 20% mica and 10% clay. Presumably both the clay and mica contribute to the creep in the Wilmington sand.

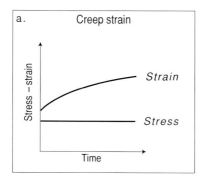

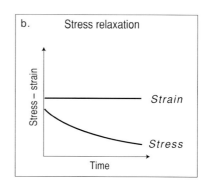

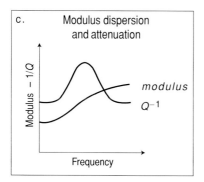

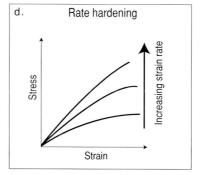

Figure 3.10. Time-dependent deformation in a viscoelastic material is most commonly observed as (a) creep strain or (b) stress relaxation. (c) Linear viscoelasticity theory predicts frequency- and rate-dependence for materials that exhibit time-dependence. Specifically, elastic modulus and attenuation (Q^{-1}) should vary with loading frequency, and (d) stiffness should increase with strain rate. From Hagin and Zoback (2004b).

Figure 3.9 (after Hagin and Zoback 2004b) illustrates a set of experiments that illustrate just how important creep strain is in this type of reservoir sand. Note that after the initial loading step to 10 MPa, the creep strain that follows each loading step is comparable in magnitude to the strain that occurs instantaneously (Figure 3.9a). The cumulative strain (Figure 3.10b) demonstrates that the creep strain accumulates linearly with pressure.

Figure 3.10 summarizes four different ways in which viscoelastic deformation manifests itself in laboratory testing, and presumably in nature. As already noted, a viscous material strains as a function of time in response to an applied stress (Figure 3.10a), and differential stress relaxes at constant strain (Figure 3.10b). In addition, the elastic moduli are frequency dependent (the seismic velocity of the formation is said to be *dispersive*) and there is marked inelastic attenuation. Q is defined as the seismic *quality factor* such that inelastic attenuation is defined as Q^{-1} (Figure 3.10c). Finally, a stress–strain test (such as illustrated in Figure 3.2) is dependent on strain rate (Figure 3.10d) such that the material seems to be both stiffer and stronger when deformed at higher rates.

The type of behavior schematically illustrated in Figure 3.10a is shown for Wilmington sand in Figure 3.8a,b and the type of behavior schematically illustrated in Figure 3.10b (stress relaxation at constant strain) is shown for Wilmington sand in Figure 3.11a. A sample was loaded hydrostatically to 3 MPa before an additional axial stress of 27 MPa was applied to the sample in a conventional triaxial apparatus (see Chapter 4). After loading, the length of the sample (the axial strain) was kept constant. Note that as a result of creep, the axial stress relaxed from 30 MPa to 10 MPa over a period of $\sim$10 hours. An implication of this behavior for unconsolidated sand reservoirs *in situ* is that very small differences between principal stresses are likely to exist. Even in an area of active tectonic activity, applied horizontal forces will dissipate due to creep in unconsolidated formations.

The type of viscous behavior is illustrated schematically in Figure 3.10d, and the rate dependence of the stress–strain behavior is illustrated in Figure 3.11b for Wilmington sand (after Hagin and Zoback 2004b). As expected, the sample is stiffer at a confining pressure of 50 MPa than it is at 15 MPa and at each confining pressure, the samples are stiffer and stronger at a strain rate of 10^{-5} sec^{-1}, than at 10^{-7} sec^{-1}.

The dispersive behavior illustrated in Figure 3.10c can be seen for dry Wilmington sand in Figure 3.11c (Hagin and Zoback 2004b). The data shown comes from a test run at 22.5 MPa hydrostatic pressure with a 5 MPa pressure oscillation. Note the dramatic dependence of the normalized bulk modulus with frequency. At the frequencies of seismic waves (10–100 Hz) and higher sonic logging frequencies of 10^4 Hz and ultrasonic lab frequencies of $\sim$$10^6$ Hz, a constant stiffness is observed. However, when deformed at very low frequencies (especially at $<10^{-3}$ Hz), the stiffness is dramatically lower. Had there been fluids present in the sample, the bulk modulus at ultrasonic frequencies would have been even higher than that at seismic frequencies. The bulk modulus increases to the Gassmann static limit at approximately 0.1 Hz and then stays constant as frequency is increased through 1 MHz. The Gassmann static limit is explained by Mavko, Mukerjii *et al.* (1998). While our experiments were conducted on dry samples, we have included the effects of poroelasticity in this diagram by including the predicted behavior of oil-saturated samples according to SQRT theory (Dvorkin, Mavko *et al.* 1995).

Because viscous deformation manifests itself in many ways, and because it is important to be able to predict the behavior of an unconsolidated reservoir sand over decades of depletion utilizing laboratory measurements made over periods of hours to days, it would be extremely useful to have a constitutive law that accurately describes the long-term formation behavior. Hagin and Zoback (2004c) discuss a variety of idealized viscous constitutive laws in terms of their respective creep responses at constant stress, the modulus dispersion and attenuation. The types of idealized models they considered

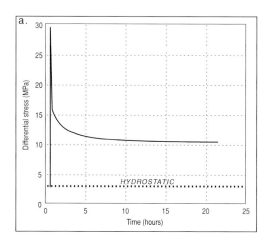

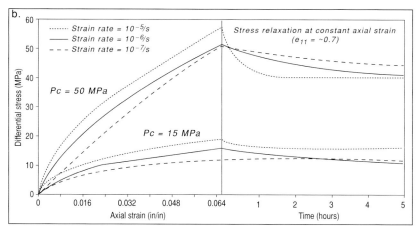

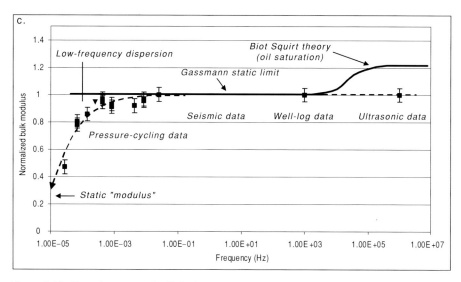

Figure 3.11. Experiments on dry Wilmington sand that illustrate different kinds of viscous deformation (modified from Hagin and Zoback 2004b). (a) Relaxation of stress after application of a constant strain step (similar to what is shown schematically in Figure 3.10b). (b) The strain rate dependence of sample stiffness (as illustrated in Figure 3.10d). (c) Normalized bulk modulus of dry Wilmington sand as a function of frequency spanning 10 decades (as shown schematically in Figure 3.10c).

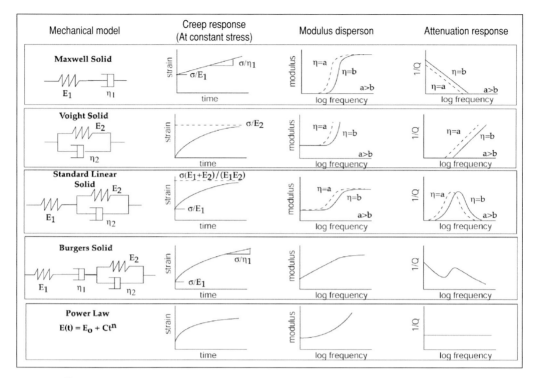

Figure 3.12. Conceptual relationships between creep, elastic stiffness, and attenuation for different idealized viscoelastic materials. Note that the creep strain curves are all similar functions of time, but the attenuation and elastic stiffness curves vary considerably as functions of frequency. From Hagin and Zoback (2004b).

are illustrated in Figure 3.12. It is important to note that if one were simply trying to fit the creep behavior of an unconsolidated sand such as shown in Figure 3.8b, four of the constitutive laws shown in Figure 3.12 have the same general behavior and could be adjusted to fit the data.

Hagin and Zoback (2004b) independently measured dispersion and attenuation and thus showed that a power-law constitutive law (the last idealized model illustrated in Figure 3.12) appears to be most appropriate. Figure 3.13a shows their dispersion measurements for unconsolidated Wilmington sand (shown previously in Figure 3.11c) as fit by three different constitutive laws. All three models fit the dispersion data at intermediate frequencies, although the Burger's model implies zero stiffness under static conditions, which is not physically plausible. Figure 3.13b shows the fit of various constitutive laws to the measured attenuation data. Note that attenuation is ~ 0.1 ($Q \sim 10$) over almost three orders of frequency and only the power-law rheology fits the essentially constant attenuation over the frequency range measured. More importantly, the power-law constitituve law fits the dispersion data, and its static value (about 40%

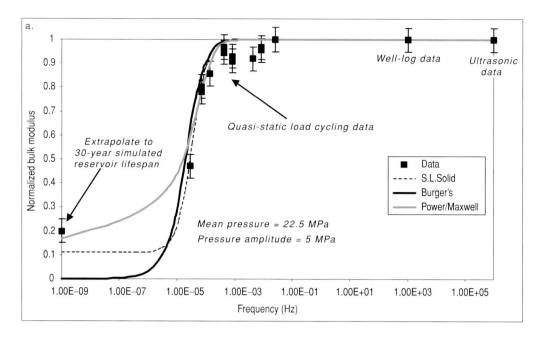

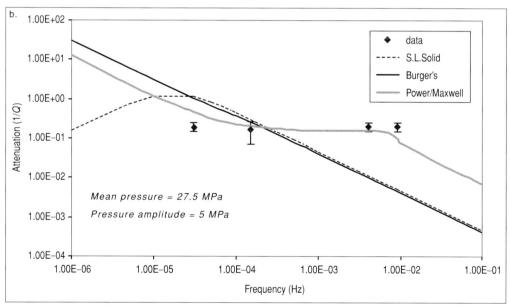

Figure 3.13. (a) Low-frequency bulk modulus dispersion predicted using parameters derived from fitting the creep strain curves compared with experimental results. The standard linear solid and Burger's solid models provide the best fit to the quasi-static data, but fail to predict reasonable results at time-periods associated with the life of a reservoir. The power law–Maxwell mixed model reproduces the overall trend of the data, and predicts a physically reasonable value of stiffness at long time-periods. The error bars represent the inherent uncertainty of the measurements (from Hagin and Zoback (2004c). (b) Low-frequency attenuation predicted using the parameters derived from the creep strain data compared with experimental results.

of the high-frequency limit) matches the compaction observed in the field. Assuming complete depletion of the producing reservoir over its ~30 year history (prior to water flooding and pressure support) results in a predicted total vertical compaction of 1.5%. This value matches closely with well-casing shortening data from the reservoir, which indicates a total vertical compaction of 2% (Kosloff and Scott 1980).

The power-law model that fits the viscous deformation data best has the form

$$\varepsilon(P_c, t) = \varepsilon_0 \left(1 + \frac{P_c}{15} t^{0.1} \right) \tag{3.12}$$

where ε_0 is the instantaneous volumetric strain, P_c is the confining pressure, and t is the time in hours. In order to complete the constitutive law, we need to combine our model for the time-dependent deformation with a model for the instantaneous deformation. Hagin and Zoback (2004) show that the instantaneous volumetric strain is also a power-law function of confining pressure, and can be described empirically with the following equation:

$$\varepsilon_0 = 0.0083 P_c^{0.54} \tag{3.13}$$

Combining the two equations results in a constitutive equation for Wilmington sand in which the volumetric strain depends on both pressure and time:

$$\varepsilon(P_c, t) = 0.0083 P_c^{0.54} \left(1 + \frac{P_c}{15} t^{0.1} \right) \tag{3.14}$$

Hence, this is a dual power-law constitutive law. Strain is a function of both confining pressure raised to an empirically determined exponent (0.54 for Wilmington sand) and time raised to another empirically determined exponent (0.1 for Wilmington sand). Other workers have also reached the conclusion that a power-law constitutive law best describes the deformation behavior of these types of materials (de Waal and Smits 1988; Dudley, Meyers *et al.* 1994).

By ignoring the quasi-static and higher frequency data, these dual power-law constitutive laws can be simplified, and the terms in the model needed to model the quasi-static data in Figure 3.13a can be eliminated (Hagin and Zoback 2007). In fact, by focusing on long-term depletion, the seven-parameter best-fitting model for Wilmington sand proposed by Hagin and Zoback (2004c) can be simplified to a three-parameter model without any loss of accuracy when considering long-term effects.

The model assumes that the total deformation of unconsolidated sands can be decoupled in terms of time. Thus, the instantaneous (time-independent) elastic–plastic component of deformation is described by a power-law function of pressure, and the viscous (time-dependent) component is described by a power-law function of time. The proposed model has the following form (written in terms of porosity for simplicity):

$$\phi(P_c, t) = \phi_i - (P_c/A) t^b \tag{3.15}$$

where the second term describes creep compaction normalized by the pressure and the first term describes the instantaneous compaction:

$$\phi_i = \phi_0 P_c^d \tag{3.16}$$

which leaves three unknown constants, A, b, and d where ϕ_0 is the initial porosity.

Apart from mathematical simplicity, this model holds other advantages over the model proposed by Hagin and Zoback (2004c). While the previous model required data from an extensive set of laboratory experiments in order to determine all of the unknown parameters, the proposed model requires data from only two experiments. The instantaneous parameters can be solved for by conducting a single constant strain-rate test, while the viscous parameters can be derived from a single creep strain test conducted at a pressure that exceeds the maximum *in situ* stress in the field (Hagin and Zoback 2007).

Predicting the long-term compaction of the reservoir from which the samples were taken can now be accomplished using the following equation:

$$\phi(P_c, t) = 0.272 P_c^{-0.046} - \left(\frac{P_c}{5410} t^{0.164} \right) \tag{3.17}$$

The first term of equation (3.17) represents the instantaneous porosity as a function of effective pressure, with ϕ_0 equal to 0.27107 and the parameter d equal to -0.046. The second term describes creep compaction normalized by effective pressure, with the parameters A equal to 5410 and b equal to 0.164.

Hagin and Zoback (2007) tested an uncemented sand from the Gulf of Mexico using experimental conditions very similar to those for the Wilmington sample, except that in this case the effective pressure was increased to 30 MPa to reflect the maximum *in situ* effective stress in the reservoir. They found that the GOM sand constant strain-rate data could be fit with the following function:

$$\phi_i = 0.2456 P_c^{-0.1518} \tag{3.18}$$

where ϕ_i is the instantaneous porosity, P_c is effective pressure, and the intercept of the equation is taken to be the initial porosity (measured gravimetrically).

The creep compaction experimental procedure used to determine the time-dependent model parameters for the GOM sand sample was also similar to that used for the Wilmington sand. Hagin and Zoback (2007) found that the creep compaction data are described quite well by the following equation:

$$\phi(P_c, t) = 0.0045105 t^{0.2318} \tag{3.19}$$

where ϕ is the porosity, P_c is the effective pressure, and t is the time in days. For details on how to determine the appropriate length of time for observing creep compaction, see Hagin and Zoback (2007).

The total deformation for this Gulf of Mexico sample can now be described in terms of porosity, effective pressure and time by combining equations (3.18) and (3.19) into the following:

$$\phi(P_c, t) = 0.2456 P_c^{-0.1518} - (P_c/6666.7)t^{0.2318} \tag{3.20}$$

As before, the first term of equation (3.20) represents the instantaneous component of deformation, with ϕ_0 equal to 0.2456 and the parameter d equal to -0.1518. The second term describes the time-dependent component of deformation, with the parameters A equal to 6667 and b equal to 0.2318. Assuming complete drawdown of the producing reservoir and an approximately 30 year history results in a predicted total vertical compaction of nearly 10%.

Table 3.2 summarizes the fitting parameters obtained from the creep strain tests and constant strain rate tests described earlier. The parameters to make note of are the exponent parameters, b and d. The apparent viscosity of a reservoir sand is captured in

Table 3.2. *Creep parameters for two uncemented sands*

Reservoir sand	A (creep)	b (creep)	ϕ_0 (instant)	d (instant)	Notes
Wilmington	5410.3	0.1644	0.271	−0.046	Stiffer and more viscous
GOM – Field X	6666.7	0.2318	0.246	−0.152	Softer and less viscous

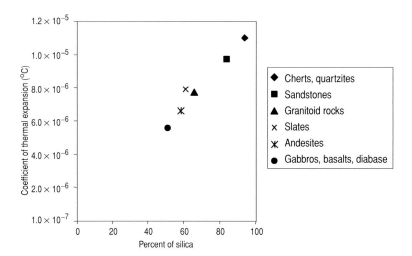

Figure 3.14. Measurements of the coefficient of linear thermal expansion for a variety of rocks as a function of the percentage of silica (data from Griffith 1936). As the coefficient of thermal expansion of silica ($\sim 10^{-5}$ °C^{-1}) is an order of magnitude higher than that of most other rock forming minerals ($\sim 10^{-6}$ °C^{-1}), the coefficient of thermal expansion ranges between those two amounts, depending on the percentage of silica.

the *b* parameter, and smaller values of *b* represent greater viscosities. Thus, Wilmington sand is more viscous than the sand from the Gulf of Mexico. The effective pressure exponent *d* represents compliance, with smaller values being stiffer. Thus, Wilmington sand is stiffer than the GOM sand. Note that stiffness is not related to *d* in a linear way, because strain and stress are related via a power law.

We will revisit the subject of viscoplastic compaction in weak sand reservoirs in Chapter 12 and relate this phenomenon to the porosity change accompanying compaction of the Wilmington reservoir in southern California and a field in the Gulf of Mexico.

Thermoporoelasticity

Because thermoporoelastic theory considers the effects of both pore fluids and temperature changes on the mechanical behavior of rock, it could be utilized as a generalized theory that might be applied generally to geomechanical problems. For most of the problems considered in this book, this application is not necessary as thermal effects are of relatively minor importance. However, as will be noted in Chapters 6 and 7, theromoporoelastic effects are sometimes important when considering wellbore failure in compression and tension and we will consider it briefly in that context.

Fundamentally, thermoporoelastic theory allows one to consider the effect of temperature changes on stress and strain. To consider the effect of temperature on stress, equation (3.21) is the equivalent of equations (3.11) where the final term represents the manner in which a temperature change, ΔT, induces stress in a poroelastic body:

$$S_{ij} = \lambda \delta_{ij} \varepsilon_{00} + 2G\varepsilon_{ij} - \alpha_T \delta_{ij} P_0 - K\alpha_T \delta_{ij} \Delta T \tag{3.21}$$

where $\alpha_T = \dfrac{1}{L}\dfrac{\delta L}{\delta T}$ is the coefficient of linear thermal expansion and defines the change in length, L, of a sample in response to a change in temperature δT.

Figure 3.14 shows the magnitude of α_T for different rocks as a function of quartz content (Griffith 1936). Because quartz has a much larger coefficient of thermal expansion than other common rock forming minerals, the coefficient of expansion of a given rock is proportional to the amount of quartz. To put this in a quantitative perspective, changes in temperature of several tens of °C can occur around wellbores during drilling in many reservoirs (much more in geothermal reservoirs or steam floods, of course), which has non-negligible stress changes around wellbores and implications for wellbore failure as discussed quantitatively in Chapters 6 and 7.

4 Rock failure in compression, tension and shear

In this chapter I review a number of fundamental principles of rock failure in compression, tension and shear that provide a foundation for many of the topics addressed in the chapters that follow. The first subject addressed below is the classical subject of rock strength in compression. While much has been written about this, it is important to review basic types of strength tests, the use of Mohr failure envelopes to represent rock failure as a function of confining stress and the ranges of strength values found for the rock types of interest here. I also discuss the relationship between rock strength and effective stress as well as a number of the strength criteria that have been proposed over the years to describe rock strength under different loading conditions. I briefly consider rock strength anisotropy resulting from the presence of weak bedding planes in rock, which can be an important factor when addressing problems of wellbore instability. This is discussed in the context of two specific case studies in Chapter 10.

In this chapter I also discuss empirical techniques for estimating rock strength from elastic moduli and porosity data obtained from geophysical logs. In practice, this is often the only way to estimate strength in many situations due to the absence of core for laboratory tests. This topic will be of appreciable interest in Chapter 10 when I address issues related to wellbore stability during drilling. I also discuss a specialized form of compressive failure – that associated with pore collapse, sometimes referred to as *shear-enhanced compaction* or *end-cap* failure. This form of rock failure will be revisited in Chapter 12 when compaction associated with depletion is addressed in some detail. I then go on to discuss rock strength in tension. Because the tensile strength of rock is quite low (and because tensile stress only acts as short-term transients in the earth at depth), this subject is principally of interest in hydraulic fracturing in formations and drilling-induced tensile wellbore failures that form in the wellbore wall (Chapter 6).

The final subjects I discuss in this chapter are related to shear sliding and the frictional strength of faults. These topics are important in a number of ways. First, I show how the shear strength of pre-existing faults constrains *in situ* stress magnitudes in the crust. These constraints will be further refined by combining with direct measurements of the least principal stress through some form of hydraulic fracturing in Chapter 6 and with observations of compressive and tensile wellbore failure in Chapters 7 and 8.

Not many years ago, the concept of a critically stressed crust with stress magnitudes controlled by the frictional strength of the crust (as discussed in this chapter) may have seemed unrelated to hydrocarbon development. It is clear today, however, that fault slip can be induced during both fluid injection and production, slip on faults may be responsible for damage to cased production wells in some fields (requiring many wells to be redrilled) and it is becoming common practice in some areas to try to intentionally induce micro-seismicity to enhance permeability in low-permeability formations. Hence, knowing that relatively small perturbations of stress may induce fault slip is important in field development in many regions. Second, slip on faults can be a source of wellbore instability (Chapters 10 and 12) and third, active faults provide efficient conduits for fluid flow in fractured reservoirs and influence the seal capacity of reservoir bounding faults (Chapter 11).

Before discussing rock strength, it is helpful to start with a clear definition of terms and common test procedures as illustrated in Figure 4.1.

- Hydrostatic compression tests ($S_1 = S_2 = S_3 = S_0$) are those in which a sample is subjected to a uniform confining pressure, S_0. Such tests yield information about rock compressibility and the pressure at which pore collapse (irreversible porosity loss) occurs. Use of an impermeable membrane around the sample isolates it from the liquid confining fluid and allows one to independently control the pore pressure in the sample as long as $P_p < S_0$.

- Uniaxial compressive tests ($S_1 > 0$, $S_2 = S_3 = 0$) are those in which one simply compresses a sample axially (with no radial stress) until it fails at a value defined as the unconfined compressive strength (usually termed either the UCS or C_0). As sample splitting (and failure on pre-existing fractures and faults) can occur in such tests, an alternative method for determining UCS is described below.

- Uniaxial tension tests ($S_1 < 0$, $S_2 = S_3 = 0$) are not shown in Figure 4.1 as they are fairly uncommon. Tensile strength ($T = -S_1$ at failure) is generally quite low in sedimentary rocks (see below) and this type of test procedure tends to promote failure along pre-existing fractures or flaws that might be present in a sample.

- Triaxial compression tests ($S_1 > S_2 = S_3 = S_0$) are the most common way of measuring rock strength under conditions that presumably simulate those at depth. Such tests are unfortunately named *triaxial* as there are only two different stresses of interest, the confining pressure S_0 and the differential stress $S_1 - S_0$. The strength of the sample at a given confining pressure is the differential stress at which it fails. The confining pressure is held constant as the sample is loaded. As in the case of hydrostatic compression tests, it is relatively straightforward to include the effects of pore pressure in such tests.

- Triaxial extension tests ($S_1 = S_2 > S_3$ where S_3 acts in an axial direction) can also be used to measure the compressive rock strength but are typically carried out only as part of specialized rock mechanical testing programs. One advantage of such tests is that they are useful for studying strength at low effective stress. A combination

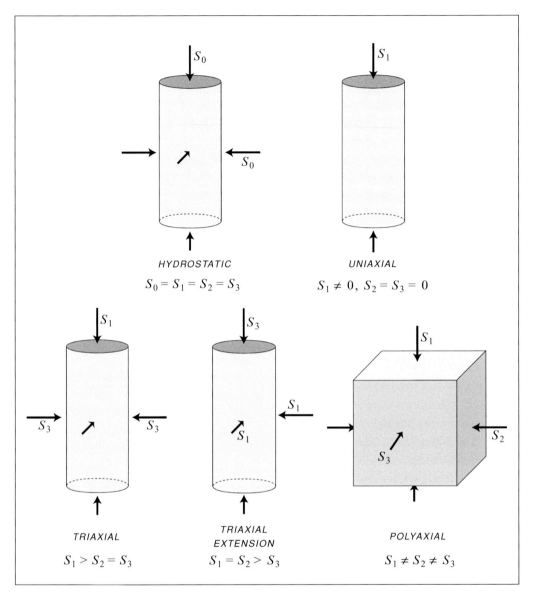

Figure 4.1. The most common types of rock mechanics tests. While it is common for petrophysical properties to be measured as a function of hydrostatic pressure, strength is typically measured via either uniaxial or triaxial tests. As discussed in the text, triaxial extension and polyaxial tests are rare. Pore pressure is frequently accommodated as an independent variable in these tests by using a flexible, impermeable sleeve outside the sample. For obvious reasons, pore pressure is not used in uniaxial tests. It is also not used in polyaxial tests because of the experimental difficulty of sealing pressure within the samples.

of triaxial compression and triaxial extension tests can be used to determine the importance of the intermediate principal stress, S_2, on failure. As discussed below, the importance of S_2 is often ignored.

- Polyaxial, or *true triaxial*, tests ($S_1 > S_2 > S_3$) are the only tests in which the three principal stresses are different. While these tests can most accurately replicate *in situ* conditions, such tests are extremely hard to conduct for several reasons: the test apparatus is somewhat complicated and difficult to use, it is nearly impossible to include the effects of pore pressure, and sample preparation is quite difficult.

Not shown in Figure 4.1 are thick-walled cylinder tests. In these tests a small axial hole is drilled along the axis of a cylindrical sample. These tests are done to determine the approximate strain around the axial hole at which failure is first noted. Such tests are done to support sand production studies such as those briefly discussed at the end of Chapter 10.

Rock strength in compression

The failure of rock in compression is a complex process that involves microscopic failures manifest as the creation of small tensile cracks and frictional sliding on grain boundaries (Brace, Paulding *et al.* 1966). Eventually, as illustrated in Figure 4.2a, there is a coalescence of these microscopic failures into a through-going shear plane (Lockner, Byerlee *et al.* 1991). In a brittle rock (with stress–strain curves like that shown in Figure 3.2) this loss occurs catastrophically, with the material essentially losing all of its strength when a through-going shear fault forms. In more ductile materials (such as poorly cemented sands) failure is more gradual. The strength is defined as the peak stress level reached during a deformation test after which the sample is said to *strain soften*, which simply means that it weakens (*i.e.* deforms at lower stresses) as it continues to deform. Simply put, rock failure in compression occurs when the stresses acting on a rock mass exceed its compressive strength. Compressive rock failure involves all of the stresses acting on the rock (including, as discussed below, the pore pressure). By rock strength we typically mean the value of the maximum principal stress at which a sample loses its ability to support applied stress.

The strength of rock depends on how it is confined. For the time being we will restrict discussion of rock strength to triaxial compression tests with non-zero pore pressure (with effective stresses $\sigma_1 > \sigma_2 = \sigma_3$). It is universally observed in such tests that sample strength is seen to increase monotonically with effective confining pressure (*e.g.* Jaeger and Cook 1979). Because of this, it is common to present strength test results using Mohr circles and Mohr failure envelopes (Figure 4.2b,c).

The basis for the Mohr circle construction is that it is possible to evaluate graphically the shear stress, τ_f, and effective normal stress ($\sigma_n = S_n - P_p$) on the fault that forms

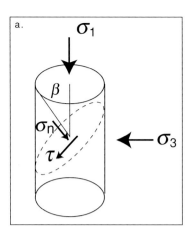

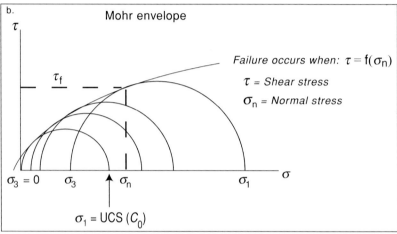

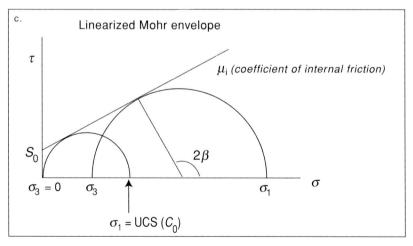

Figure 4.2. (a) In triaxial strength tests, at a finite effective confining pressure σ_3 (S_3-P_0), samples typically fail in compression when a through-going fault develops. The angle at which the fault develops is described by β, the angle between the fault normal and the maximum compressive stress, σ_1. (b) A series of triaxial strength tests at different effective confining pressures defines the Mohr failure envelope which typically flattens as confining pressure increases. (c) The linear simplification of the Mohr failure envelope is usually referred to as Mohr–Coulomb failure.

during the failure process in terms of the applied effective principal stresses σ_1 and σ_3,

$$\tau_f = 0.5(\sigma_1 - \sigma_3)\sin 2\beta \tag{4.1}$$

$$\sigma_n = 0.5(\sigma_1 + \sigma_3) + 0.5(\sigma_1 - \sigma_3)\cos 2\beta \tag{4.2}$$

where β is the angle between the fault normal and σ_1 (Figure 4.2a).

Conducting a series of triaxial tests defines an empirical Mohr–Coulomb failure envelope that describes failure of the rock at different confining pressures (Figure 4.2b). Allowable stress states (as described by Mohr circles) are those that do not intersect the Mohr–Coulomb failure envelope. Stress states that describe a rock just at the failure point "touch" the failure envelope. Stress states corresponding to Mohr circles which exceed the failure line are not allowed because failure of the rock would have occurred prior to the rock having achieved such a stress state.

The slope of the Mohr failure envelopes for most rocks decreases as confining pressure increases, as shown schematically in Figure 4.2b and for a sandstone in Figure 4.3a. However, for most rocks it is possible to consider the change of strength with confining pressure in terms of a linearized Mohr–Coulomb failure envelope (Figures 4.2c and 4.3a) defined by two parameters: μ_i, the slope of the failure line, termed the coefficient of *internal* friction, and the unconfined compressive strength (termed the UCS or C_0). One could also describe the linear Mohr failure line in terms of its intercept when $\sigma_3 = 0$ which is called the cohesive strength (or cohesion), S_0, as is common in soil mechanics. In this case, the linearized Mohr failure line can be written as

$$\tau = S_0 + \sigma_n \mu_i \tag{4.3}$$

As cohesion is not a physically measurable parameter, it is more common to express rock strength in terms of C_0. The relationship between S_0 and C_0 is:

$$C_0 = 2S_0 \left[\left(\mu_i^2 + 1 \right)^{1/2} + \mu_i \right] \tag{4.4}$$

While uniaxial tests are obviously the easiest way to measure C_0, it is preferable to determine C_0 by conducting a series of triaxial tests to avoid the axial splitting of the samples that frequently occurs during uniaxial tests and the test results are sensitive to the presence of pre-existing flaws in the samples. Once a Mohr envelope has been obtained through a series of tests, one can find C_0 by either fitting the envelope with a linear Mohr failure line and determining the uniaxial compressive strength graphically, or simply by measuring strength at many pressures and plotting the data as shown in Figure 4.3b, for Darley Dale sandstone (after Murrell 1965). As shown, C_0 is the intercept in the resultant plot (94.3 MPa) and μ_i is found to be 0.83 from the relationship

$$\mu_i = \frac{n-1}{2\sqrt{n}} \tag{4.5}$$

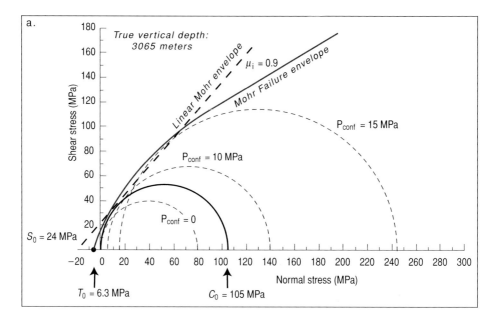

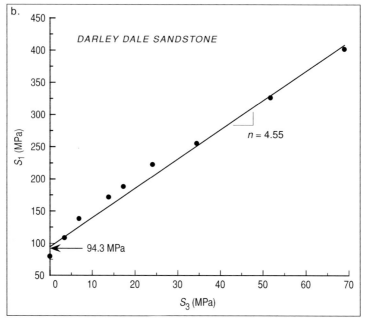

Figure 4.3. (a) Laboratory strength tests at 0, 10 and 15 MPa confining pressure defines the Mohr envelope for a sandstone sample recovered from a depth of 3065 m in southeast Asia. A linearized Mohr–Coulomb failure envelope and resultant values of S_0, C_0 and μ_i are shown. (b) Illustration of how strength values determined from a series of triaxial tests can be used to extrapolate a value of C_0. As explained in the text, the slope of this line can be used to determine μ_i.

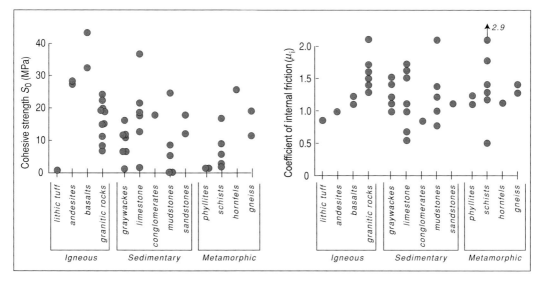

Figure 4.4. Cohesion and internal friction data for a variety of rocks (data replotted from the compilation of Carmichael 1982). Note that weak rocks with low cohesive strength still have a significant coefficient of internal friction.

where n is the slope of failure line when the stress at failure, S_1, is plotted as a function of the confining pressure, S_3, as shown in Figure 4.3b.

The fact that the test data can be fairly well fitted by a straight line in Figure 4.3b illustrates that using a linearized Mohr failure envelope for these rocks is a reasonable approximation. An important concept to keep in mind when considering rock strength is that while strong rocks have high cohesion and weak rocks have low cohesion, nearly all rocks have relatively high coefficients of internal friction. In other words, the rocks with low cohesion (or low compressive strength) are weak at low mean stresses but increase in strength as the mean stress increases. This is shown in the compilation shown in Figures 4.4a,b (data from Carmichael 1982). For sedimentary rocks, cohesive strengths are as low as 1 MPa and as high as several tens of MPa. Regardless, coefficients of internal friction range from about 0.5 to 2.0 with a median value of about 1.2. One exception to this is shales, which tend to have a somewhat lower value of μ_i. This is discussed below in the section discussing how rock strength is derived from geophysical logs.

A simple, but very important illustration of the importance of cohesion on wellbore stability is illustrated in Figure 4.5. Linearized Mohr envelopes are shown schematically for a strong rock (high cohesive strength) and weak rock (low cohesive strength) with the same μ_i. As discussed in detail in Chapter 6, when one considers the stresses at the wall of a vertical wellbore that might cause compressive rock failure, the least principal stress, σ_3, is usually the radial stress, σ_{rr}, which is equal to the difference between

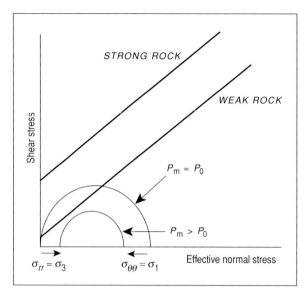

Figure 4.5. Schematic illustration of how raising mud weight helps stabilize a wellbore. The Mohr circle is drawn for a point around the wellbore. For weak rocks (low cohesion), when mud weight and pore pressure are equal, the wellbore wall fails in compression as the radial stress, σ_{rr} is equal to 0. Raising mud weight increases σ_{rr} and decreases $\sigma_{\theta\theta}$, the *hoop stress* acting around the wellbore. This stabilizes the wellbore by reducing the size of the Mohr circle all around the circumference of the well.

the mud weight, P_m, and the pore pressure, P_0. The maximum principal stress driving failure of the wellbore wall is $\sigma_{\theta\theta}$, the *hoop stress* acting parallel to the wellbore wall in a circumferential manner (Figure 6.1). Note that if the cohesive strength of the rock is quite low, when the mud weight is exactly equal to the pore pressure (*i.e.* the mud weight is exactly *balanced* with the pore pressure), $\sigma_{\theta\theta}$ does not have to be very large to exceed the strength of the rock at the wellbore wall and cause wellbore failure because $\sigma_{rr} = 0$. However, if the mud weight exceeds the pore pressure, σ_{rr} increases and $\sigma_{\theta\theta}$ decreases, thus resulting in a more stable wellbore. This is discussed more thoroughly in Chapter 6. Of course, drillers learned this lesson empirically a century ago as the use of mud weight to stabilize wellbores is one of a number of considerations which are discussed at some length in Chapter 10.

Compressive strength criteria

Over the years, many different failure criteria have been proposed to describe rock strength under different stress conditions based on the different types of laboratory tests illustrated in Figure 4.1 (as well as other types of tests). While somewhat complicated, these criteria are intended to better utilize laboratory strength data in actual case studies. It is obvious that the loading conditions common to laboratory tests are not very indicative of rock failure in cases of practical importance (such as wellbore stability). However, while it is possible in principle to utilize relatively complex failure criteria, it is often impractical to do so because core is so rarely available for comprehensive laboratory testing (most particularly in overburden rocks where many wellbore stability problems are encountered). Moreover, because the stresses acting in the earth at depth are strongly concentrated around wellbores (as discussed in Chapter 6), it is usually more important to estimate the magnitudes of *in situ* stresses correctly than to have a precise value of rock strength (which would require exhuming core samples for extensive rock strength tests) in order to address practical problems (as demonstrated in Chapter 10).

In this section, we will consider five different criteria that have been proposed to describe the value of the maximum stress, σ_1, at the point of rock failure as a function of the other two principal stresses, σ_2 and σ_3. Two commonly used rock strength criteria (the Mohr–Coulomb and the Hoek–Brown criteria), ignore the influence of the intermediate principal stress, σ_2, and are thus derivable from conventional triaxial test data ($\sigma_1 > \sigma_2 = \sigma_3$). We also consider three true triaxial, or polyaxial criteria (modified Wiebols–Cook, modified Lade, and Drucker–Prager), which consider the influence of the intermediate principal stress in polyaxial strength tests ($\sigma_1 > \sigma_2 > \sigma_3$). We illustrate below how well these criteria describe the strength of five rocks: amphibolite from the KTB site, Dunham dolomite, Solenhofen limestone, Shirahama sandstone and Yuubari shale as discussed in more detail by Colmenares and Zoback (2002).

Linearized Mohr–Coulomb

The linearized form of the Mohr failure criterion may be generally written as

$$\sigma_1 = C_0 + q\sigma_3 \tag{4.6}$$

where C_0 is solved-for as a fitting parameter,

$$q = \left[\left(\mu_i^2 + 1\right)^{1/2} + \mu_i\right]^2 = \tan^2(\pi/4 + \phi/2) \tag{4.7}$$

and

$$\phi = \tan^{-1}(\mu_i) \tag{4.8}$$

This failure criterion assumes that the intermediate principal stress has no influence on failure.

As viewed in σ_1, σ_2, σ_3 space, the yield surface of the linearized Mohr–Coulomb criterion is a right hexagonal pyramid equally inclined to the principal stress axes. The intersection of this yield surface with the π-plane is a hexagon. The π-plane is the plane

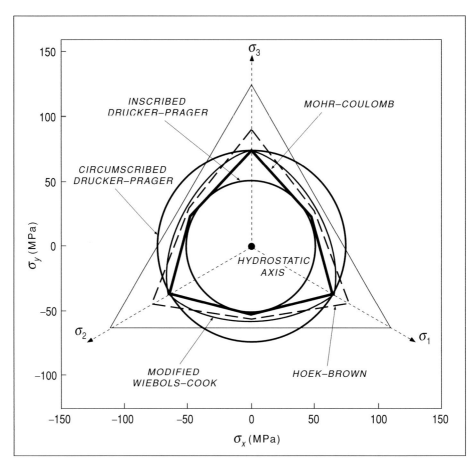

Figure 4.6. Yield envelopes projected in the π-plane for the Mohr–Coulomb criterion, the Hoek–Brown criterion, the modified Wiebols–Cook criterion and the circumscribed and inscribed Drucker–Prager criteria. After Colmenares and Zoback (2002). *Reprinted with permission of Elsevier.*

perpendicular to the straight line $\sigma_1 = \sigma_2 = \sigma_3$. Figure 4.6 shows the yield surface of the linearized Mohr–Coulomb criterion is hexagonal in the π-plane. Figure 4.7a shows the representation of the linearized Mohr–Coulomb criterion in $\sigma_1 - \sigma_2$ space for $C_0 = 60$ MPa and $\mu_i = 0.6$. In this figure (and Figures 4.8, 4.9 and 4.10b below), σ_1 at failure is shown as a function of σ_2 for experiments done at different values of σ_3.

Hoek– Brown criterion

This empirical criterion uses the uniaxial compressive strength of the intact rock material as a scaling parameter, and introduces two dimensionless strength parameters,

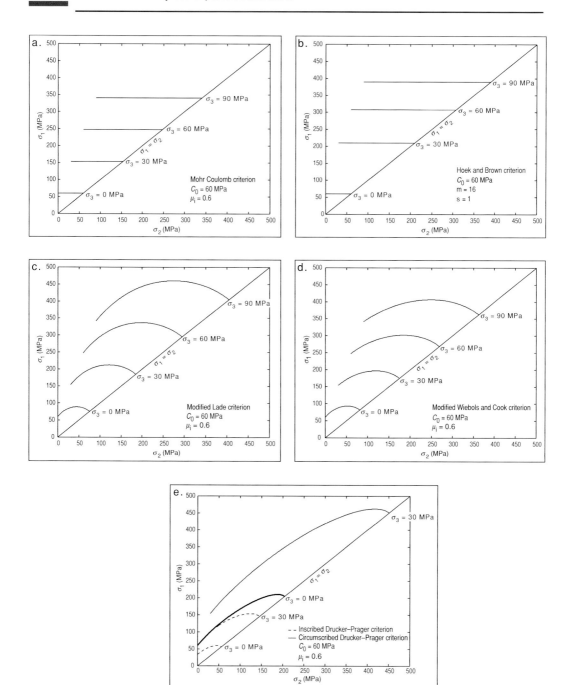

Figure 4.7. To observe how different compressive failure criteria define the importance of the intermediate principal stress, σ_2, on rock strength, for $\sigma_3 = 0, 30, 60$ and 90 MPa and $C_0 = 60$ MPa and $\mu_i = 0.6$, we show the curves corresponding to (a) linearized Mohr–Coulomb criterion; (b) Hoek–Brown criterion ($m = 16$ and $s = 1$); (c) modified Lade criteria; (d) modified Wiebols–Cook criterion; (e) inscribed and circumscribed Drucker–Prager criteria (shown for only for $\sigma_3 = 0$ and 30 MPa for simplicity). After Colmenares and Zoback (2002). *Reprinted with permission of Elsevier.*

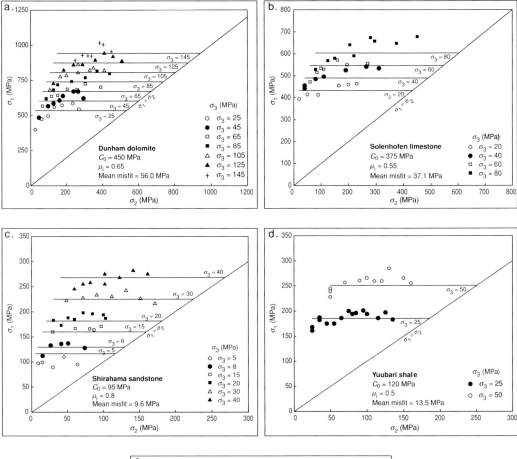

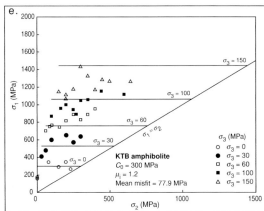

Figure 4.8. Best-fitting solution for all the rocks using the Mohr–Coulomb criterion. (a) Dunham dolomite. (b) Solenhofen limestone. (c) Shirahama sandstone. (d) Yuubari shale. (e) KTB amphibolite. After Colmenares and Zoback (2002). *Reprinted with permission of Elsevier.*

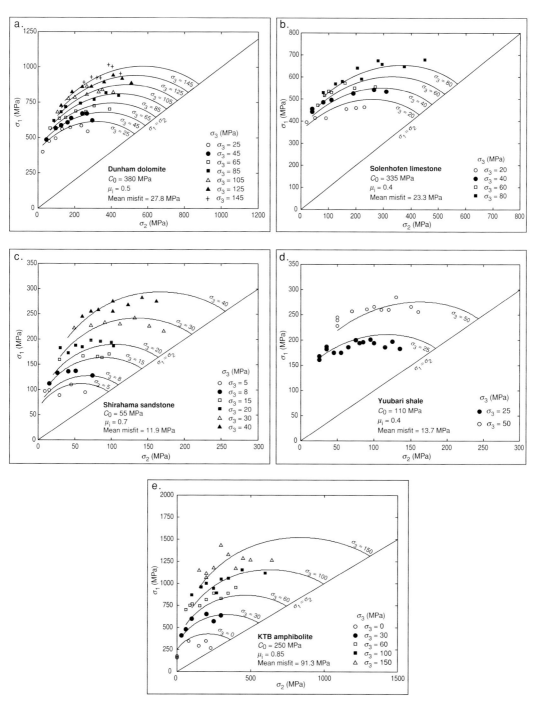

Figure 4.9. Best-fitting solution for all the rocks using the modified Lade criterion; (a) Dunham dolomite; (b) Solenhofen limestone; (c) Shirahama sandstone; (d) Yuubari shale; (e) KTB amphibolite. After Colmenares and Zoback (2002). *Reprinted with permission of Elsevier.*

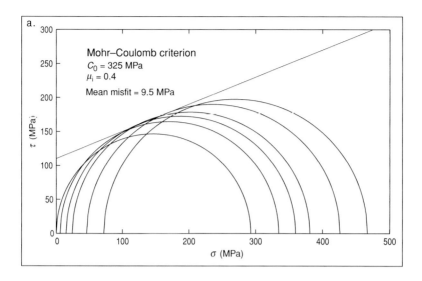

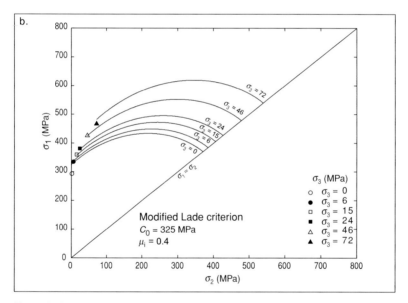

Figure 4.10. Best-fitting solution for the Solenhofen limestone using the triaxial test data; (a) Mohr–Coulomb criterion; (b) modified Lade criterion. After Colmenares and Zoback (2002). *Reprinted with permission of Elsevier.*

m and s. After studying a wide range of experimental data, Hoek and Brown (1980) proposed that the maximum principal stress at failure is given by

$$\sigma_1 = \sigma_3 + C_0 \sqrt{m \frac{\sigma_3}{C_0} + s} \qquad (4.9)$$

where m and s are constants that depend on the properties of the rock and on the extent to which it had been broken before being tested. Note that this form of the failure law results in a curved (parabolic) Mohr envelope, similar to that shown in Figures 4.2b and 4.3a. The Hoek–Brown failure criterion was originally developed for estimating the strength of rock masses for application to excavation design.

According to Hoek and Brown (1997), m depends on rock type and s depends on the characteristics of the rock mass such that:

- $5 < m < 8$: carbonate rocks with well-developed crystal cleavage (dolomite, limestone, marble).
- $4 < m < 10$: lithified argillaceous rocks (mudstone, siltstone, shale, slate).
- $15 < m < 24$: arenaceous rocks with strong crystals and poorly developed crystal cleavage (sandstone, quartzite).
- $16 < m < 19$: fine-grained polyminerallic igneous crystalline rocks (andesite, dolerite, diabase, rhyolite).
- $22 < m < 33$: coarse-grained polyminerallic igneous and metamorphic rocks (amphibolite, gabbro, gneiss, granite, norite, quartz-diorite).

While these values of m obtained from lab tests on intact rock are intended to represent a good estimate when laboratory tests are not available, we will compare them with the values obtained for the five rocks studied. For intact rock materials, s is equal to one. For a completely granulated specimen or a rock aggregate, s is equal to zero.

Figure 4.6 shows that the intersection of the Hoek–Brown yield surface with the π-plane is approximately a hexagon. The sides of the Hoek–Brown failure cone are not planar, as is the case for the Mohr–Coulomb criterion but, in the example shown, the curvature is so small that the sides look like straight lines. Figure 4.7b shows this criterion in σ_1–σ_2 space for $C_0 = 60$ MPa, $m = 16$ and $s = 1$. The Hoek–Brown criterion is independent of σ_2, like the Mohr–Coulomb criterion. One practical disadvantage of the Hoek–Brown criterion, discussed later, is that correlations are not readily available in the published literature to relate m to commonly measured with geophysical well logs, nor are there relationships to relate m to the more commonly defined angle of internal friction.

Modified Lade criterion

The Lade criterion (Lade 1977) is a three-dimensional failure criterion that was originally developed for frictional materials without effective cohesion (such as granular soils). It was developed for soils with curved failure envelopes. This criterion is given by

$$\left(\frac{I_1^3}{I_3} - 27\right)\left(\frac{I_1}{p_a}\right)^{m'} = \eta_1 \tag{4.10}$$

where I_1 and I_3 are the first and third invariants of the stress tensor

$$I_1 = S_1 + S_2 + S_3 \tag{4.11}$$

$$I_3 = S_1 \cdot S_2 \cdot S_3 \tag{4.12}$$

p_a is atmospheric pressure expressed in the same units as the stresses, and m' and η_1 are material constants.

In the modified Lade criterion developed by Ewy (1999) m' was set equal to zero in order to obtain a criterion which is able to predict a linear shear strength increase with increasing mean stress, $I_1/3$. For considering materials with cohesion, Ewy (1999) included pore pressure as a necessary parameter and introduced the parameters S and η as material constants. The parameter S is related to the cohesion of the rock, while the parameter η represents the internal friction.

Doing all the modifications and defining appropriate stress invariants, the following failure criterion was obtained:

$$\frac{(I_1')^3}{I_3'} = 27 + \eta \tag{4.13}$$

where

$$I_1' = (\sigma_1 + S) + (\sigma_2 + S) + (\sigma_3 + S) \tag{4.14}$$

and

$$I_3' = (\sigma_1 + S)(\sigma_2 + S)(\sigma_3 + S) \tag{4.15}$$

S and η can be derived directly from the Mohr–Coulomb cohesion S_0 and internal friction angle ϕ by

$$S = \frac{S_0}{\tan \phi} \tag{4.16}$$

$$\eta = \frac{4(\tan \phi)^2 (9 - 7\sin \phi)}{(1 - \sin \phi)} \tag{4.17}$$

where $\tan \phi = \mu_i$ and $S_0 = C_0/(2\,q^{1/2})$ with q as defined in equation (4.7).

The modified Lade criterion predicts a strengthening effect with increasing intermediate principal stress, σ_2, followed by a slight reduction in strength as σ_2 increases. It should be noted that the equations above allow one to employ this criterion using the two parameters most frequently obtained in laboratory strength tests, C_0 and μ_i. This makes this criterion easy to use, and potentially more generally descriptive of rock failure, when considering problems such as wellbore stability. The modified Lade criterion can be observed in Figure 4.7c where it has been plotted in σ_1–σ_2 space for $C_0 = 60$ MPa and $\mu_i = 0.6$, the same parameters used for the Mohr–Coulomb criterion in Figure 4.7a.

Modified Wiebols–Cook criterion

Wiebols and Cook (1968) proposed an effective strain energy criterion for rock failure that depends on all three principal stresses. Zhou (1994) presented a failure criterion with features similar to the Wiebols–Cook criterion which is an extension of the circumscribed Drucker–Prager criterion (described below).

The failure criterion proposed by Zhou predicts that a rock fails if

$$J_2^{1/2} = A + B J_1 + C J_1^2 \tag{4.18}$$

where

$$J_1 = \frac{1}{3}(\sigma_1 + \sigma_2 + \sigma_3) \tag{4.19}$$

and

$$J_2^{1/2} = \sqrt{\frac{1}{6}\left[(\sigma_1 - \sigma_2)^2 + (\sigma_1 - \sigma_3)^2 + (\sigma_2 - \sigma_3)^2\right]} \tag{4.20}$$

J_1 is the mean effective confining stress and, for reference, $J_2^{1/2}$ is equal to $(3/2)^{1/2}\tau_{\text{oct}}$, where τ_{oct} is the octahedral shear stress

$$\tau_{\text{oct}} = \frac{1}{3}\sqrt{(\sigma_1 - \sigma_2)^2 + (\sigma_2 - \sigma_3)^2 + (\sigma_2 - \sigma_1)^2} \tag{4.21}$$

The parameters A, B, and C are determined such that equation (4.18) is constrained by rock strengths under triaxial ($\sigma_2 = \sigma_3$) and triaxial extension ($\sigma_1 = \sigma_2$) conditions (Figure 4.1). Substituting the given conditions plus the uniaxial rock strength ($\sigma_1 = C_0$, $\sigma_2 = \sigma_3 = 0$) into equation (4.18), it is found that

$$C = \frac{\sqrt{27}}{2C_1 + (q-1)\sigma_3 - C_0}\left(\frac{C_1 + (q-1)\sigma_3 - C_0}{2C_1 + (2q+1)\sigma_3 - C_0} - \frac{q-1}{q+2}\right) \tag{4.22}$$

with $C_1 = (1 + 0.6\,\mu_i)C_0$ and q given by equation (4.7),

$$B = \frac{\sqrt{3}(q-1)}{q+2} - \frac{C}{3}[2C_0 + (q+2)\sigma_3] \tag{4.23}$$

and

$$A = \frac{C_0}{\sqrt{3}} - \frac{C_0}{3}B - \frac{C_0^2}{9}C \tag{4.24}$$

The rock strength predictions produced using equation (4.18) are similar to those of Wiebols and Cook and thus it is referred to as the modified Wiebols–Cook criterion. For polyaxial states of stress, the strength predictions made by this criterion are slightly higher than those found using the linearized Mohr–Coulomb criterion. This can be seen in Figure 4.6 because the failure cone of the modified Wiebols–Cook criterion just coincides with the outer apices of the Mohr–Coulomb hexagon. This criterion is plotted in σ_1–σ_2 space in Figure 4.7d. Note its similarity to the modified Lade criterion.

Drucker–Prager criterion

The extended von Mises yield criterion, or Drucker–Prager criterion, was originally developed for soil mechanics (Drucker and Prager 1952). The von Mises criterion may be written in the following way

$$J_2 = k^2 \tag{4.25}$$

where k is an empirical constant. The yield surface of the modified von Mises criterion in principal stress space is a right circular cone equally inclined to the principal stress axes. The intersection of the π-plane with this yield surface is a circle. The yield function used by Drucker and Prager to describe the cone in applying the limit theorems to perfectly plastic soils has the form:

$$J_2^{1/2} = k + \alpha J_1 \tag{4.26}$$

where α and k are material constants. The material parameters α and k can be determined from the slope and the intercept of the failure envelope plotted in the J_1 and $(J_2)^{1/2}$ space. α is related to the internal friction of the material and k to the cohesion of the material. In this way, the Drucker–Prager criterion can be compared to the Mohr–Coulomb criterion. When α is equal to zero, equation (4.26) reduces to the von Mises criterion.

The Drucker–Prager criteria can be divided into an outer bound criterion (or circumscribed Drucker–Prager) and an inner bound criterion (or inscribed Drucker–Prager). These two versions of the Drucker–Prager criterion come from comparing the Drucker–Prager criterion with the Mohr–Coulomb criterion. In Figure 4.6 the two Drucker–Prager criteria are plotted in the π-plane. The inner Drucker–Prager circle only touches the inside of the Mohr–Coulomb criterion and the outer Drucker-Prager circle coincides with the outer apices of the Mohr–Coulomb hexagon.

The inscribed Drucker–Prager criterion is obtained when (Veeken, Walters *et al.* 1989; McLean and Addis 1990)

$$\alpha = \frac{3\sin\phi}{\sqrt{9 + 3\sin^2\phi}} \tag{4.27}$$

and

$$k = \frac{3C_0\cos\phi}{2\sqrt{q}\sqrt{9 + 3\sin^2\phi}} \tag{4.28}$$

where ϕ is the angle of internal friction, as defined above.

The circumscribed Drucker–Prager criterion (McLean and Addis 1990; Zhou 1994) is obtained when

$$\alpha = \frac{6\sin\phi}{\sqrt{3}\,(3 - \sin\phi)} \tag{4.29}$$

and

$$k = \frac{\sqrt{3}C_0 \cos\phi}{\sqrt{q}\,(3 - \sin\phi)} \tag{4.30}$$

As equations (4.29) and (4.30) show, α only depends on ϕ which means that α has an upper bound in both cases; 0.866 in the inscribed Drucker–Prager case and 1.732 in the circumscribed Drucker–Prager case.

In Figure 4.7e we show the Drucker–Prager criteria for $C_0 = 60$ MPa and $\mu_i = 0.6$ in comparison with other failure criteria. As shown in Figure 4.7e, for the same values of C_0 and μ_i, the inscribed Drucker–Prager criterion predicts failure at much lower stresses as a function of σ_2 than the circumscribed Drucker–Prager criterion.

As mentioned above, Colmenares and Zoback (2002) considered these failure criterion for five rock types: amphibolite from the KTB site in Germany (Chang and Haimson 2000), Dunham dolomite and Solenhofen limestone (Mogi 1971) and Shirahama sandstone and Yuubari shale (Takahashi and Koide 1989).

Figure 4.8 presents all the results for the Mohr–Coulomb criterion with the best-fitting parameters for each rock type. As the Mohr–Coulomb does not take into account the influence of σ_2, the best fit would be the horizontal line that goes through the middle of the data for each σ_3. The smallest misfits associated with the Mohr–Coulomb criterion were obtained for the Shirahama sandstone and the Yuubari shale. The largest misfits were for Dunham dolomite, Solenhofen limestone and KTB amphibolite, which are rocks showing the greatest influence of the intermediate principal stress on failure.

The modified Lade criterion (Figure 4.9) works well for the rocks with a high σ_2-dependence of failure such as Dunham dolomite and Solenhofen limestone. For the KTB amphibolite, this criterion reasonably reproduces the trend of the experimental data but not as well as for the Dunham dolomite. We see a similar result for the Yuubari shale. The fit to the Shirahama sandstone data does not reproduce the trends of the data very well.

We now briefly explore the possibility of using triaxial test data to predict the σ_2-dependence using the modified Lade failure criterion. The reason for doing this is to be able to characterize rock strength with relatively simple triaxial tests, but to allow all three principal stresses to be considered when addressing problems such as wellbore failure. We utilize only the triaxial test data for Solenhofen limestone (Figure 4.8b) which would not have detected the fact that the strength is moderately dependent on α_2. As shown by Colmenares and Zoback (2002), by using only triaxial test data (shown in Figure 4.10a), we obtain a value of C_0 as a function of α_2 (Figure 4.10b) that is within $\pm 3\%$ of that obtained had polyaxial test data been collected.

Because the subject of rock strength can appear to be quite complex, it might seem quite difficult to know how to characterize the strength of a given rock and to utilize this knowledge effectively. In practice, however, the size of the failure envelope (Figure 4.6) is ultimately more important than its exact shape. When applied to problems of wellbore stability (Chapter 10), for example, practical experience has shown that for relatively strong rocks, either the Mohr–Coulomb criterion or the Hoek–Brown criterion yield reliable results. In fact, when using these data to fit the polyaxial strength data shown in Figure 4.8, the two criteria worked equally well (Colmenares and Zoback 2002). However, because the value for the parameter m in the Hoek–Brown criterion is rarely measured, it is usually most practical to use the Mohr–Coulomb criterion when considering the strength of relatively strong rocks. Similarly, in weaker rocks, both the modified Lade and the modified Wiebols–Cook criteria, both polyaxial criteria, seem to work well and yield very similar fits to the data shown in Figure 4.8 (Colmenares and Zoback 2002). The modified Lade criterion is easily implemented in practice as it is used with the two parameters most commonly measured in laboratory tests, μ_i and C_0.

Strength and pore pressure

As mentioned in Chapter 3, pore pressure has a profound effect on many rock properties, including rock strength. Figure 4.11 shows conventional triaxial strength tests on Berea sandstone and Mariana limestone by Handin, Hager *et al.* (1963). In Figures 4.11a and c, the strength tests are shown without pore pressure in the manner of Figure 4.3b where the strength at failure, S_1, is shown as a function of confining pressure, S_3. As discussed above, S_1 depends linearly on S_3 such that

$$S_1 = C_0 + nS_3 \tag{4.31}$$

where C_0, n and μ_i are 62.8 MPa, 2.82 and 0.54 for Berea sandstone and 40.8 MPa, 3.01 and 0.58 for Marianna limestone, respectively. Rearrangement of equation (4.31) yields the following

$$S_1 - S_3 = C_0 + (1 - n)P_p - (1 - n)S_3 \tag{4.32}$$

Assuming that it is valid to replace S_1 with $(S_1 - P_p)$ and S_3 with $(S_3 - P_p)$ in equation (4.31), it would mean that strength is a function of the simple form of effective stress (equation 3.8). Figures 4.11b,c show that the straight lines predicted by equation (4.32) fit the data exactly for the various combinations of confining pressures and pore pressures at which the tests were conducted. In other words, the effect of pore pressure on rock strength is described very well by the simple (or Terzaghi) form of the effective stress law in these and most rocks. One important *proviso*, however, is that this has not been tested in a comprehensive way in the context of wellbore failure, a topic of considerable interest. Hence, more research on this topic is needed and points to a clear need for investigating the strength of a variety of rocks (of different strength, stiffness, permeability, etc.) at range of conditions (different loading rates, effective confining pressures, etc.).

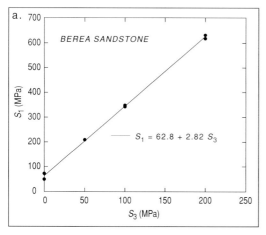
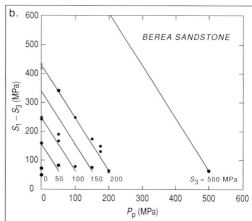
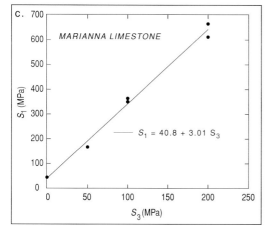
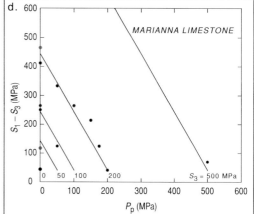

Figure 4.11. (a) Dependence of rock strength on confining pressure in the absence of pore pressure for Berea sandstone. (b) Dependence of strength on confining pressure and pore pressure assuming the simple Terzaghi effective stress law (equation 3.8) is valid (straight diagonal lines). (c) and (d) show similar data for Marianna limestone. Data derived from Handin, Hager *et al.* (1963).

Rock strength anisotropy

The presence of weak bedding planes in shaley rocks (or finely laminated sandstones or foliation planes in metamorphic rocks) can sometimes have a marked effect on rock strength. In Chapter 10 we will investigate several cases that illustrate the importance of slip on weak bedding planes on wellbore stability when wells are drilled at an oblique angle to the bedding planes.

The influence of weak bedding planes on rock strength is referred to as strength anisotropy. The importance of this depends both on the relative weakness of the bedding

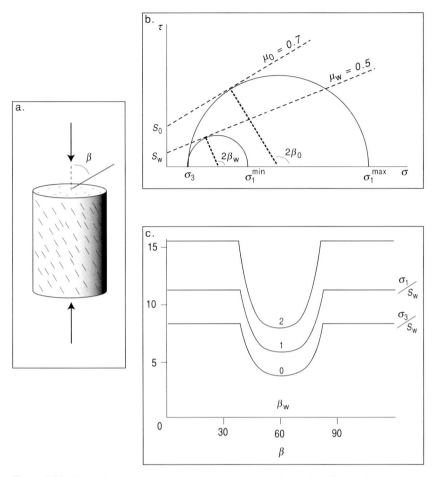

Figure 4.12. Dependence of rock strength on the angle of weak bedding or foliation planes. (a) Rock samples can be tested with the orientation of weak planes at different angles, β, to the maximum principal stress, σ_1. (b) The strength can be defined in terms of the intact rock strength (when the weak planes do not affect failure) and the strength of the weak planes. (c) Prediction of rock strength (normalized by the cohesion of bedding planes) as function of β. Modified from Donath (1966) and Jaeger and Cook (1979).

plane and the orientation of the plane with respect to the applied stresses. This is illustrated in Figure 4.12a, for strength tests with bedding planes whose normal is at an angle, β, to the applied maximum stress. Intuitively, one can see that when $\beta \sim 0°$ or $90°$, the bedding planes will have relatively little influence on rock strength. However, when $\beta \sim 60°$, slip on a weak bedding plane would occur at a markedly lower stress level than that required to form a new fault. To be more quantitative, one could view a rock as having two strengths (illustrated in the Mohr diagram in Figure 4.12b). The intact rock would have its *normal* strength which would control failure when slip on bedding planes did not occur and a lower strength, defined by the cohesion, S_w, and internal friction, μ_w, of the weak bedding planes which would apply. These parameters are only relevant, of course, when slip occurs along the pre-existing planes of weakness and affects rock strength.

Mathematically, it is possible to estimate the degree to which bedding planes lower rock strength using a theory developed by Donath (1966) and Jaeger and Cook (1979). The maximum stress at which failure will occur, σ_1, will depend on σ_3, S_w, and σ_w by

$$\sigma_1 = \sigma_3 \frac{2(S_w + \mu_w \sigma_3)}{(1 - \mu_w \cot \beta_w)\sin 2\beta} \qquad (4.33)$$

This is shown in Figure 4.12c. At high and low β, the intact rock strength (shown normalized by S_w) is unaffected by the presence of the bedding planes. At $\beta \sim 60°$, the strength is markedly lower. Using

$$\tan 2\beta_w = \frac{1}{\mu_w}$$

it can be shown that the minimum strength is given by

$$\sigma_1^{\min} = \sigma_3 + 2(S_w + \mu_w \sigma_3)\left[\left(\mu_w^2\right)^{\frac{1}{2}} + \mu_w\right] \qquad (4.34)$$

As shown in the schematic example in Figure 4.12c, the rock strength is reduced markedly. Strength tests performed on a granitic muscovite gneiss (a metamorphic rock) from the KTB scientific research borehole in Germany with pronounced foliation planes (Vernik, Lockner *et al.* 1992) are fitted by this theory extremely well (Figure 4.13). Note that when the foliation planes have their maximum effect on rock strength (at $\beta = 60°$), strength is reduced approximately by half and the dependence of strength on β_w is well described by equation (4.33).

The importance of weak bedding planes in shale is quite important in two case studies presented in Chapter 10. In northern South America, shales which have a UCS of about 10,000 psi (when measured normal to bedding) have very weak bedding planes with $S_w = 300$ psi and $\mu_w = 0.5$ which greatly affects wellbore stability for wells at some trajectories to the steeply dipping bedding planes. Off the eastern coast of Canada, a formation with a UCS of about 8000 psi (when measured normal to bedding) has weak bedding planes with $S_w = 700$ psi and $\mu_w = 0.2$. In this case, bedding is nearly horizontal and horizontal stress magnitudes are relatively low, but long reach-deviated wells are severely affected by the formation's low bedding-plane strength.

Estimating rock strength from geophysical log data

As alluded to above, many geomechanical problems associated with drilling must be addressed when core samples are unavailable for laboratory testing. In fact, core samples

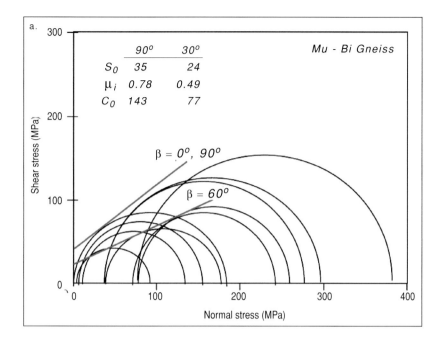

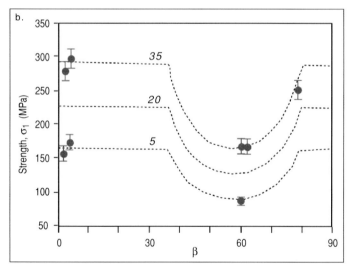

Figure 4.13. Fit of compressive strength tests to the theory illustrated in Figure 4.12 and defined by equation (4.33). Modified from Vernik, Lockner *et al.* (1992).

of overburden formations (where many wellbore instability problems are encountered) are almost never available for testing. To address this, numerous relations have been proposed that relate rock strength to parameters measurable with geophysical well logs. The basis for these relations is the fact that many of the same factors that affect rock strength also affect elastic moduli and other parameters, such as porosity.

Nearly all proposed formulae for determination of rock strength from geophysical logs utilize either:

- P-wave velocity, V_p, or equivalently, the travel time of compressional waves along the wellbore wall, Δt ($\Delta t = V_p^{-1}$), expressed as slowness, typically μs/ft,
- Young's modulus, E (usually derived from V_p and density data as illustrated in Table 3.1), or
- Porosity, ϕ (or density) data.

The justification for proposed relations is illustrated by the dependence of uniaxial compressive strength on these parameters as shown in Figures 4.14, 4.15 and 4.16 for sandstones, shales, and limestone and dolomite, respectively utilizing data from Lama and Vutukuri (1978), Carmichael (1982), Jizba (1991), Wong, David *et al.* (1997), Horsrud (2001) and Kwasniewski (1989). In each of the figures, the origin of the laboratory data is indicated by the symbol used. Despite the considerable scatter in the data, for each rock type, there is marked increase of strength with V_p and E and a marked decrease in strength with increased porosity.

Table 4.1 presents a number of relations for different sandstones from different geological settings for predicting rock strength from log data studied (Chang, Zoback *et al.* 2006). Equations (1)–(3) use P-wave velocity, V_p (or equivalently as Δt) measurements obtained from well logs. Equations (5)–(7) utilize both density and V_p data, and equation (4) utilizes V_p, density, Poisson's ratio (requiring V_s measurements) and clay volume (from gamma ray logs). Equation (8) utilizes Young's modulus, E, derived from V_p and V_s, and equations (9) and (10) utilize log-derived porosity measurements to estimate rock strength. Because of the considerable scatter in Figure 4.14, it is obvious that it would be impossible for any of the relations in Table 4.1 to fit all of the data shown. It also needs to be remembered that P-wave velocity in the lab is measured at ultrasonic frequencies (typically ~1 MHz) in a direction that is frequently orthogonal to bedding and typically on the most intact samples available that may not be representative of weaker rocks responsibility for wellbore failure. As discussed in Chapter 3, sonic velocities used in geophysical well logs operate at much lower frequencies. Moreover, such measurements are made parallel to the wellbore axis, which frequently is not perpendicular to bedding. Hence, there are significant experimental differences between field measurements and laboratory calibrations that need to be taken into consideration.

For the relations that are based on V_p (Figure 4.14a), it is noteworthy that except for equations (1) and (6) (derived for relatively strong rocks), all of the relations predict extremely low strengths for very slow velocities, or high travel times ($\Delta t \geq 100\ \mu$s/ft), and appear to badly underpredict the data. Such velocities are characteristic of weak sands such as found in the Gulf of Mexico, and a number of the relations in Table 4.1 were derived for these data. However, while equation (6) appears to fit the reported values better than the other equations, one needs to keep in mind that there are essentially no very weak GOM sands represented in the strength data available in the three studies represented in this compilation. Similarly, for fast, high-strength rocks, equation

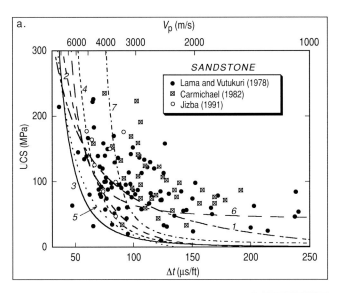

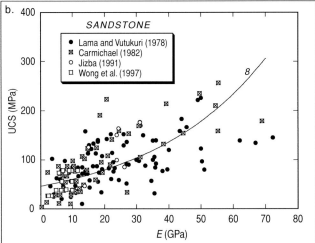

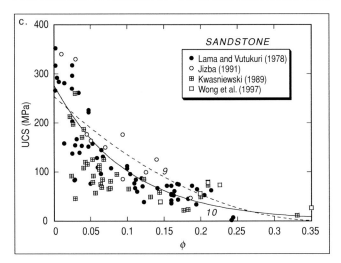

Figure 4.14. Comparison between different empirical laws (Table 4.1) for the dependence of the strength of sandstones on (a) compressional wave velocity, (b) Young's modulus and (c) porosity. After Chang, Zoback *et al.* (2006). *Reprinted with permission of Elsevier.*

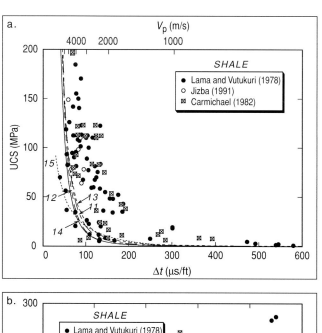

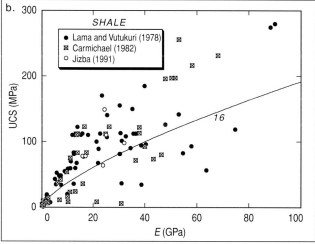

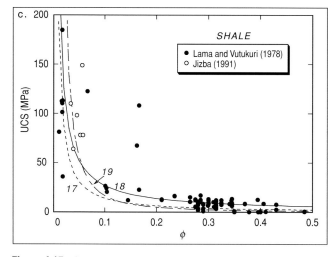

Figure 4.15. Comparison between different empirical laws (Table 4.2) for the dependence of the strength of shale on (a) compressional wave velocity, (b) Young's modulus and (c) porosity. After Chang, Zoback *et al.* (2006). *Reprinted with permission of Elsevier.*

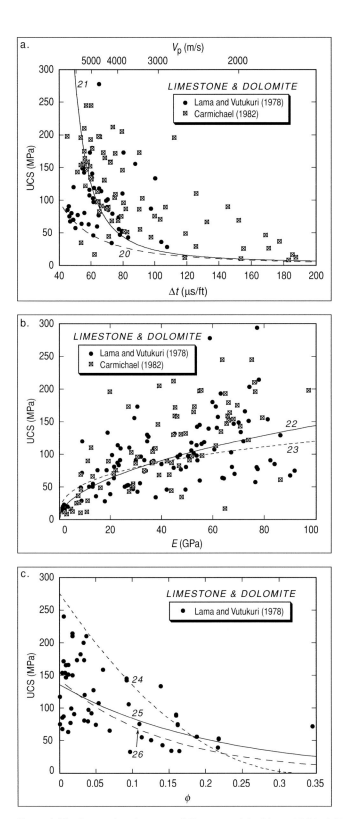

Figure 4.16. Comparison between different empirical laws (Table 4.3) for the dependence of the strength of carbonate rocks on (a) compressional wave velocity, (b) Young's modulus and (c) porosity. After Chang, Zoback *et al.* (2006). *Reprinted with permission of Elsevier.*

Table 4.1. *Empirical relationships between UCS and other physical properties in sandstones. After Chang, Zoback et al. (2006). Reprinted with permission of Elsevier*

Equation No.	UCS, MPa	Region where developed	General comments	Reference
1	$0.035\,V_p - 31.5$	Thuringia, Germany	–	(Freyburg 1972)
2	$1200\exp(-0.036\Delta t)$	Bowen Basin, Australia	Fine grained, both consolidated and unconsolidated sandstones with wide porosity range	(McNally 1987)
3	$1.4138 \times 10^7\,\Delta t^{-3}$	Gulf Coast	Weak and unconsolidated sandstones	Unpublished
4	$3.3 \times 10^{-20}\,\rho^2 V_p^2\,[(1+\nu)/(1-\nu)]^2(1-2\nu)[1+0.78V_{\text{clay}}]$	Gulf Coast	Applicable to sandstones with UCS >30 MPa	(Fjaer, Holt et al. 1992)
5	$1.745 \times 10^{-9}\,\rho V_p^2 - 21$	Cook Inlet, Alaska	Coarse grained sands and conglomerates	(Moos, Zoback et al. 1999)
6	$42.1\exp(1.9 \times 10^{-11}\,\rho V_p^2)$	Australia	Consolidated sandstones with $0.05 < \phi < 0.12$ and UCS > 80MPa	Unpublished
7	$3.87\exp(1.14 \times 10^{-10}\,\rho V_p^2)$	Gulf of Mexico	–	Unpublished
8	$46.2\exp(0.000027E)$	–	–	Unpublished
9	$A\,(1-B\phi)^2$	Sedimentary basins worldwide	Very clean, well consolidated sandstones with $\phi < 0.30$	(Vernik, Bruno et al. 1993)
10	$277\exp(-10\phi)$	–	Sandstones with $2 <$ UCS $< 360\,\text{MPa}$ and $0.002 < \phi < 0.33$	Unpublished

Units used: V_p (m/s), Δt (μs/ft), ρ (kg/m^3), V_{clay} (fraction), E (MPa), ϕ (fraction)

Table 4.2. *Empirical relationships between UCS and other physical properties in shale. After Chang, Zoback* et al. *(2006).* Reprinted with permission of Elsevier

	UCS, MPa	Region where developed	General comments	Reference
11	$0.77 (304.8/\Delta t)^{2.93}$	North Sea	Mostly high porosity Tertiary shales	(Horsrud 2001)
12	$0.43 (304.8/\Delta t)^{3.2}$	Gulf of Mexico	Pliocene and younger	Unpublished
13	$1.35 (304.8/\Delta t)^{2.6}$	Globally	–	Unpublished
14	$0.5 (304.8/\Delta t)^{3}$	Gulf of Mexico	–	Unpublished
15	$10 (304.8/\Delta t - 1)$	North Sea	Mostly high porosity Tertiary shales	(Lal 1999)
16	$0.0528 E^{0.712}$	–	Strong and compacted shales	Unpublished
17	$1.001 \phi^{-1.143}$	–	Low porosity ($\phi < 0.1$), high strength shales	(Lashkaripour and Dusseault 1993)
18	$2.922 \phi^{-0.96}$	North Sea	Mostly high porosity Tertiary shales	(Horsrud 2001)
19	$0.286 \phi^{-1.762}$	–	High porosity ($\phi > 0.27$) shales	Unpublished

Units used: Δt (μs/ft), E (MPa), ϕ (fraction)

(3) (derived for low-strength rocks) does a particularly poor job of fitting the data. The single equation derived using Young's modulus, (8), fits the available data reasonably well, but there is considerable scatter at any given value of E. Both of the porosity relations in Table 4.1 seem to generally overestimate strength, except for the very lowest porosities.

Overall, it is reasonable to conclude that none of the equations in Table 4.1 seem to do a very good job of fitting the data in Figure 4.14. That said, it is important to keep in mind that the validity of any of these relations is best judged in terms of how well it would work for the rocks for which it was originally derived. Thus, calibration is extremely important before utilizing any of the relations shown. Equation (5), for example, seems to systematically underpredict all the data in Figure 4.14a, yet worked very well for the clean, coarse-grained sands and conglomerates for which it was derived (Moos, Zoback *et al.* 1999). It is also important to emphasize that relations that accurately capture the lower bound of the strength data (such as equations 2–5) can be used to take a conservative approach toward wellbore stability. While the strength may be larger than predicted (and thus the wellbore more stable) it is not likely to be lower.

Considering now the empirical relations describing the strength of shales (Table 4.2), equations (11)–(15) seem to provide a lower bound for the data in Figure 4.15a. While it might be prudent to underestimate strength, the difference between these relations and the measured strengths is quite marked, as much as 50 MPa for high-velocity rocks ($\Delta t < 150$ μs/ft). For low-velocity rocks ($\Delta t > 200$ μs/ft), the relations under-predict strengths by 10–20 MPa. However, the porosity relations (equations 17–19) seem to

Table 4.3. *Empirical relationships between UCS and other physical properties in limestone and dolomite. After Chang, Zoback* et al. *(2006).* Reprinted with permission of Elsevier

	UCS, MPa	Region where developed	General comments	Reference
20	$(7682/\Delta t)^{1.82} / 145$	–	–	(Militzer 1973)
21	$10^{(2.44 + 109.14/(t)} / 145$	–	–	(Golubev and Rabinovich 1976)
22	$0.4067 \, E^{0.51}$	–	Limestone with 10 < UCS < 300 MPa	Unpublished
23	$2.4 \, E^{0.34}$	–	Dolomite with 60 < UCS < 100 MPa	Unpublished
24	$C \, (1-D\phi)^2$	Korobcheyev deposit, Russia	C is reference strength for zero porosity (250 < C < 300 MPa). D ranges between 2 and 5 depending on pore shape	(Rzhevsky and Novick 1971)
25	$143.8 \, \exp(-6.95\phi)$	Middle East	Low to moderate porosity (0.05 < ϕ < 0.2) and high UCS (30 < UCS < 150 MPa)	Unpublished
26	$135.9 \, \exp(-4.8\phi)$	–	Representing low to moderate porosity (0 < ϕ < 0.2) and high UCS (10 < UCS < 300 MPa)	Unpublished

Units used: Δt (μs/ft), E (MPa), ϕ (fraction)

fit the available data for shales quite well, especially high-porosity Tertiary shales. It should be noted that in the context of these equations, porosity is defined as the porosity that one would derive from well logs.

Another type of correlation is sometimes useful for predicting shale strength is its dependence on shaliness. This arises in a case study considered in Chapter 10, where the stability of wells drilled through shales is highly variable because of strong variations of rock strength. In very relatively low gamma shales (<80 API), the UCS is quite high ($\tilde{1}25$ MPa) whereas in shalier formations (<100 API) the UCS is only about 50 MPa.

Empirical relations relating the strength of carbonate rocks to geophysical parameters are presented in Table 4.3 and do a fairly poor job whether considering velocity, modulus or porosity data (Figure 4.16). One of the reasons for this is that there is a very large variation in strength of any given rock type. For example, strong carbonate rocks of low porosity, high velocity and high stiffness show strength values that vary by almost a factor of 4. The same is true for high porosity and low velocity, very large fluctuations in observed strength are observed. All of this emphasizes the importance of being able to calibrate strength relations in any particular case.

There have been relatively few attempts to find relationships between the angle of internal friction, Φ and geophysical measurements, in part because of the fact that even weak rocks have relatively high Φ (Figure 4.4), and there are relatively complex

Table 4.4. *Empirical relationships between* Φ *and other logged measurements. After Chang, Zoback* et al. *(2006).* Reprinted with permission of Elsevier

Eq. No.	Φ, degree	General Comments	Reference
27	$\sin^{-1}((V_p - 1000) / (V_p + 1000))$	Shale	(Lal 1999)
28	$70 - 0.417\,GR$	Shales with $60 < GR < 120$	Unpublished
29	$\tan^{-1}\left(\dfrac{78 - 0.4GR}{60}\right)$	Shaly sedimentary rocks	Unpublished

Units used: V_p (m/s), GR (API)

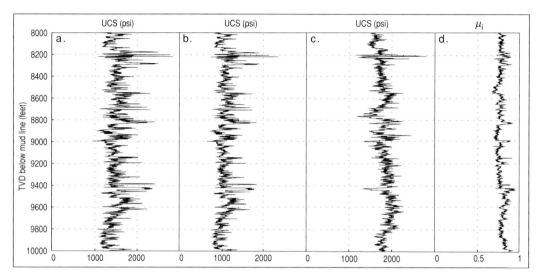

Figure 4.17. Utilization of equations (11) (a), (12) (b) and (18) (c) from Table 4.2, to predict rock strength for a shale section of a well drilled in the Gulf of Mexico. (d) The coefficient of internal friction is from equation (28) in Table 4.4. After Chang, Zoback *et al.* (2006). *Reprinted with permission of Elsevier.*

relationships between Φ and micro-mechanical features of rock such as a rock's stiffness, which largely depends on cementation and porosity. Nonetheless, some experimental evidence shows that shale with higher Young's modulus generally tends to possess a higher Φ (Lama and Vutukuri 1978). Two relationships relating Φ to rock properties for shale and shaly sedimentary rocks are listed in Table 4.4. It is relatively straightforward to show that the importance of Φ in wellbore stability analysis is much less significant than UCS.

An example illustrating how rock strength is determined from geophysical logs using three of the empirical relations in Table 4.2 is illustrated in Figures 4.17 and 4.18 for a shale section in a vertical well in the Gulf of Mexico. We focus on the interval from 8000 to 10,000 ft where there are logging data available that include compressional wave

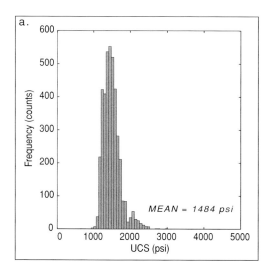

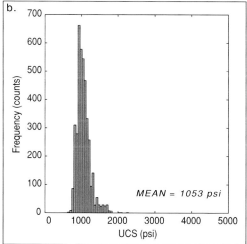

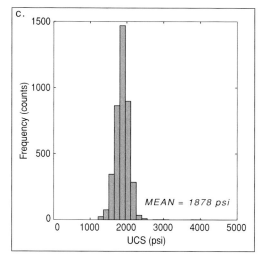

Figure 4.18. Histogram of shale strengths for the log-derived values shown in Figure 4.17: (a), (b) and (c) correspond to equations (11), (12) and (18) in Table 4.2, respectively. Note that the mean strength varies considerably, depending on which empirical relation is chosen. After Chang, Zoback *et al.* (2006). *Reprinted with permission of Elsevier.*

velocity, gamma ray and density. The coefficient of internal friction was determined using the gamma relation, equation (28) in Table 4.4. Although this interval is comprised of almost 100% shale, the value of μ_i obtained using equation (28) ranges between 0.7 and 0.84. Using the velocity data, the UCS was determined using equations (11) and (12) (Table 4.2, Figure 4.17a, b). While the overall shape of the two strength logs is approximately the same (as both are derived from the V_p data), the mean vertically averaged strength derived using equations (11) is 1484 ± 233 psi (Figure 4.18a) whereas that derived with equations (12) has a strength of 1053 ± 182 psi (Figure 4.18b). Porosity was derived from the density log assuming a matrix density of 2.65 g/cm^3 and a fluid density of 1.1 g/cm^3. The porosity-derived UCS shown in Figure 4.17c with equation (18) indicates an overall strength of 1878 ± 191 psi (Figure 4.18c). It is noteworthy in this single example that there is an almost factor of 2 variation in mean strength. However, as equation (12) was derived for the Gulf of Mexico region, it is probably more representative of actual rock strength at depth as it is was derived for formations of that particular region.

Shear-enhanced compaction

Another form of compressional rock failure of particular interest in porous rocks is sometimes referred to as shear-enhanced compaction. It refers to the fact that there will be irreversible deformation (*i.e.* plasticity) characterized by the loss of porosity due to pore collapse as confining pressure and/or shear stress increases beyond a limiting value. To represent these ductile yielding behaviors of rocks, *end-caps* (or yield surfaces of constant porosity) are used. These end-caps represent the locus of points that have reached the same volumetric plastic strain and their position (and exact shape) depends on the properties of the specific rock being considered.

A theoretical formalism known as the *Cambridge Clay* (or Cam-Clay) model is useful for describing much laboratory end-cap data (Desai and Siriwardane 1984). In this case, failure envelopes are determined by relatively simple laboratory experiments and are commonly represented in p–q space where p is the mean effective stress and q is the differential stress. Mathematically, the three principal stresses and p–q space are related as follows:

$$p = \frac{1}{3}J_1 = \frac{1}{3}(\sigma_1 + \sigma_2 + \sigma_3)$$

$$p = \frac{1}{3}(S_1 + S_2 + S_3) - P_P$$

(4.35)

$$q = \sqrt{3J_{2D}}$$

$$q^2 = \frac{1}{2}[(S_1 - S_2)^2 + (S_2 - S_3)^2 + (S_1 - S_3)^2]$$

(4.36)

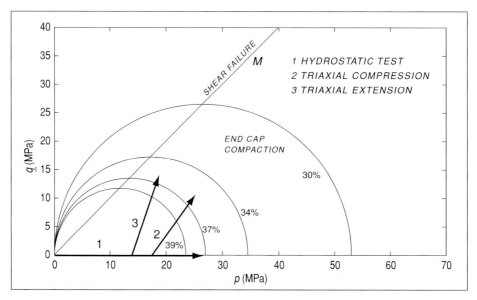

Figure 4.19. The Cam–Clay model of rock deformation is presented in p–q space as modified by Chan and Zoback (2002) following Desai and Siriwardane (1984) which allows one to define how inelastic porosity loss accompanies deformation. The contours defined by different porosities are sometimes called *end-caps*. Loading paths consistent with hydrostatic compression, triaxial compression and triaxial extension tests are shown. © *2002 Society Petroleum Engineers*

J_1 and J_{2D} are the first and the second invariant of the deviatoric stress tensor respectively. The equation of the yield loci shown in Figure 4.19 for the Cam-Clay model is given by Desai and Siriwardane (1984) as:

$$M^2 p^2 - M^2 p_0 p + q^2 = 0 \tag{4.37}$$

where M is known as the critical state line and can be expressed as $M = q/p$.

The Cam-Clay model in p–q space is illustrated in Figure 4.19 from Chan and Zoback (2002). Note that the shape of the yield surface as described by equation (4.37) in the Cam-Clay model is elliptical. If the *in situ* stress state in the reservoir is within the domain bounded by the failure envelope in p–q space, the formation is not likely to undergo plastic deformation. The intersection of the yielding locus and the p-axis is defined as p_0 (also known as the preconsolidation pressure) and each end-cap has its own unique p_0 that defines the hardening behavior of the rock sample. The value of p_0 can be determined easily from a series of hydrostatic compression tests in which porosity is measured as a function of confining pressure. Conceptually, it is easy to see why the end-caps should be roughly elliptical. Because shear stress facilitates the process of compaction and porosity loss, the mean confining pressure at which a certain end-cap is reached will decrease as shear stress increases.

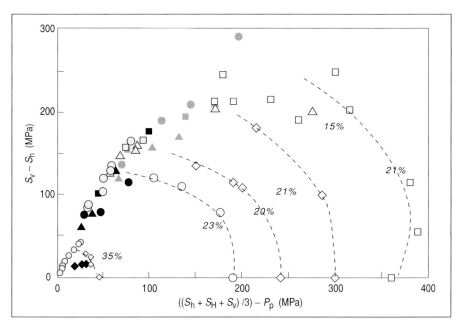

Figure 4.20. Compilation rock strength data for a wide variety of sandstones (different symbols) define the overall trend of irreversible porosity loss and confirms the general curvature of the end-caps to be similar to that predicted by the Cam-Clay model. After Schutjens, Hanssen *et al.* (2001). © *2001 Society Petroleum Engineers*

Three different loading paths are shown in Figure 4.19. Path 1 corresponds to hydrostatic loading to 26 MPa ($\underline{q} = 0$), Path 2 corresponds to a triaxial compression test ($\underline{q}/p = 3$) after loading to an initially hydrostatic pressure of 14 MPa, and Path 3 corresponds to triaxial extension ($\underline{q}/p = 3/2$) after hydrostatic loading to 18 MPa.

In weak formations such as weakly cemented sand, porous chalk or diatomite, once loading reaches an end-cap, compaction and grain rearrangement (and eventually grain crushing and pore collapse) will be the dominant deformation modes. If the loading path reaches the shear failure line, M, slip on a pre-existing fault will occur.

An example of end-cap deformation is illustrated in Figure 4.20 for a compilation of lab tests on a wide variety of sandstones (Schutjens, Hanssen *et al.* 2001). The contour lines show the end-caps, which demonstrate how porosity is irreversibly lost at shear stresses less than that required to cause shear failure (see Desai and Siriwardane 1984 and Wood 1990). Note that even in the absence of shear stress (*i.e.* moving just along the abscissa) porosity would be irreversibly lost as the mean stress increases from initial porosities greater than 35% to as low as 21% at $p \sim 360$ MPa. With increased $\underline{q}$, the contours that define the end-caps curve back toward the ordinate because the confining pressure required to cause a given reduction in porosity decreases a moderate amount.

In weak formations, such as chalks, grain crushing can occur at much lower pressures than those for sandstones (Teufel, Rhett *et al.* 1991).

   Because reservoir compaction associated with depletion is an important process in many reservoirs, inelastic compaction is discussed in detail in Chapter 12. In addition to porosity loss, there can be substantial permeability loss in compacting reservoirs as well as the possibility of surface subsidence and production-induced faulting in normal faulting environments. The degree to which these processes are manifest depends on the properties of the reservoir (compaction will be an important factor in weak formations such as chalks and highly compressible uncemented sands), the depth and thickness of the reservoir, the initial stress state and pore pressure and the reservoir stress path, or change in horizontal stress with depletion (as described in Chapter 3). Wong, David *et al.* (1997) demonstrated that the onset of grain crushing and pore collapse in sand reservoirs depends roughly on the product of the porosity times the grain radius. However, in uncemented or poorly cemented sand reservoirs, there will also be inelastic compaction due to grain rearrangement, which can be appreciable (Chapter 12).

Tensile rock failure

Compared to the compressional strength of rock (as discussed above) and the frictional strength of fractures and faults in earth's crust (as discussed below), the tensile strength of rock is relatively unimportant. The reasons for this are multifold: First, the tensile strength of essentially all rocks is quite low, on the order of just a few MPa (Lockner 1995) and when pre-existing flaws exist in rock (as is the case when considering any appreciable rock volume), tensile strength would be expected to be near zero. Second, as argued in Chapter 1, *in situ* stress at depth is never tensile. As discussed in Chapter 6, tensile fractures can occur around wellbores in some stress states because of the stress concentration at the wellbore wall. Hydraulic fracturing is a form of tensile failure that occurs when fluid pressure exceeds the local least principal stress. This can be a natural process, leading to the formation of joints in rock (opening-mode, planar fractures) as illustrated in the inset of Figure 4.21. While joints are relatively ubiquitous in nature, they are unlikely to have a significant effect on reservoir properties (such as bulk permeability) at depth because they are essentially closed at any finite effective stress. Because fracture permeability is highly dependent on the width of any open fracture at depth, small tensile micro-fractures will have little influence on flow. The extension of a tensile fracture also occurs during hydraulic fracturing operations when fluid pressure is intentionally raised above the least principal stress to propagate a fracture which is then filled with sand or another material as a propant to increase formation permeability (Chapter 6).

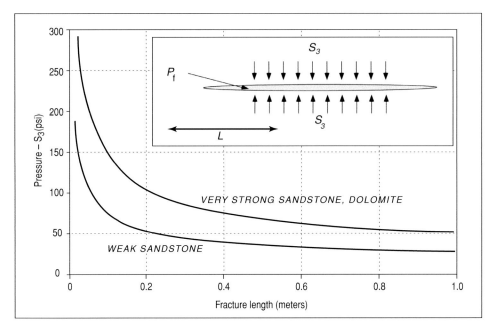

Figure 4.21. The difference between internal fluid pressure and the least principal stress as a function of fracture length for propagation of a Mode I fracture (see inset) for rocks with extremely high fracture toughness (such as very strong sandstone or dolomite) and very low fracture toughness (weakly cemented sandstone).

In the case of hydraulic fracture propagation, it is quite straightforward to demonstrate that rock strength in tension is essentially unimportant in the fracture extension process. In terms of fracture mechanics, the stress intensity at the tip of an opening mode planar fracture (referred to as a Mode I fracture), is given by

$$K_{\mathrm{i}} = (P_{\mathrm{f}} - S_3)\pi L^{1/2} \tag{4.38}$$

where K_{i} is the stress intensity factor, P_{f} is the pressure within the fracture (taken to be uniform for simplicity), L is the length of the fracture and S_3 is the least principal stress. Fracture propagation will occur when the stress intensity factor K_{i} exceeds K_{ic}, the critical stress intensity, or fracture toughness. Figure 4.21 shows the value of $(P_{\mathrm{f}} - S_3)$ required to cause failure as a function of fracture length L, for a rock with a high fracture toughness, such as a very strong, low-porosity sandstone or a strong dolomite, and a rock with a very low fracture toughness, such as a poorly cemented sandstone (Rummel and Winter 1983). It is clear that while the fracture toughness is important to initiate and initially extend a fracture, once a fracture reaches a length of a few tens of cm, extremely small pressures in excess of the least principal stress are required to make the fracture grow, regardless of the rock's fracture toughness. This means, of course, that the principal control on fracture propagation is that P_{f} exceed S_3 by only a small amount. Once the Mode I fracture starts to grow, the strength of the rock in tension is irrelevant.

Unlike compressional strength, tensile strength does not seem to be dependent on simple effective stress, especially in low-porosity/low-permeability rocks. Schmitt and Zoback (1992) demonstrated in the laboratory with granitic rocks that pore pressure acting in rocks had less of an effect on reducing the tensile stress at which failure would be expected. They attributed this effect to *dilatancy hardening* – as failure was approached the creation of micro-cracks in the incipient failure zone causes pore pressure to locally drop, thereby negating its effect on strength. It is not known how significant this effect is in higher-porosity rocks.

Shear failure and the frictional strength of rocks

Slip on faults is important in a number of geomechanical contexts. Slip on faults can shear well casings and it is well known that fluid injection associated with water flooding operations can induce earthquakes, for reasons explained below. As will be discussed in Chapter 12, some stress paths associated with reservoir depletion can induce normal faulting. Chapter 11 discusses fluid flow along active shear faults at a variety of scales. In this chapter we discuss the frictional strength of faults in order to provide constraints on the magnitudes of principal stresses at depth.

Friction experiments were first carried out by Leonardo da Vinci, whose work was later translated and expanded upon by Amontons. Da Vinci found that frictional sliding on a plane will occur when the ratio of shear to normal stress reaches a material property of the material, μ, the coefficient of friction. This is known as Amontons' law

$$\frac{\tau}{\sigma_n} = \mu \tag{4.39}$$

where τ is the shear stress resolved onto the sliding plane. The role of pore pressure on frictional sliding is introduced via σ_n, the effective normal stress, defined as $(S_n - P_p)$, where S_n is the normal stress resolved onto the sliding plane. Thus, raising the pore pressure on a fault (through fluid injection, for example) could cause fault slip by reducing the effective normal stress (Hubbert and Rubey 1959). The coefficient of friction, μ, is not to be confused with the coefficient of internal friction μ_i, defined above in the context of the linearized Mohr–Coulomb criterion. In fact, equation (4.39) appears to be the same as equation (4.3), with the cohesion set to zero. It is important to remember, however, that μ in equation (4.39) describes slip on a pre-existing fault whereas μ_i is defined to describe the increase in strength of intact rock with pressure (*i.e.* the slope of the failure line on a Mohr diagram) in the context of failure of an initially intact rock mass using the linearized Mohr–Coulomb failure criterion.

Because of his extensive research on friction (Coulomb 1773), equation (4.39) is sometimes referred to as the Coulomb criterion. One can define the Coulomb failure function (CFF) as

$$\text{CFF} = \tau - \mu\sigma_n \tag{4.40}$$

When the Coulomb failure function is negative, a fault is stable as the shear stress is insufficient to overcome the resistance to sliding, $\mu\sigma_n$. However, as CFF approaches zero, frictional sliding will occur on a pre-existing fault plane as there is sufficient shear stress to overcome the effective normal stress on the fault plane. Again, the CFF in this manner presupposes that the cohesive strength of a fault is very small compared to the shear and normal stresses acting upon it. As will be illustrated below, this assumption appears to be quite reasonable.

As mentioned above, equation (4.39) predicts that raising pore pressure would tend to de-stabilize faults and encourage slip to take place by raising the ratio of shear to normal stress on any pre-existing fault. While there have been many examples of seismicity apparently induced by fluid injection in oil fields (see the review by Grasso 1992), two experiments in the 1960s and 1970s in Colorado first drew attention to this phenomenon (Figure 4.22) and provided implicit support for the applicability of Amontons' law/Coulomb failure to crustal faulting. A consulting geologist in Denver named David Evans pointed out an apparent correlation between the number of earthquakes occurring at the Rocky Mountain Arsenal and the volume of waste fluid being injected into the fractured basement rocks at 3.7 km depth. Subsequently, Healy, Rubey *et al.* (1968) showed there to be a close correlation between the downhole pressure during injection and the number of earthquakes (Figure 4.22a). The focal mechanisms of the earthquakes were later shown to be normal faulting events. This enabled Zoback and Healy (1984) to demonstrate that the magnitudes of the vertical stress, least principal stress and pore pressure during injection were such that equation (4.39) was satisfied and induced seismicity was to be expected for a coefficient of friction of about 0.6 (see below). A similar study was carried out only a few years later at Rangeley, Colorado (Figure 4.22b) where water was being injected at high pressure in an attempt to improve production from the extremely low permeability Weber sandstone (Raleigh, Healy *et al.* 1976). In this case, it could be seen that a downhole pressure of 3700 psi (25.5 MPa) was required to induce slip on pre-existing faults in the area, as predicted by equation (4.39) (Zoback and Healy 1984).

As mentioned above, friction is a material property of a fault and Byerlee (1978) summarized numerous laboratory experiments on a wide variety of faults in different types of rock. He considered natural faults in rock, faults induced in triaxial compression tests and artificial faults (*i.e.* sawcuts in rock) of different roughness. His work (and that of many others) is summarized in Figure 4.23 (modified from Byerlee 1978). Note that for an extremely wide variety of rock types, Byerlee showed that at elevated effective normal stress ($\geq \sim 10$ MPa), friction on faults is independent of surface roughness,

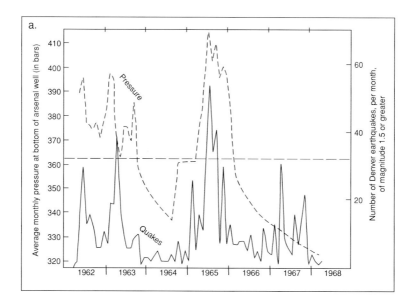

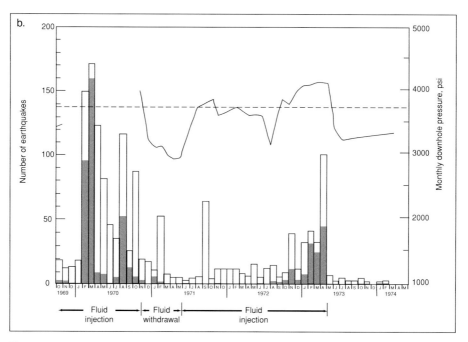

Figure 4.22. (a) Correlation between downhole pressure and earthquake occurrence during periods of fluid injection and seismicity at the Rocky Mountain Arsenal. Modified from Healy, Rubey *et al.* (1968). (b) Correlation between downhole pressure and earthquake occurrence triggered by fluid injection at the Rangely oil field in Colorado. After Raleigh, Healy *et al.* (1976).

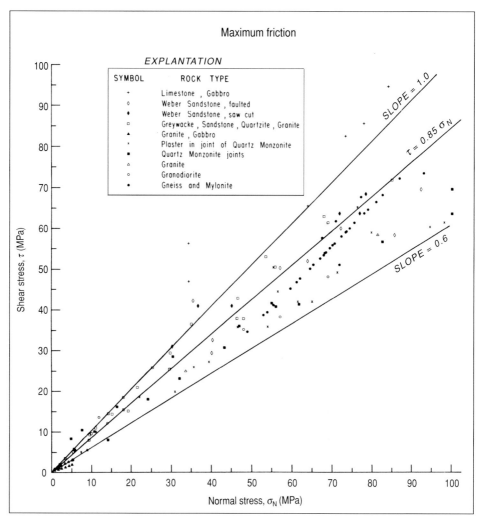

Figure 4.23. Rock mechanics tests on wide range of rocks (and plaster in a rock joint) demonstrating that the coefficient of friction (the ratio of shear to effective normal stress) ranges between 0.6 and 1.0 at effective confining pressures of interest here. Modified after Byerlee (1978).

normal stress, rate of slip, etc. such that the coefficient of friction is found to be within a relatively small range:

$$0.6 \leq \mu \leq 1.0 \tag{4.41}$$

This relation is sometimes known as Byerlee's law. In fact, John Jaeger, perhaps the leading figure in rock mechanics of the twentieth century, once said: *There are only two things you need to know about friction. It is always 0.6, and it will always make a monkey out of you.*

The critically stressed crust

Three independent lines of evidence indicate that intraplate continental crust is generally in a state of incipient, albeit slow, frictional failure: (i) the widespread occurrence of seismicity induced by either reservoir impoundment or fluid injection (Healy, Rubey *et al.* 1968; Raleigh, Healy *et al.* 1972; Pine, Jupe *et al.* 1990; Zoback and Harjes 1997); (ii) earthquakes triggered by small stress changes associated with other earthquakes (Stein, King *et al.* 1992); and (iii) *in situ* stress measurements in deep wells and boreholes (see the review by Townend and Zoback 2000). The *in situ* stress measurements further demonstrate that the stress magnitudes derived from Coulomb failure theory utilizing laboratory-derived frictional coefficients of 0.6–1.0 are consistent with measured stress magnitudes. This is well illustrated in Figure 4.24 by stress magnitude data collected in the KTB borehole to ~8 km depth. Measured stresses are quite high and consistent with the frictional faulting theory with a frictional coefficient of ~0.7 (Zoback, Apel *et al.* 1993; Brudy, Zoback *et al.* 1997). Further evidence for such a *frictional failure* stress state is provided by the fact that a series of earthquakes could be triggered at ~9 km depth in rock surrounding the KTB borehole with extremely low perturbations of the ambient, approximately hydrostatic pore pressure (Zoback and Harjes 1997). In Chapter 9 we evaluate stress magnitude data from a variety of sedimentary basins around the world that illustrate how stress magnitudes are in equilibrium with frictional strength in normal, strike-slip and reverse faulting environments.

That the state of stress in the crust is generally in a state of incipient frictional failure might seem surprising, especially for relatively stable intraplate areas. However, a reason for this can be easily visualized in terms of a simple cartoon as shown in Figure 4.25 (after Zoback and Townend 2001). The lithosphere as a whole (shown simply in Figure 4.25 as three distinct layers – the brittle upper crust, the ductile lower crust and the ductile uppermost mantle) must support plate driving forces. The figure indicates a power-law creep law (*e.g.* Brace and Kohlstedt 1980) typically used to characterize the ductile deformation of the lower crust and upper mantle. Because the applied force to the lithosphere will result in steady-state creep in the lower crust and upper mantle, as long as the "three-layer" lithosphere is coupled, stress will build up in the upper brittle layer due to the creep deformation in the layers below. Stress in the upper crust builds over time, eventually to the point of failure. The fact that intraplate earthquakes are relatively infrequent simply means that the ductile strain rate is low in the lower crust and upper mantle (Zoback, Townend *et al.* 2002). Zoback and Townend (2001) discuss the fact that at the relatively low strain rates characterizing intraplate regions, sufficient plate-driving force is available to maintain a "strong" brittle crust in a state of frictional failure equilibrium.

Stress measurements in many parts of the world indicate that earth's crust is in a state of frictional failure equilibrium as described by equation (4.39) and coefficients of

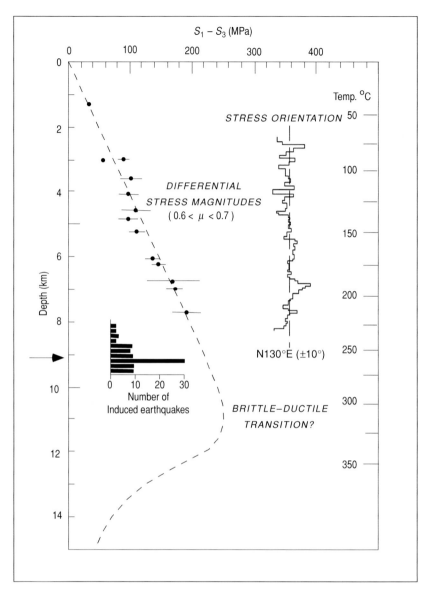

Figure 4.24. Stress measurements in the KTB scientific research well indicate a *strong* crust, in a state of failure equilibrium as predicted by Coulomb theory and laboratory-derived coefficients of friction of 0.6–0.7 (after Zoback and Harjes 1997). The arrow at 9.2 km depth indicates where the fluid injection experiment occurred.

friction consistent with equation (4.41). Figure 4.26, from Zoback and Townend (2001), shows a compilation of stress measurements in relatively deep wells and boreholes in various parts of the world. As shown, the ratio of the maximum and minimum effective stresses corresponds to a crust in frictional failure equilibrium with a coefficient of friction ranging between 0.6 and 1.0. A similar conclusion was reached by Zoback and

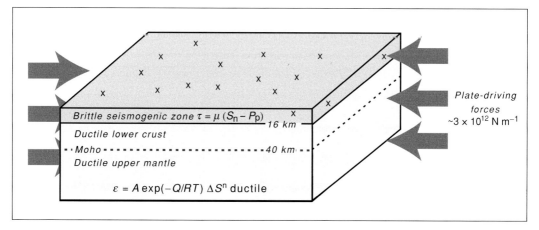

Figure 4.25. Schematic illustration of how the forces acting on the lithosphere keep the brittle crust in frictional equilibrium through creep in the lower crust and upper mantle (after Zoback and Townend 2001). *Reprinted with permission of Elsevier.*

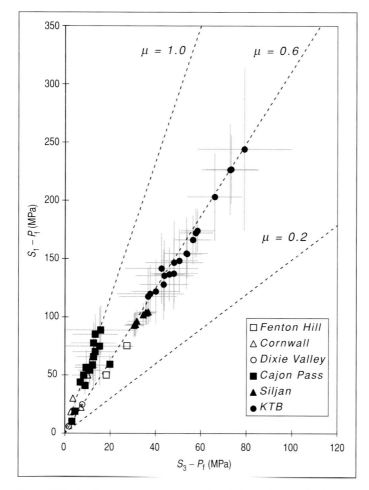

Figure 4.26. *In situ* stress measurements in relatively deep wells in crystalline rock indicate that stress magnitudes seem to be controlled by the frictional strength of faults with coefficients of friction between 0.6 and 1.0. After Zoback and Townend (2001). *Reprinted with permission of Elsevier.*

Healy (1984) based on stress measurements from much shallower depths. It should be noted that Townend (2003) pointed out that the uncertainty estimates in Figure 4.26 are likely significantly smaller than those shown.

There are two implications of the data shown in Figure 4.26. First, Byerlee's law (equation 4.41), defined on the basis of hundreds of laboratory experiments, appears to correspond to faults *in situ*. This is a rather amazing result when one considers the huge difference between the size of samples used for friction experiments in the lab and the size of real faults *in situ*, the variability of roughness of the sliding surface, the idealized conditions under which laboratory experiments are conducted, etc. Second, everywhere that stress magnitudes have been measured at appreciable depth, they indicate that they are controlled by the frictional strength of pre-existing faults in the crust. In other words, the earth's crust appears to be in a state of failure equilibrium and the law that describes that state is simple Coulomb friction, or Amontons' law as defined in equation (4.39). In fact, we will find that this is the case in many sedimentary basins around the world (Chapters 9–12).

In shaley rocks, it is widely suspected that the coefficient of friction may be significantly lower than 0.6, especially at low effective pressure. In fact, Byerlee pointed out that due to water layers within its crystallographic structure, montmorillonite has unusually lower frictional strength because intracrystalline pore pressure develops as it is being deformed. This manifests itself as a low friction. In recent drained laboratory tests Ewy, Stankowich *et al.* (2003) tested a deep clay and three shale samples and found coefficients of friction that range between 0.2 and 0.3. The subject of the frictional strength of shaley rocks is complicated, not only by the issue of pore pressure but by the fact that many tests reveal that clays that have low frictional strength at low effective pressure have higher frictional strength at higher effective pressures (Morrow, Radney *et al.* 1992; Moore and Lockner 2006).

Limits on *in situ* stress from the frictional strength of faults

Because earth's crust contains widely distributed faults, fractures, and planar discontinuities at many different scales and orientations, it is self-evident that stress magnitudes at depth (specifically, the differences in magnitude between the maximum and minimum principal stresses) will be limited by the frictional strength of these planar discontinuities. Building upon the arguments of the previous section, we demonstrate below how the frictional strength of pre-existing faults in the crust limits the possible range of stress magnitudes at any given depth and pore pressure. While the observation that the stress magnitudes in the crust in many areas are in equilibrium with its frictional strength (see Chapter 1) enables us to make specific predictions of stress magnitudes at depth, we will not assume that this is always the case. Rather, we will

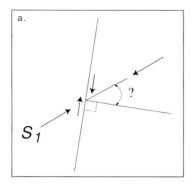
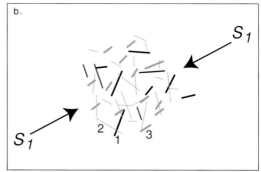

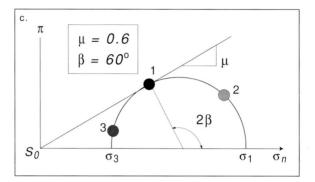

Figure 4.27. (a) Frictional sliding on an optimally oriented fault in two dimensions. (b) One can consider the Earth's crust as containing many faults at various orientations, only some of which are optimally oriented for frictional sliding. (c) Mohr diagram corresponding to faults of different orientations. The faults shown by black lines in (b) are optimally oriented for failure (labeled 1 in b and c), those shown in light gray in (b) (and labeled 2 in b and c) in (b) trend more perpendicular to S_{Hmax}, and have appreciable normal stress and little shear stress. The faults shown by heavy gray lines and labeled 3 in (b) are more parallel to S_{Hmax} have significantly less shear stress and less normal stress than optimally oriented faults as shown in (c).

simply assume that stresses in the Earth cannot be such that they exceed the frictional strength of pre-existing faults. This concept is schematically illustrated in Figure 4.27.

We first consider a single fault in two dimensions (Figure 4.27a) and ignore the magnitude of the intermediate principal effective stress because it is in the plane of the fault. The shear and normal stresses acting on a fault whose normal makes and an angle β with respect to the direction of maximum horizontal compression, S_1, was given by equations (4.1) and (4.2). Hence, the shear and normal stresses acting on the fault depend on the magnitudes of the principal stresses, pore pressure and the orientation of the fault with respect to the principal stresses.

It is clear in the Mohr diagram shown in Figure 4.27c that for any given value of σ_3 there is a maximum value of σ_1 established by the frictional strength of the pre-existing fault (the Mohr circle cannot exceed the maximum frictional strength). If the fault is critically oriented, that is, at the optimal angle for frictional sliding,

$$\beta = \frac{\pi}{4} + \frac{1}{2}\tan^{-1}\mu \tag{4.42}$$

Combining this relation with the principles of Anderson's classification scheme (see also Figure 5.1) it is straightforward to see (assuming $\mu \approx 0.6$):

- Normal faults are expected to form in conjugate pairs that dip $\sim 60°$ and strike parallel to the direction of S_{Hmax}.
- Strike-slip faults are expected to be vertical and form in conjugate pairs that strike $\sim 30°$ from the direction of S_{Hmax}.
- Reverse faults are expected to dip $\sim 30°$ and form in conjugate pairs that strike normal to the direction of S_{Hmax}.

Jaeger and Cook (1979) showed that the values of σ_1 and σ_3 (and hence S_1 and S_3) that corresponds to the situation where a critically oriented fault is at the frictional limit (*i.e.* equation 4.39 is satisfied) are given by:

$$\frac{\sigma_1}{\sigma_3} = \frac{S_1 - P_p}{S_3 - P_p} = [(\mu^2 + 1)^{1/2} + \mu]^2 \tag{4.43}$$

such that for $\mu = 0.6$ (see Figure 4.26),

$$\frac{\sigma_1}{\sigma_3} = 3.1 \tag{4.44}$$

In Figure 4.27c, we generalize this concept and illustrate the shear and normal stresses acting on faults with three different orientations. As this is a two-dimensional illustration, it is easiest to consider this sketch as a map view of vertical strike-slip faults in which $\sigma_2 = \sigma_v$ is in the plane of the faults (although this certainly need not be the case). In this case, the difference between σ_{Hmax} (defined as $S_{Hmax} - P_p$) and σ_{hmin} (defined as $S_{hmin} - P_p$), the maximum and minimum principal effective stresses for the case of strike-slip faulting, is limited by the frictional strength of these pre-existing faults as defined in equation (4.43). In other words, as S_{Hmax} increases with respect to S_{hmin}, the most well-oriented pre-existing faults begin to slip as soon as their frictional strength is reached (those shown by heavy black lines and labeled 1). As soon as these faults start to slip, further stress increases of S_{Hmax} with respect to S_{hmin} cannot occur. We refer to this subset of faults *in situ* (those subparallel to set 1) as *critically stressed* (*i.e.* to be just on the verge of slipping), whereas faults of other orientations are not (Figure 4.27b,c). The faults that are oriented almost orthogonally to S_{Hmax} have too much normal stress and not enough shear stress to slip (those shown by thin gray lines and labeled set 2) whereas those striking sub parallel to S_{Hmax} have low normal stress and low shear stress (those shown by thick gray lines and labeled set 3).

We can use equation (4.43) to estimate an upper bound for the ratio of the maximum and minimum effective stresses and use Anderson's faulting theory (Chapter 1) to determine which principal stress (*i.e.* S_{Hmax}, S_{hmin}, or S_v) corresponds to S_1, S_2 and S_3, respectively. This depends, of course, on whether it is a normal, strike-slip, or reverse faulting environment. In other words:

$$\text{Normal faulting } \frac{\sigma_1}{\sigma_3} = \frac{S_\text{v} - P_\text{p}}{S_\text{hmin} - P_\text{p}} \leq [(\mu^2 + 1)^{1/2} + \mu]^2 \qquad (4.45)$$

$$\text{Strike-slip faulting } \frac{\sigma_1}{\sigma_3} = \frac{S_\text{Hmax} - P_\text{p}}{S_\text{hmin} - P_\text{p}} \leq [(\mu^2 + 1)^{1/2} + \mu]^2 \qquad (4.46)$$

$$\text{Reverse faulting } \frac{\sigma_1}{\sigma_3} = \frac{S_{H\text{max}} - P_\text{p}}{S_v - P_\text{p}} \leq [(\mu^2 + 1)^{1/2} + \mu]^2 \qquad (4.47)$$

As referred to above, the limiting ratio of principal effective stress magnitudes defined in equations (4.45)–(4.47) is 3.1 for $\mu = 0.6$, regardless of whether one considers normal, strike-slip or reverse faulting regime. However, it should be obvious from these equations that stress magnitudes will increase with depth (as S_v increases with depth). The magnitude of pore pressure will affect stress magnitudes as will whether one is in a normal, strike-slip, or reverse faulting environment. This is illustrated in Figures 4.28 and 4.29, which are similar to Figure 1.4 except that we now include the limiting values of *in situ* principal stress differences at depth for both hydrostatic and overpressure conditions utilizing equations (4.45)–(4.47). In a normal faulting environment in which pore pressure is hydrostatic (Figure 4.28a), equation (4.45) defines the lowest value of the minimum principal stress with depth. It is straightforward to show that in an area of critically stressed normal faults, when pore pressure is hydrostatic, the lower bound value of the least principal stress $S_\text{hmin} \sim 0.6 S_\text{v}$, as illustrated by the heavy dashed line in Figure 4.28a. The magnitude of the least principal stress cannot be lower than this value because well-oriented normal faults would slip. Or in other words, the inequality in equation (4.45) would be violated. In the case of strike-slip faulting and hydrostatic pore pressure (Figure 4.28b), the maximum value of S_Hmax (as given by equation 4.46) depends on the magnitude of the minimum horizontal stress, S_hmin. If the value of the minimum principal stress is known (from extended leak-off tests or hydraulic fracturing, as discussed in Chapter 6), equation (4.46) can be used to put an upper bound on S_Hmax. The position of the heavy dashed line in Figure 4.28b shows the maximum value of S_Hmax for the S_hmin values shown by the tick marks. Finally, for reverse faulting (equation 4.47 and Figure 4.27c), because the least principal stress is the vertical stress, S_v, it is clear that the limiting value for S_Hmax (heavy dashed line) is very high. In fact, the limiting case for the value of S_Hmax is $\sim 2.2 S_\text{v}$ for hydrostatic pore pressure and $\mu = 0.6$.

Many regions around the world are characterized by a combination of normal and strike-slip faulting (such as western Europe) and reverse and strike-slip faulting (such as the coast ranges of western California). It is clear how these types of stress states come about. In an extensional environment, if S_hmin is near its lower limit ($\sim 0.6 S_\text{v}$) and S_Hmax near its upper limit S_v (such that $S_1 \approx S_2$), the equalities in equations (4.45) and (4.46) could both be met and both normal and strike-slip faults would be potentially active. In a

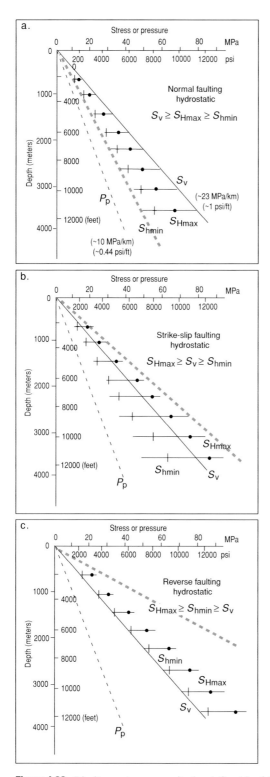

Figure 4.28. Limits on stress magnitudes defined by frictional faulting theory in normal (a), strike-slip and (b) reverse faulting (c) regimes assuming hydrostatic pore pressure. The heavy line in (a) shows the minimum value of the least principal stress, S_{hmin}, in normal faulting environments, in (b) the maximum value of S_{Hmax} for the values of S_{hmin} shown by the ticks and (c) the maximum value of S_{Hmax} for reverse faulting regimes where the least principal stress is the vertical stress S_v.

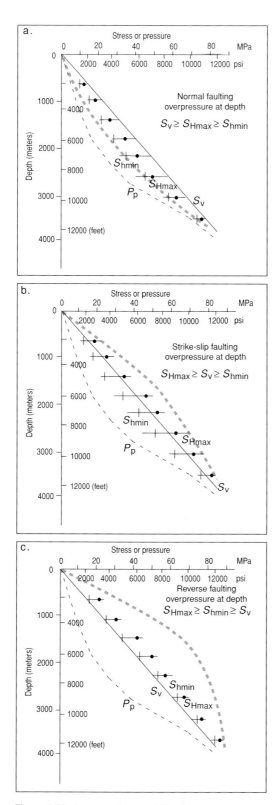

Figure 4.29. Same as Figure 4.28 when overpressure develops at depth as shown. Note that in all three stress states, when pore pressure is nearly lithostatic, all three principal stresses are also close to S_v.

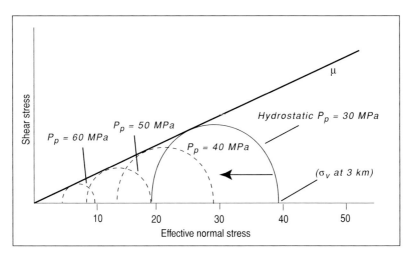

Figure 4.30. In terms of frictional faulting theory, as pore pressure increases (and effective stress decreases), the difference between the maximum and minimum effective principal stress (which defines the size of the Mohr circle) decreases with increasing pore pressure at the same depth.

compressional environment, if S_{hmin} is approximately equal S_v (such that $S_2 \approx S_3$), and S_{Hmax} much larger, it would correspond to a state in which both strike-slip and reverse faults were potentially active.

Figures 4.29a–c illustrate the limiting values of stress magnitudes when pore pressure increases markedly with depth in a manner similar to cases like that illustrated in Figure 1.4. As pore pressure approaches S_v at great depth, as is the case in some sedimentary basins, the limiting stress magnitudes (heavy dashed lines) are not significantly different from the vertical stress, regardless of whether it is a normal, strike-slip or reverse faulting environment. While this might seem counter-intuitive, when pore pressure is extremely high, fault slip will occur on well-oriented faults when there is only a small difference between the maximum and minimum principal stresses.

The way in which the difference in principal stresses is affected by pore pressure is illustrated by the Mohr circles in Figure 4.30. When stress magnitudes are controlled by the frictional strength of faults, as pore pressure increases, the maximum size of the Mohr circle decreases. In other words, as pore pressure gets higher and higher, faulting occurs with smaller and smaller differences between the maximum and minimum effective principal stress. Hence, the Mohr circle gets smaller as pore pressures increases.

While Figure 4.30 may seem obvious, there are two issues to draw attention to because it is most commonly assumed that for given values of S_1, S_2 and S_3, changing pore pressure simply shifts the position of a Mohr circle along the abscissa. The first point worth emphasizing is that when the state of stress at depth is limited by the frictional strength of pre-existing faults, the ratio of effective stresses remains the same (in accord with equations 4.45, 4.46 and 4.47) as pore pressure changes (as illustrated). But this is not true of the ratios (or differences) in the absolute stress magnitudes (as shown in Figure 4.29) such that the higher the pore pressure, the lower the principal stress differences. At extremely high pore pressure, relatively small stress perturbations are sufficient to change the style of faulting from one stress regime to the other (for example, to go from normal faulting to reverse faulting). This is dramatically different from the case in which pore pressure is hydrostatic. The second point to note is that perturbations of pore pressure associated with depletion (or injection) will also affect stress magnitudes through the types of poroelastic effects discussed in Chapter 3. Hence, the size and position of the Mohr circle is affected by the change in pore pressure. This can have an important influence on reservoir behavior, especially in normal faulting regions (Chapter 12).

Stress polygon

For reasons that will become clear when we start to utilize observations of wellbore failure to constrain stress magnitudes in Chapters 7 and 8, it is convenient to be able to simply estimate the range of possible stress states at any given depth and pore pressure given that stress in the crust is limited by the frictional strength of faults. Figure 4.31 illustrates the range of allowable values for horizontal principal stresses in the earth's crust for normal, strike-slip and reverse faulting environments using equations (4.45), (4.46) and (4.47) and E. M. Anderson's stress and faulting classification system discussed in Chapter 1. Such figures, introduced by Zoback, Mastin *et al.* (1987) and Moos and Zoback (1990), allow one to illustrate the range of possible magnitudes of S_{hmin} and S_{Hmax} at a particular depth for a given pore pressure and assumed coefficient of friction (here taken to be 0.6). Figure 4.31 is illustrated for a depth of 3 km assuming an average overburden density of 2.3 g/cm^3. Allowable stress states are shown for hydrostatic pore pressure (Figure 4.31a) and significant overpressure (Figure 4.31b). The contruction of such figures is straightforward. The fact that $S_{Hmax} \geq S_{hmin}$ requires all possible stress states to be above a diagonal line of unit slope. The vertical and horizontal lines intersecting at $S_{Hmax} = S_{hmin} = S_v$ separate the stress fields associated with normal (N), strike-slip (SS) and reverse (RF) faulting stress environments as defined by Anderson. The vertical line in the lower left of the polygon indicates the lowest value of S_{hmin} possible in a normal faulting environment as predicted using equation (4.45). In other words, for the value of S_{hmin} shown by this line, a Mohr circle would exactly touch a frictional failure envelope with a slope of 0.6. Similarly, the horizontal line defining the top of the polygon corresponds the value of S_{Hmax} at which reverse faulting would occur (equation 4.47). The diagonal line bounding the polygon on the upper left corresponds to the value of S_{Hmax} at which strike-slip faulting would occur for a given value of S_{hmin} (equation 4.46). Thus, in every case, the stress at depth must be somewhere within

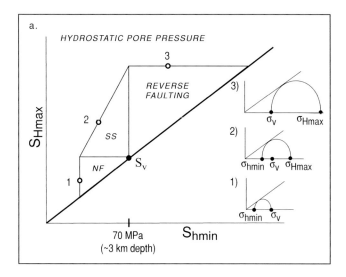

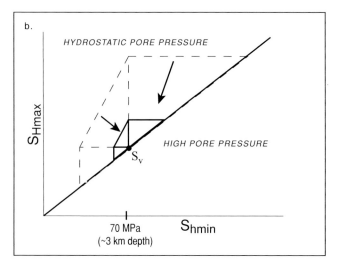

Figure 4.31. Polygons which define possible stress magnitudes at a given depth are shown for a depth of 3 km for (a) hydrostatic pore pressure and (b) pore pressure equal to 80% of the overburden. After Zoback, Mastin *et al.* (1987) and Moos and Zoback (1990).

the stress polygon. If the state of stress is in frictional failure equilibrium, the state of stress falls on the outer periphery of the polygon, depending, of course, on whether the stress state is normal, strike-slip or reverse faulting. As demonstrated in Chapter 9, *in situ* stress measurements from sedimentary basins around the world confirm the fact that the differences in stress magnitudes are frequently limited by the frictional strength of pre-existing faults that are well-oriented for slip in the current stress field and coefficients of friction of 0.6–0.7 seem to work quite well. In terms of Figure 4.31, this means that the stress state *in situ* is often found to lie around the periphery of the figure.

Figure 4.31b again illustrates the fact that elevated pore pressure reduces the difference between principal stresses at depth as shown previously in Figure 4.30. When pore pressure is elevated, all three principal stresses are close in magnitude to the vertical stress and relatively small changes in the stress field can cause a transition from one style of faulting to another. Moos and Zoback (1993) hypothesize that because of elevated pore pressure at depth in the vicinity of Long Valley caldera, the style of faulting goes from NF/SS faulting on one side of the caldera to RF/SS faulting on the other side as the direction of the horizontal principal stresses change.

The stress polygon shown in Figure 4.31a permits a very wide range of stress values at depth and would not seem to be of much practical use in limiting stress magnitudes. However, as extended leak-off tests or hydraulic fracturing tests are often available to provide a good estimate of the least principal stress (Chapter 6), the polygon is useful for estimating the possible range of values of S_{Hmax}. As noted above, we will illustrate in Chapters 7 and 8 that if one also has information about the existence of either compressive or tensile wellbore failures, one can often put relatively narrow (and hence, useful) bounds on possible stress states at depth. In other words, by combining the constraints on stress magnitudes obtained from the frictional strength of the crust, measurements of the least principal stress from leak-off tests and observations of wellbore failure place strong constraints on the *in situ* stress state (Chapters 6–8) which can be used to address the range of problems encountered in reservoir geomechanics addressed in Chapters 10–12.

5 Faults and fractures at depth

In this chapter I consider a number of topics related to faults and fractures in rock. Faults and fractures exist in essentially all rocks at depth and can have a profound effect on fluid transport, mechanical properties and wellbore stability. As discussed in Chapter 4 (and later demonstrated through a number of case studies), frictional slip along pre-existing fractures and faults limits *in situ* stress magnitudes in a predictable – and useful – way.

To begin this chapter, I distinguish between opening mode (Mode I) fractures and faults, and briefly discuss the importance of faults in influencing fluid flow in low permeability rock. The influence of faults on permeability is discussed at length in Chapter 11, as well as the sealing (or leakage) potential of reservoir-bounding faults. The manner in which slip along weak bedding planes can affect wellbore stability is discussed in Chapter 10. I briefly discuss wellbore imaging devices as it is now routine to use such devices to map fractures and faults in reservoirs, and then discuss common techniques for representing fracture orientation data, including stereonets and three-dimensional Mohr diagrams, when the state of stress is known. Faulting in three dimensions is revisited in Chapter 11. I conclude this chapter by briefly discussing earthquake focal mechanisms and their use in determining approximate stress orientations and relative stress magnitudes.

There are a number of books and collections of scientific papers on the subject of fractures and faults in rock. Of particular note are the compilations by Long *et al.* (1996); Jones, Fisher *et al.* (1998) Hoak, Klawitter *et al.* (1997) and the collections of papers on the mechanical involvement of fluids in faulting (Hickman, Sibson *et al.* 1995; Haneberg, Mozley *et al.* 1999; Faybishenko, Witherspoon *et al.* 2000; Vigneresse 2001; Jones, Fisher *et al.* 1998; Davies and Handschy 2003). Hence, the purpose of this chapter is not to provide a comprehensive review of this subject. Rather, my goal is to cover a number of basic principles about the nature of fractures and faults at depth and provide a basis for concepts discussed in subsequent chapters. For reasons that will soon be apparent, we will be principally concerned with faults; planar discontinuities associated with shear deformation. As the topic of this book is geomechanics, I consider only mechanical discontinuities at depth and not those associated with chemical processes – dissolution features, stylolites, etc. – which are encountered

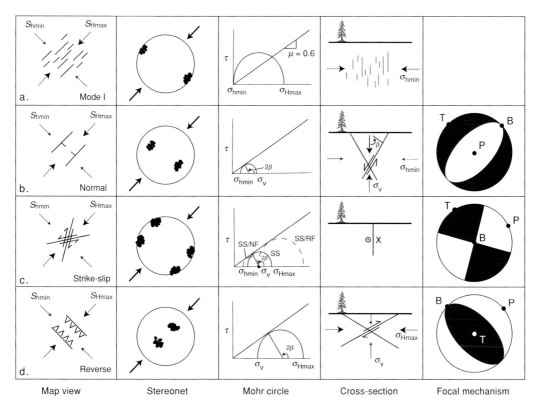

| Map view | Stereonet | Mohr circle | Cross-section | Focal mechanism |

Figure 5.1. Schematic illustration of the orientation of various types of fractures and faults with respect to the orientation of S_{Hmax} and S_{hmin}. (a) Mode I fractures and joints are expected to form parallel to S_{Hmax} and normal to S_{hmin}. (b) Conjugate strike-slip faults are expected to be vertical and strike $\sim 30°$ from the direction of S_{Hmax} (for $\mu \sim 0.6$). (c) Reverse faults are expected to dip $\sim 30°$ (for $\mu \sim 0.6$) and strike normal to the direction of S_{Hmax}. (d) Conjugate normal faults are expected to dip $\sim 60°$ (for $\mu \sim 0.6$) and strike parallel to the direction of S_{Hmax}. Because fractures and faults are introduced during multiple deformational episodes (depending on the age and geologic history of the formation) it is common for formations to contain numerous fractures at a variety of orientations.

in carbonate rocks, although such features may play a role in localizing subsequent shear deformation.

The relationship between the *in situ* state of stress and the orientation of hydraulically conductive fractures is frequently viewed in the context of Mode I fractures – extensional fractures oriented perpendicular to the least principal stress (Secor 1965; du Rouchet 1981; Nur and Walder 1990). There are a number of excellent papers on joints and Mode I fractures in rock (see the review by Pollard and Aydin 1988) and a number of papers on the application of the theory of fracture mechanics to rock (including utilizing shear fracture Modes 2 and 3 representations of faults) is presented by Atkinson (1987). As illustrated in Figure 5.1a, if the least principal stress is S_{hmin} (as is true in normal and strike-slip faulting regimes), Mode I fractures would be expected to form in the

S_Hmax–S_v plane. If such fractures were to form in the current stress field and have an appreciable effect on fluid flow in otherwise low permeability reservoirs, it would result in a simple relationship between fracture orientation, stress orientation and permeability anisotropy. Moreover, the simplistic cartoon shown in Figure 5.1a has straightforward implications for using geophysical techniques such as seismic velocity anisotropy, shear-wave splitting, and amplitude versus offset (AVO) to identify *in situ* directions of permeability anisotropy (*e.g.* Crampin 1985; Winterstein and Meadows 1995). The subject of the relationships among freacture orientation, stress orientation and shear velocity will be revisited at the end of Chapter 8.

Faults, fractures and fluid flow

While Mode I features are ubiquitous in some outcrops (*e.g.* Engelder 1987; Lorenz, Teufel *et al.* 1991) and can be seen as micro-cracks in core (e.g. Laubach 1997), it is unlikely that they contribute appreciably to fluid flow at depth where appreciable stresses exist. To consider flow through a fracture, we begin by considering a parallel plate approximation for fluid flow through a planar fracture. For a given fluid viscosity, η, the volumetric flow rate, Q, resulting from a pressure gradient, ∇P, is dependent on the cube of the separation between the plates, b,

$$Q = \frac{b^3}{12\eta} \nabla P \tag{5.1}$$

To make this more relevant to flow through a Mode I fracture, consider flow through a long crack of length L, with elliptical cross-section (such as that shown in the inset of Figure 4.21). The maximum separation aperture of the fracture at its midpoint is given by

$$b_\text{max} = \frac{2(P_\text{f} - S_3)L(1 - v^2)}{E} \tag{5.2}$$

where P_f is the fluid pressure in the fracture, v is Poisson's ratio and E is Young's modulus. This results in a flow rate given by

$$Q = \frac{\pi}{8\eta} \left(\frac{b_\text{max}}{2} \right)^3 \nabla P \tag{5.3}$$

which yields

$$Q = \frac{\pi}{8\eta} \left[\frac{L(1 - v^2)(P_\text{f} - S_3)}{E} \right]^3 \nabla P \tag{5.4}$$

Hence, the flow rate through a fracture in response to a pressure gradient will be proportional to the cube of the product of the length times the difference between the fluid pressure inside the fracture (acting to open it) and the least principal stress normal to it (acting to close it). This said, it is not likely that Mode I fractures affect fluid flow because they cannot have significant aperture at depth. As illustrated in Figure 4.21, when P_f slightly exceeds S_3, fractures of any appreciable length would be expected to propagate, thereby dropping P_f and causing the fracture to close. In fact, because the static case in the earth is that $(P_f-S_3) < 0$, transient high fluid pressures are required to initiate Mode I fractures (as natural hydrofracs), but following initiation, the pressure is expected to drop and the fractures to close. Hence, only extremely small fracture apertures would be expected, having little effect on flow. Let us consider 0.3 MPa as a reasonable upper bound for $P_f - S_3$ in a one meter long Mode I fracture because of the relative ease with the fracture would propagate (Figure 4.21). For reasonable values of v and E, equation (5.2) demonstrates that the maximum aperture of a Mode I fracture would be on the order of 0.01 mm. Obviously, considering a fracture to be only 1 m long is arbitrary (especially because b_{max} increases as L increases), but as L increases, the maximum value of $P_f - S_3$ decreases thereby limiting b_{max} (equation 5.2). Of course, real Mode I fractures in rock will not have perfectly smooth surfaces so that even when they are closed, a finite aperture will remain (Brown and Scholz 1986) such that in rocks with almost no matrix permeability, closed Mode 1 fractures can enhance flow to a some extent.

Faults (Mode 2 or 3 fractures that have appreciable shear deformation) are likely to be much better conduits for flow than Mode I fractures. Figure 5.2a (modified from Dholakia, Aydin *et al.* 1998) schematically illustrates how faults evolve from initially planar Mode I fractures, sometimes called joints, or in some cases, bedding planes. After the passage of time and rotation of principal stresses, shear stress acting on a planar discontinuity can cause slip to occur. In cemented rocks, shearing will cause brecciation (fragmentation and grain breakage) along the fault surface (as well as dilatancy associated with shear) as well as damage to the rocks adjacent to the fault plane. Both processes enable the fault to maintain permeability even if considerable effective normal stress acts across the fault at depth. For this reason, faults that are active in the current stress field can have significant effects on fluid flow in many reservoirs (Barton, Zoback *et al.* 1995). This will be discussed at greater length in Chapter 11. It should be pointed out that the terms *fractures* and *faults* are used somewhat informally in this and the chapters that follow. It should be emphasized that it is likely that with the exception of bedding planes, the majority of planar features observed in image logs (next section) that will have the greatest effect on the flow properties of formations at depth are, in fact, faults – planar discontinuities with a finite amount of shear deformation.

The photographs in Figure 5.2a,b (also from Dholakia, Aydin *et al.* 1998) illustrates the principle of fault-controlled permeability in the Monterey formation of western California at two different scales. The Monterey is a Miocene age siliceous shale with extremely low matrix permeability. It is both the source rock and reservoir for many oil fields in the region. The porosity created in fault-related breccia zones encountered in

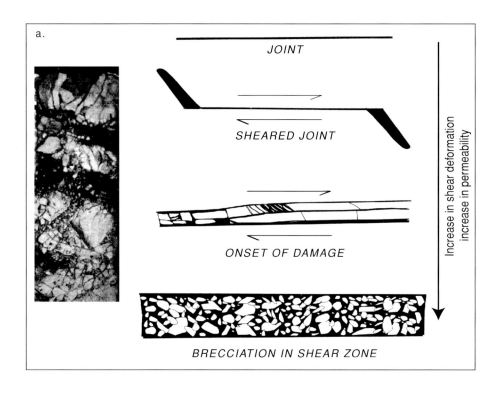

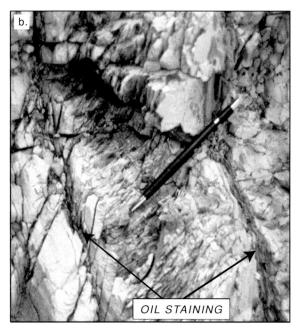

Figure 5.2. Schematic illustration of the evolution of a fault from a joint (after Dholakia, Aydin *et al*. 1998). As shear deformation occurs, brecciation results in interconnected porosity thus enhancing formation permeability. In the Monterey formation of California, oil migration is strongly influenced by the porosity generated by brecciation accompanying shear deformation on faults. This can be observed at various scales in core (a) and outcrop (b). *AAPG© 1998 reprinted by permission of the AAPG whose permission is required for futher use.*

core samples of the Antelope shale from the Buena Vista Hills field of the San Joaquin basin (left side of Figure 5.2a), as well as outcrops along the coastline (Figure 5.2b), are clearly associated with the presence of hydrocarbons. Thus, the enhancement of permeability resulting from the presence of faults in the Monterey is critically important for hydrocarbon production.

Figures 5.1b–d illustrates the idealized relationships between conjugate sets of normal, strike-slip and reverse faults and the horizontal principal stress (as well as the corresponding Mohr circles). Recalling subjects first mentioned in Chapters 1 and 4 related to Andersonian faulting theory and Mohr–Coulomb failure, respectively, Figures 5.1b–d illustrate the orientation of shear faults with respect to the horizontal and vertical principal stresses, associated Mohr circles and earthquake focal plane mechanisms associated with normal, strike-slip and reverse faulting. Lower hemisphere stereonets (second column) and earthquake focal plane mechanisms (fifth column) are described below. The first and fourth columns (map views and cross-sections) illustrate the geometrical relations discussed in Chapter 4 in the context of equation (4.43). For a coefficient of friction of 0.6, active normal faults (Figure 5.1b) are expected to strike nearly parallel to the direction of S_{Hmax} and conjugate fault sets are expected to be active that dip $\sim 60°$ from horizontal in the direction of S_{hmin}. Strike-slip faults (Figure 5.1c) are expected to be nearly vertical and form in conjugate directions approximately $30°$ from the direction of S_{Hmax}. Reverse faults (Figure 5.1d) are expected to strike in a direction nearly parallel to the direction of S_{hmin} and dip approximately $30°$ in the S_{Hmax} direction. The Mohr circles associated with each of these stress states (middle column) simply illustrate the relative magnitudes of the three principal stresses (shown as effective stresses) associated with each stress state.

There are three points that need to be remembered about the idealized relationships illustrated in Figure 5.1. First, these figures illustrate the relationship between potentially active faults and the stress state that caused them. In reality, many fractures and faults (of quite variable orientation) may be present *in situ* that have been introduced by various deformational episodes throughout the history of a given formation. It is likely that many of these faults may be inactive (*dead*) in the current stress field. Because currently active faults seem most capable of affecting permeability and reservoir performance (see also Chapter 11), it will be the subset of all faults *in situ* that are currently active today that will be of primary interest. The second point to note is that in many parts of the world a *transitional* stress state is observed. That is, a stress state associated with concurrent strike-slip and normal faulting ($S_{Hmax} \sim S_v > S_{hmin}$) in which both strike-slip and normal faults are potentially active or strike-slip and reverse faulting regime ($S_{Hmax} > S_v \sim S_{hmin}$) in which strike-slip and reverse faults are potentially active. Examples of these stress states will be seen in Chapters 9–11. Finally, while the concept of conjugate fault sets is theoretically valid, in nature one set of faults is usually dominant such that the simple symmetry seen in Figure 5.1 is a reasonable idealization, but is rarely seen.

Wellbore imaging

Wellbore imaging devices make it possible to obtain direct information on the distribution and orientation of fractures and faults encountered at depth. One family of wellbore image tools is collectively known as the ultrasonic borehole televiewer (BHTV). Such tools scan the wellbore wall with a centralized rotating ultrasonic (several hundred kilohertz to ~1 megahertz) transducer that is oriented with respect to magnetic north (Figure 5.3a from Zemanek, Glenn *et al.* 1970). The amplitude of the reflected pulse is diminished when the wellbore wall is rough (such as where a fracture or bedding plane intersects the well) and the travel time increases when the wellbore radius is enlarged by features such as wellbore breakouts (Chapter 6). These devices provide an image of both the acoustic reflectance and the radius of the wellbore such that it is possible to construct three-dimensional images of a wellbore (Figure 5.3b after Barton, Tessler *et al.* 1991). The reflectance depends on both the stiffness of the formation and the smoothness of the wellbore wall. Figure 5.3d illustrates an *unwrapped* image of the wellbore wall in which position around the well (with respect to north in this case) is shown on the abscissa and depth is shown on the ordinate. The amplitude of the reflected pulse is displayed as brightness. In such a display, planar fractures (or bedding planes) cutting the wellbore (Figure 5.3c) have a sinusoidal appearance (Figure 5.3d) resulting from the low amplitude of the reflected acoustic pulse along the intersection of the fault plane and wellbore wall. The dip direction is obvious in the unwrapped image (the direction of the lowest point where the fracture leaves the wellbore) and the amount of dip is determined from

$$\text{Dip} = \tan^{-1}(h/d) \tag{5.5}$$

where h is the height of the fracture as measured at the top and bottom of its intersection with the wellbore and d is the diameter of the well. In Chapter 6, we will demonstrate another important application of borehole televiewer data in the context of analysis of stress-induced compressional wellbore failures (or breakouts) as the time it takes the pulse to travel to/from the wellbore wall (and knowledge of the acoustic velocity of the wellbore fluid) enables one to reconstruct the detailed cross-sectional shape of the wellbore wall. Ultrasonic wellbore imaging is now available from a number of geophysical logging companies. While the details of operation of these types of instruments are slightly different (such as the number of pulses per rotation, the frequency of the ultrasonic transducer and the way in which the transducer beam is focused on the wellbore wall), the fundamental operation of all such tools is quite similar.

Figure 5.4b is an unwrapped image of wellbore wall made with the other type of image data used widely in the oil and gas industry, an electrical imaging device that uses arrays of electrodes on pads mounted on arms that press against the wellbore wall. The imaging device (Figure 5.4a from Ekstrom, Dahan *et al.* 1987) monitors the contact resistance

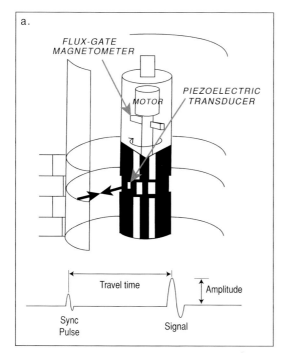

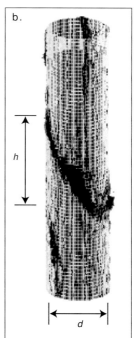

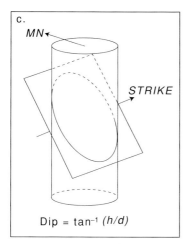

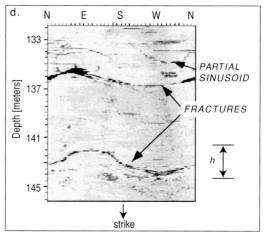

Figure 5.3. The principles of operation of an ultrasonic borehole televiewer (after Zemanek, Glenn *et al.* 1970). (a) An ultrasonic transducer is mounted on a rotating shaft. The transducer emits a high-frequency pulse that is transmitted through the wellbore fluid, reflected off the wellbore wall and returned to the transducer. Typically, several hundred pulses are emitted per rotation. A magnetometer in the tool allows the orientation of the transducer to be known with respect to magnetic North. (b) The amplitude data can be displayed in three-dimensional views that illustrate how faults or bedding planes cut across the wellbore. (c) Schematic view of a plane cutting through a wellbore. (d) An *unwrapped* view of a wellbore image with depth on the ordinate and azimuth (or position around the wellbore) on the ordinate. In this case planar features scutting through the wellbore such as bedding planes or fractures appear as sinusoids.

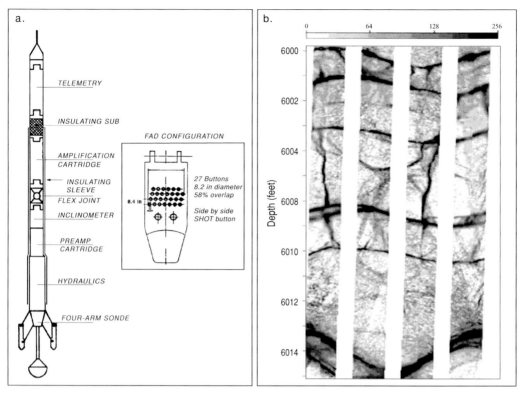

Figure 5.4. The principles of electrical imaging devices (after Ekstrom, Dahan *et al.* 1987).
(a) Arrays of electrodes are deployed on pads mounted on four or six caliper arms and pushed
against the side of the well. The entire pad is kept at constant voltage with respect to a reference
electrode, and the current needed to maintain a constant voltage at each electrode is an indication of
the contact resistance, which depends on the smoothness of the wellbore wall. (b) It is most
common to display these data as unwrapped images of the wellbore wall.

of an array of electrodes which are depth shifted as the tool is pulled up the hole so as to
achieve an extremely small effective spacing between measurement points. Thus, these
types of tools create a fine-scale map of the smoothness of the wellbore wall revealing
with great precision features such as bedding planes, fractures and features such as
drilling-induced tensile wall fractures (Chapter 6). Because the arrays of electrodes
are in direct contact with the wellbore wall, they tend to be capable of imaging finer
scale fractures than borehole televiewers, but provide less useful information about the
size and shape of the well. As with televiewers, wellbore imaging with these types of
tools is now widely available commercially. Some companies operate tools with four
pads, others with six, which cover various fractions of the wellbore circumference.
Nonetheless, the principles of operation are quite similar. The gaps in Figure 5.4b
represent the areas between electrode arrays on the four pads of this tool where no data
are collected.

Although wellbore image data only provide information on the apparent aperture of faults where they are intersected by a wellbore, empirical relations have been proposed that attempt to relate fault length to aperture (*e.g.* Gudmundsson 2000) and enable one to develop ideas about fracture networks from wellbore image data. However, it should be pointed out that the aperture at the wellbore wall is not the actual width of the fault plane. As discussed by Barton, Tessler *et al.* (1991), image logs are basically measuring the smoothness of the wellbore wall, such that the logs are quite sensitive to any spalling at the wellbore wall that occurs as the bit penetrates the fault plane. Thus, the apparent aperture is always going to be larger than the actual aperture. Barton, Tessler *et al.* (1991) point out that it is sometimes useful to qualitatively distinguish large from small aperture faults during the analysis of image data.

One obvious limitation of imaging tools is that they will undersample fractures and faults whose planes are nearly parallel to the wellbore axis. This is most easily visualized for a vertical well – the probability of intersecting horizontal fractures in the formation is one, but the probability of intersecting vertical fractures is essentially zero. Following Hudson and Priest (1983), it is straightforward to estimate the correction to apply to a fracture population of a given dip. If we have a set of fractures of constant dip, ϕ, with an average separation D (measured normal to the fracture planes), the number of fractures observed along a length of wellbore, $L = D/\cos\phi$, $N_{obs}(\phi)$ must be corrected in order to obtain the true number of fractures that occur in the formation over a similar distance, $N_{true}(\phi)$ via

$$N_{true}(\phi) = (\cos\phi)^{-1} N_{obs}(\phi) \tag{5.6}$$

This is illustrated for densely fractured granitic rock by Barton and Zoback (1992), who also discuss sampling problems associated with truncation, the inability to detect extremely small fractures, and censoring, the fact that when an extremely large fault zone is penetrated by a wellbore, it is often difficult to identify the fault plane because the wellbore is so strongly perturbed that data quality is extremely poor.

Representation of fracture and fault data at depth

Figure 5.5 illustrates the commonly used parameters that describe the orientation of faults and fractures in three dimensions. Fault *strike* refers to the azimuth of a horizontal line in the plane of the fracture and is measured from north. Unless otherwise noted, dip is measured from a horizontal plane and is positive to the right, when *looking* in the strike direction. Dip direction is often used instead of strike to define fracture orientation. Measured from north, dip direction is the azimuth of the projection of a vector in the plane of the fault that is normal to the strike direction and points down dip. Slip on the fault is also defined by a vector, which can be thought of as the

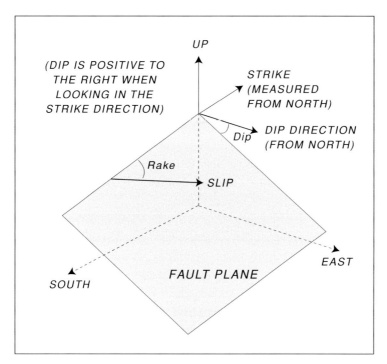

Figure 5.5. Definition of strike, dip and dip direction on an arbitrarily oriented planar feature such as a fracture or fault. Rake is the direction of slip in the plane of the fault as measured from horizontal.

scratch on the footwall, resulting from relative motion of the hanging wall. The slip direction is defined by the rake angle, which is measured in the plane of the fault from horizontal.

A variety of techniques are used to represent the orientation of fractures and faults at depth. One of the most common techniques in structural geology is the use of lower hemisphere stereographic projections as illustrated in Figure 5.6 (see detailed discussions in Twiss and Moores 1992 and Pollard and Fletcher 2005). Stereographic projections show either the trace of a fracture plane (where it intersects the lower half of the hemisphere) or the intersection of fracture poles (normals to the fracture planes) and the hemisphere (Figure 5.6a). The circular diagrams (Figure 5.6b) used to represent such projections are referred to as stereonets (Schmidt equal area stereonets).

As shown in Figure 5.6b, near-horizontal fractures dipping to the northwest have poles that plot near the center of the stereonet whereas the trace of the fractures plot near the edge of the stereonet (upper left stereonet). Conversely, near-vertical fractures striking to the southeast and dipping to the southwest have poles that plot near the edge of the figure and fracture traces that cut through the stereonet near its center (lower right stereonet). The cloud of poles shown in each figure illustrates a group of fracture or fault planes with similar, but slightly different, strikes and dips. The second column

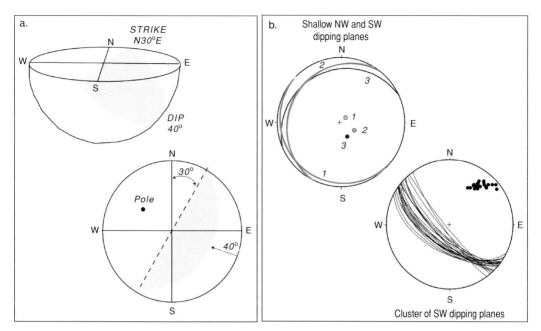

Figure 5.6. Illustration of the display of fracture and fault data using lower hemisphere stereographic projections. Either the intersection of the plane with the hemisphere can be shown or the pole to the plane. Planes which are sub-horizontal have poles that plot near the center of the stereonet whereas steeply dipping planes have poles which plot near the edge.

in Figure 5.1 illustrates the orientations of Mode I, normal, strike-slip and reverse faults in stereographic presentations.

Figure 5.7 shows the distribution of relatively small faults encountered in granitic rock in a geothermal area over the depth range of 6500–7000 feet that have been mapped with an electrical imaging device such as that shown in Figure 5.4. Figure 5.7a shows the data as a *tadpole* plot, the most common manner of portraying fracture data in the petroleum industry. The depth of each fracture is plotted as a dot with its depth along the ordinate and the amount of dip along the abscissa (ranging from 0 to 90°). The dip direction is shown by the direction that the *tail* of the tadpole points. This interval is very highly fractured (almost 2000 fractures are intersected over the 500 ft interval). The stereonet shown in Figure 5.7b illustrates the wide range of fracture orientations over the entire interval. Poles are seen at almost every location on the stereonet but the numerous poles just south of west indicate a concentration of steeply dipping fractures striking roughly north northwest and dipping steeply to the northeast. The advantages and disadvantages of the two types of representations are fairly obvious. Figure 5.7a allows one to see the exact depths at which the fractures occur as well as the dip and dip direction of individual planes, whereas Figure 5.7b provides a good overview of concentrations of fractures and faults with similar strike and dip.

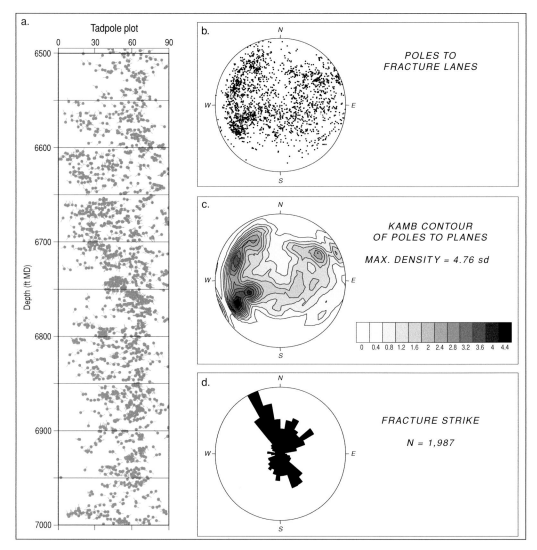

Figure 5.7. The distribution of fault data from a geothermal well drilled into granite can be displayed in various ways. (a) A *tadpole* plot, where depth is shown on the ordinate and dip on the abscissa. The dip direction is shown by the direction of the *tail* on each dot. (b) A stereographic projection shows the wide distribution of fracture orientation. (c) A contour plot of the fracture density (after Kamb 1959) to indicate statistically significant pole concentrations. (d) A *rose* diagram (circular histogram) indicating the distribution of fracture strikes.

In highly fractured intervals such as that shown in Figure 5.7b, it is not straightforward to characterize the statistical significance of concentrations of fractures at any given orientation. Figure 5.7c illustrates the method of Kamb (1959) used to contour the difference between the concentration of fracture poles with respect to a random distribution. This is expressed in terms of the number standard deviations that the observed

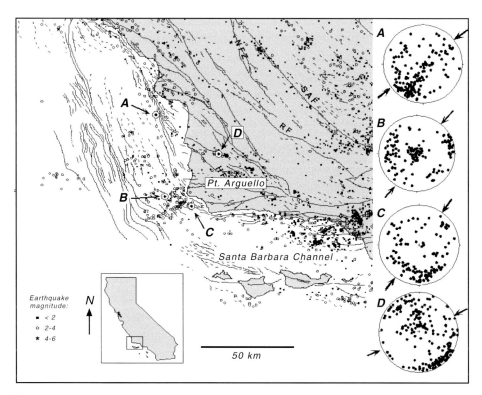

Figure 5.8. The Point Arguello area in western California is characterized by numerous earthquakes (dots), active faults and folds (thin lines). Image data from four wells drilled into the Monterey formation (A,B,C,D) illustrate the complex distribution of faults and fractures in each as shown in the stereonets (after Finkbeiner, Barton *et al.* 1997). *AAPG© 1997 reprinted by permission of the AAPG whose permission is required for futher use.*

concentration deviates from a random distribution. The areas with dark shading (>2 sd) indicate statistically significant fracture concentrations. The concentration of poles that were noted above corresponding to fractures striking north northwest and dipping steeply to the northeast is associated with a statistically significant concentration of poles.

Figure 5.7d shows a *rose diagram* (circular histogram) of fracture strikes. While the data indicate that the majority of fractures strike NNW–SSE, there are obviously many fractures with other orientations. One shortcoming of such representations is that they do not represent any information about fracture dip.

Figure 5.8 illustrates a case study in the Monterey formation in western California. Finkbeiner, Barton *et al.* (1998) studied four wells penetrating the highly folded, fractured and faulted Monterey formation at the sites shown. As discussed further in Chapter 11, the presence of these fractures and faults in the Monterey is essential for there being sufficient permeability to produce hydrocarbons. In fact, the outcrop photograph shown in Figure 5.2b was taken just west of well C in Figure 5.8. Numerous earthquakes occur in this region (shown by small dots), and are associated with slip on numerous reverse and strike-slip faults throughout the area. The trends of active faults and fold axes are shown on the map. The distribution of fractures and faults from analysis of ultrasonic borehole televiewer data in wells A–D are presented in the four stereonets shown. The direction of maximum horizontal compression in each well was determined from analysis of wellbore breakouts (explained in Chapter 6) and is northeast–southwest compression (as indicated by the arrows on stereonets). Note that while the direction of maximum horizontal compression is relatively uniform (and consistent with both the trend of the active geologic structures and earthquake focal mechanisms in the region as discussed later in this chapter), the distribution of faults and fractures in each of these wells is quite different. The reason for this has to do with the geologic history of each site. The highly idealized relationships between stress directions and Mode I fractures and conjugate normal, reverse and strike-slip faults (illustrated in Figure 5.1) are not observed. Rather, complex fault and fracture distributions are often seen which reflect not only the current stress field, but deformational episodes that have occurred throughout the geologic history of the formation. Thus, the ability to actually map the distribution of fractures and faults using image logs is essential to actually knowing what features are present *in situ* and thus which fracture and faults are important in controlling fluid flow at depth (Chapter 11).

Three-dimensional Mohr diagrams

In a number of applications and case studies discussed in the chapters that follow it will be necessary to calculate the shear and normal stress acting on arbitrarily oriented faults in three dimensions. One classical way to do this is to utilize three-dimensional Mohr diagrams as illustrated in Figure 5.9a (again see detailed discussions in Twiss and Moores 1992 and Pollard and Fletcher 2005). As shown in the figure, the values of the three principal stresses σ_1, σ_2 and σ_3 are used to define three Mohr circles. All planar features can be represented by a point, P, located in the space between the two smaller Mohr circles (defined by the differences between σ_1 and σ_2, and σ_2 and σ_3, respectively), and the big Mohr circle (defined by the difference between σ_1 and σ_3). As in the case of two-dimensional Mohr circles (*e.g.* Figure 4.2), the position of point P defines the shear and normal stress on the plane. The figure illustrates that critically oriented faults (*i.e.* those capable of sliding in the ambient stress field) plot in the region shown in Figure 5.9a as corresponding to coefficients of friction between 0.6 and 1.0.

Graphically, the position of point P in the three-dimensional Mohr circle is found using two angles, β_1 and β_3, that define the angles between the normal to the fault and the S_1 and S_3 axes, respectively (Figure 5.9b). As shown in Figure 5.9a, to find the

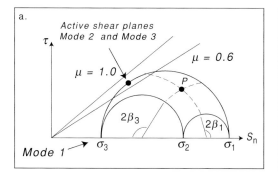 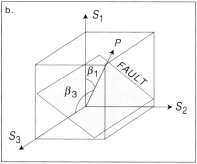

Figure 5.9. Representation of the shear and effective normal stress on an arbitrarily oriented fault can be accomplished with a three-dimensional Mohr circle (a). Although the exact position of point P can be determined with angles β_1 and β_3 measured between the fault normal and S_1 and S_3 directions (b) utilizing the graphical reconstruction shown, it is typical to calculate this mathematically (see text) and utilize the three-dimensional Mohr diagram for representation of the data.

location of P graphically, one utilizes the angles $2\beta_1$ and $2\beta_3$ to find points on the two small circles, and by constructing arcs drawn from the center of the other Mohr circle, P is determined as the intersection of the two arcs. It is obvious that a Mode I plane (normal to the least principal stress) plots at the position of σ_3 in the Mohr diagram.

Of course, it is not necessary these days to use graphical techniques alone for determining the shear and effective normal stress on arbitrarily oriented planes, but three-dimensional Mohr diagrams remain quite useful for representing fault data, as will be illustrated in the chapters that follow.

There are two common methods for calculating the magnitude of shear and normal stress on an arbitrarily oriented plane. The first technique defines the shear and normal stress in terms of the effective principal stresses and the orientation of the fault plane to the stress field. The shear and effective normal stresses are given by

$$\tau = a_{11}a_{12}\sigma_1 + a_{12}a_{22}\sigma_2 + a_{13}a_{23}\sigma_3 \tag{5.7}$$
$$\sigma_n = a_{11}^2\sigma_1 + a_{12}^2\sigma_2 + a_{13}^2\sigma_3 \tag{5.8}$$

where a_{ij} are the direction cosines (Jaeger and Cook 1971):

$$A = \begin{bmatrix} \cos\gamma\cos\lambda & \cos\gamma\sin\lambda & -\sin\gamma \\ -\sin\lambda & \cos\lambda & 0 \\ \sin\gamma\cos\lambda & \sin\gamma\sin\lambda & \cos\gamma \end{bmatrix} \tag{5.9}$$

where γ is the angle between the fault normal and S_3, and λ is the angle between the projection of the fault strike direction and S_1 in the S_1–S_2 plane.

Alternatively, one can determine the shear and normal stress via tensor transformation. If principal stresses at depth are represented by

$$S = \begin{bmatrix} S_1 & 0 & 0 \\ 0 & S_2 & 0 \\ 0 & 0 & S_3 \end{bmatrix}$$

we can express stress in a geographical coordinate system with the transform

$$S_{\mathrm{g}} = R_1' S R_1 \qquad (5.10)$$

where

$$R_1 = \begin{bmatrix} \cos a \cos b & \sin a \cos b & -\sin b \\ \cos a \sin b \sin c - \sin a \cos c & \sin a \sin b \sin c + \cos a \cos c & \cos b \sin c \\ \cos a \sin b \cos c + \sin a \sin c & \sin a \sin b \cos c - \cos a \sin c & \cos b \cos c \end{bmatrix}$$

$$(5.11)$$

and the Euler (rotation) angles that define the stress coordinate system in terms of geographic coordinates are as follows:

$a =$ trend of S_1
$b = -$plunge of S_1
$c =$ rake S_2.

However, if S_1 is vertical (normal faulting), these angles are defined as

$a =$ trend of $S_{\mathrm{Hmax}} - \pi/2$
$b = -$trend of S_1
$c = 0$.

Using the geographical coordinate system, it is possible to project the stress tensor on to an arbitrarily oriented fault plane. To calculate the stress tensor in a fault plane coordinate system, S_{f}, we once again use the principles of tensor transformation such that

$$S_{\mathrm{f}} = R_2 S_{\mathrm{g}} R_2' \qquad (5.12)$$

where

$$R_2 = \begin{bmatrix} \cos(\mathrm{str}) & \sin(\mathrm{str}) & 0 \\ \sin(\mathrm{str})\cos(\mathrm{dip}) & -\cos(\mathrm{str})\cos(\mathrm{dip}) & -\sin(\mathrm{dip}) \\ -\sin(\mathrm{str})\sin(\mathrm{dip}) & \cos(\mathrm{str})\sin(\mathrm{dip}) & -\cos(\mathrm{dip}) \end{bmatrix} \qquad (5.13)$$

where str is the fault strike and dip is the fault dip (positive dip if fault dips to the right when the fault is viewed in the direction of the strike). The shear stress, τ, which acts in the direction of fault slip in the fault plane, and normal stress, S_{n}, are given by

$$\tau = S_{\mathrm{r}}(3, 1) \qquad (5.14)$$

$$S_{\mathrm{n}} = S_{\mathrm{f}}(3, 3) \qquad (5.15)$$

where

$$S_r = R_3 S_f R'_3 \tag{5.16}$$

and

$$R_3 = \begin{bmatrix} \cos(\text{rake}) & \sin(\text{rake}) & 0 \\ -\sin(\text{rake}) & \cos(\text{rake}) & 0 \\ 0 & 0 & 1 \end{bmatrix} \tag{5.17}$$

Here rake of the slip vector is given by

$$\text{rake} = \arctan\left(\frac{S_f(3, 2)}{S_f(3, 1)}\right) \tag{5.18a}$$

if $S_f(3,2) > 0$ and $S_f(3,1) > 0$ or $S_f(3,2) > 0$ and $S_f(3,1) < 0$; alternatively

$$\text{rake} = 180° - \arctan\left(\frac{S_f(3, 2)}{-S_f(3, 1)}\right) \tag{5.18b}$$

if $S_f(3,2) < 0$ and $S_f(3,1) > 0$; or

$$\text{rake} = \arctan\left(\frac{-S_f(3, 2)}{-S_f(3, 1)}\right) - 180° \tag{5.18c}$$

if $S_f(3,2) < 0$ and $S_f(3,1) < 0$.

To illustrate these principles for a real data set, Figure 5.10 shows a stereonet representation of 1688 faults imaged with a borehole televiewer in crystalline rock from the Cajon Pass research borehole over a range of depths from 1750 to 3500 m depth (after Barton and Zoback 1992). Shear and normal stress were calculated using equations (5.14) and (5.15) with the magnitude and orientation of the stress tensor from Zoback and Healy (1992). We can then represent the shear and normal stress on each plane with a three-dimensional Mohr circle in the manner of Figure 5.9a. Because of the variation of stress magnitudes over this depth range, we have normalized the Mohr diagram by the vertical stress, S_V. As illustrated, most of the faults appear to be inactive in the current stress field. As this is Cretaceous age granite located only 4 km from the San Andreas fault, numerous faults have been introduced into this rock mass over tens of millions of years. However, a number of faults are oriented such that the ratio of shear to normal stress is in the range 0.6–0.9. These are active faults, which, in the context of the model shown in Figures 4.24c,d, are critically stressed and hence limit principal stress magnitudes. In Chapter 12 we show that whether a fault is active or inactive in the current stress field determines whether it is hydraulically conductive (permeable) at depth.

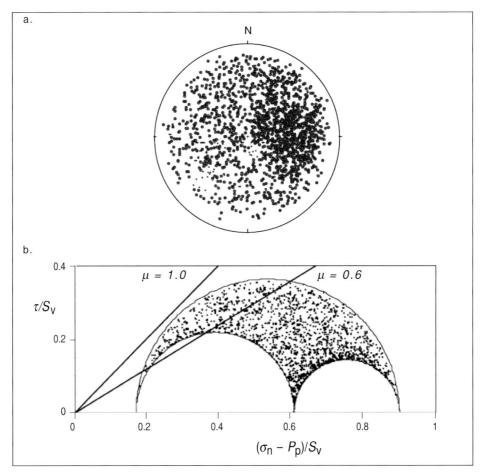

Figure 5.10. (a) Stereographic representation of fault data detected through wellbore image analysis in highly fractured granitic rock encountered in the Cajon Pass research well from 1750 to 3500 m depth (after Barton and Zoback 1992). (b) Representation of the same data utilizing a three-dimensional Mohr diagram normalized by the vertical stress. While many fractures appear to be critically stressed, most are not and thus reflect the rock's geologic history (after Barton, Zoback *et al.* 1995).

Earthquake focal mechanisms

In seismically active regions, important information can be obtained about the stress field from earthquake focal mechanisms (also known as fault plane solutions). The *beach balls* shown in Figures 1.2 and 5.1 that correspond to normal, strike-slip and reverse faulting stress regimes are based on the pattern of seismic radiation resulting from slip on a fault. A good description of earthquake focal plane mechanisms for non-experts can be found in Fowler (1990).

For simplicity, let us consider right-lateral slip on a vertical, east–west trending strike slip fault at the surface of a half-space (Figure 5.11a). When the fault slips, P-waves radiate outward with both positive and negative polarities that map onto symmetric compressional and dilatational quadrants. The four lobes shown in the figure illustrate the variation of wave amplitude with the direction of wave propagation relative to the fault plane. Note that the P-wave amplitude is zero in the direction parallel and perpendicular to the fault plane, such that these planes are referred to as *nodal* planes. If there were seismometers distributed over the surface of this half space, the orientation of the two nodal planes and the sense of motion on the planes could be determined by mapping the polarity of the first arriving waves from the earthquake. Thus, in the idealized case shown, data from a number of seismometers distributed on the surface of the half space could be used to determine both the orientation of the fault plane and the fact that right-lateral slip occurred on this plane. There is, however, a 90° ambiguity in the orientation of the fault plane as left-lateral slip on a north–south trending fault plane would produce exactly the same pattern of seismic radiation as right lateral slip on an east–west striking plane. Thus, an earthquake focal plane mechanism contains two orthogonal nodal planes, one of which is the fault plane and the other is referred to as the auxiliary plane. In the absence of additional data (such as coincidence of the earthquake hypocenter with the location of a mapped fault or the alignment of aftershocks along the fault surface), it cannot be determined which of the two planes is the actual fault.

Actual earthquakes are more complicated in several regards. First, they usually occur at depth such that seismic radiation propagates outward in all directions; it also quite common for faults to be dipping and, of course, strike-slip, reverse or normal fault slip (or a combination of strike-slip with normal or strike-slip with reverse) could occur. Figure 5.11b is a cross-section illustrating the radiation pattern for a dipping normal fault. By constructing an imaginary sphere around the hypocenter, we can portray the radiation pattern on a lower-hemisphere stereographic projection (Figure 5.11c), producing figures that look like *beach balls* where the compressional quadrants are shaded dark and the dilatational quadrants are shown in white. Thus, for the case illustrated in Figure 5.11c, we know from the dilatational arrivals in the center of the figure that it was a normal faulting event. By definition, the *P*-axis bisects the dilatational quadrant, the *T*-axis bisects the compressional quadrant and *B*-axis is orthogonal to *P* and *T*. In this simple case, the orientation of the two nodal planes trend north–south but knowing that the east dipping plane is the fault plane requires additional information, as noted above. Of course, if the seismic waves are recorded on relatively few seismographs, the planes of the focal mechanism will be poorly constrained, as will the *P*- and *T*-axes. Nonetheless, as discussed in more detail in Chapter 9 (and illustrated in the stress maps presented in Chapter 1), earthquake focal plane mechanisms prove useful for determining both the style of faulting and approximate directions of the principal stresses (see below).

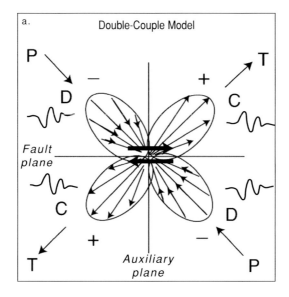

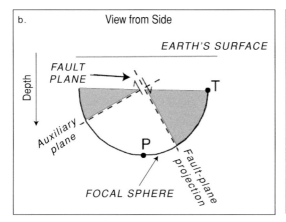
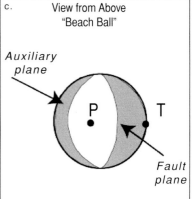

Figure 5.11. (a) Schematic illustration of the radiation pattern and force-couple associated with earthquakes as the basis earthquake focal plane mechanisms. An east–west striking, vertical right-lateral strike slip fault intersecting a half space is shown. The polarity of the P-waves defines the compressional and dilatational quadrants. (b) Cross-sectional view of the nodal planes, radiation pattern and *P*- and *T*-axes associated with an east-dipping normal fault. The radiation pattern does not uniquely distinguish the fault plane from the auxiliary plane. (c) Lower hemisphere stereonet representation of the normal faulting focal mechanism.

Earthquake focal mechanisms associated with normal, strike slip and reverse faults are illustrated in Figure 5.1. Note that while the conjugate shear faults are at angles ±30° on either side of the maximum principal stress, the focal mechanism illustrates orthogonal nodal planes, one of which is the fault. As a point of historical interest, focal plane mechanisms were instrumental in establishing the theory of plate tectonics as they illustrated that extensional normal faulting was occurring along mid-ocean spreading centers and the appropriate sense of lateral slip occurred on transform faults (see the review by Stein and Klosko 2002).

With respect to the orientation of *in situ* stress, the advantages of utilizing well-constrained earthquake focal plane mechanisms to map the stress field are fairly obvious: earthquakes record stress-induced deformation at mid-crustal depths, they sample relatively large volumes of rock and, due to the continued improvement of regional and global networks, more well-constrained focal mechanisms for mapping the stress field are available now than ever before. However, it is important to keep in mind that focal plane mechanisms record deformation and not stress. The *P*- and *T*-axes shown in Figure 5.11 are, by definition, the bisectors of the dilatational and compressional quadrants of the focal mechanism. Thus, they are not the maximum and minimum principal stress directions (as is sometimes assumed) but are the compressional and extensional strain directions for slip on either of the two possible faults. As most crustal earthquakes appear to occur on pre-existing faults (rather than resulting from new fault breaks), the slip vector is a function both of the orientation of the fault and the orientation and relative magnitude of the principal stresses, and the *P*- and *T*-axes of the focal plane mechanism do not correlate directly with principal stress directions. In an attempt to rectify this problem, Raleigh, Healy *et al.* (1972) showed that if the nodal plane of the focal mechanism corresponding to the fault is known, it is preferable not to use the *P*-axes of the focal-plane mechanism but instead to assume an angle between the maximum horizontal stress and the fault plane defined by the coefficient of friction of the rock. Because the coefficient of friction of many rocks is often about 0.6, Raleigh, Healy *et al.* (1972) suggested that the expected angle between the fault plane and the direction of maximum principal stress would be expected to be about 30°. Unfortunately, for intraplate earthquakes (those of most interest here), we usually do not know which focal plane corresponds to the fault plane. Nevertheless, in most intraplate areas, *P*-axes from well-constrained focal plane mechanisms do seem to represent a reasonable approximation of the maximum horizontal stress direction, apparently because intraplate earthquakes do not seem to occur on faults with extremely low friction (Zoback and Zoback 1980, 1989; Zoback, Zoback *et al.* 1989) and give an indication of relative stress magnitude (normal, strike-slip or reverse faulting). This will be discussed in more detail in Chapter 9.

As mentioned in Chapter 1, if the coefficient of friction of the fault is quite low, the direction of maximum compression can be anywhere in the dilatational quadrant and the *P*-axis can differ from the true maximum stress direction by as much as 45° (MacKenzie 1969). In fact, studies such as Zoback, Zoback *et al.* (1987) excluded as tectonic stress indicators right-lateral strike-slip focal plane mechanisms right on the San Andreas fault as did subsequent stress compilations at global scale as discussed in Chapter 9. In the case of the San Andreas, appreciable heat flow data collected in the vicinity of the San Andreas show no evidence of frictionally generated heat

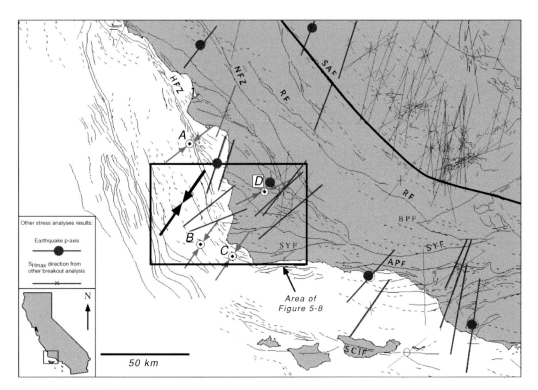

Figure 5.12. Stress map (direction of S_{Hmax}) of western California in the Point Arguello area (after Finkbeiner, Barton *et al.* 1997). Earthquake focal mechanisms within the rectangle were inverted using the technique of Gephart (1990) to obtain a direction of S_{Hmax} shown by the heavy arrow. The S_{Hmax} direction from wellbore breakouts studied with wellbore image data is shown for wells A–D discussed in the text. The *P*-axes of reverse faulting focal mechanisms are shown by the lines with a dot in the center. *AAPG© 1997 reprinted by permission of the AAPG whose permission is required for futher use.*

(Lachenbruch and Sass 1992) and appears to limit average shear stresses acting on the fault to depths of $\sim$15 km to about 20 MPa, approximately a factor of 5 below the stress levels predicted by the Coulomb criterion assuming that hydrostatic pore pressure at depth and the applicability of laboratory-derived friction coefficients of $\sim$0.6. Stress orientation data near the San Andreas fault also imply low resolved shear stresses on the fault at depth (Mount and Suppe 1987; Zoback, Zoback *et al.* 1987; Hickman 1991; Lachenbruch and Sass 1992; Townend and Zoback 2001; Hickman and Zoback 2004; Boness and Zoback 2006).

To optimize the use of focal plane mechanism data for determining stress orientations it is necessary to consider multiple events in a given region and use either the average *P*-axis direction as the maximum horizontal stress direction or, preferably, to formally invert a group of focal-plane mechanisms to determine the orientation and relative magnitude of the principal stress tensor (see Angelier 1979; Angelier 1984; Gephart and Forsyth 1984; Gephart 1990; Michael 1987).

Figure 5.12 is a map showing S_{Hmax} directions in the area shown in Figure 5.8. The lines with inward pointed arrows are derived from wellbore breakouts (Chapters 6 and 9) and those with a circle in the middle show the P-axis of reverse-faulting focal plane mechanisms. For slip on pure reverse faults, the horizontal projection of the P-axis is quite similar to the S_{Hmax} direction because the projection of the P-axis onto a horizontal plane will be the same as the S_{Hmax} direction regardless of either the choice of nodal plane or the coefficient of friction of the fault. The S_{Hmax} direction shown by the heavy arrows was obtained from inversion of earthquake focal plane mechanisms in the area enclosed by the rectangle (Finkbeiner 1998). Note that this direction compares quite well with the stress orientations obtained from wells A–D, wellbore breakouts in other wells and individual earthquake focal plane mechanisms. Because the majority of earthquakes in this region are reverse faulting events, the direction of S_{Hmax} is not greatly affected by uncertainties in knowing either the coefficient of friction of the fault or which nodal plane in the focal mechanism is the fault and which is the auxiliary plane.

Part II Measuring stress orientation and magnitude

6 Compressive and tensile failures in vertical wells

The principal topics I address in this chapter are the relationships among *in situ* stress magnitudes, rock strength and the nature of compressive and tensile failures that can result from the concentration stress around a wellbore. To establish the principles of wellbore failure with relatively simple mathematics, I consider in this chapter only vertical wells drilled parallel to the vertical principal stress, S_v. In Chapter 8 I generalize this discussion and consider deviated wells of arbitrary orientation in an arbitrarily oriented stress field.

In some ways every well that is drilled can be thought of as a rock mechanics experiment. The formation surrounding the wellbore wall is subject to a stress concentration that varies strongly with the position around the well and the distance from the wellbore wall. The way in which this formation responds to the stress concentration is a function of both the stress field and rock strength. As discussed in Chapter 7, detailed knowledge of the nature of wellbore failure (especially as revealed by wellbore imaging, as discussed in Chapter 5) allows one to estimate (or constrain) the magnitude and orientation of *in situ* stresses at depth. In some cases it also allows one to obtain direct information about rock strength *in situ*.

The concentration of stress around wellbores can lead to compressive failures known as stress-induced breakouts and/or tensile failure of the wellbore wall that we will refer to as drilling-induced tensile wall fractures. Breakouts are quite common in many wells and yield important information about both stress orientation and magnitude. However, excessive wellbore breakouts can lead to problematic (potentially catastrophic) wellbore instabilities. I address methods to analyze and mitigate such problems in Chapter 10.

In vertical wells, the occurrence of tensile fractures in a well usually implies (*i*) that S_{hmin} is the minimum principal stress and (*ii*) there are large differences between the two horizontal principal stresses, S_{Hmax} and S_{hmin}. As discussed below, the occurrence of tensile fractures is also influenced by high mud weight and cooling of the wellbore wall. The processes that control the initiation of tensile wall fractures are important for understanding the initiation of hydraulic fractures (Hubbert and Willis 1957; Haimson and Fairhurst 1967). However, *hydrofracs* are distinguished from tensile wall fractures in that they propagate from the wellbore into the *far field*, away from the wellbore stress concentration. In Chapter 7, I discuss the importance of drilling-induced tensile fractures as means of obtaining important information about stress orientation and magnitude as well as the manner in which hydraulic fractures yield extremely important information about the magnitude of the least principal stress. Of course, if hydraulic fracturing occurs unintentionally during drilling (due to excessively high mud weights), lost circulation can occur. This is another serious problem during drilling, especially in areas of severe overpressure. Options for avoiding lost circulation during drilling in overpressured areas are discussed in Chapter 8.

In the sections of this chapter that follow, I first introduce the concept of stress concentrations around a vertical well, how this stress concentration can lead to compressive and tensile wall failures and how such failures are used to determine the orientation of the horizontal principal stresses that exist *in situ*. The majority of stress orientation data shown in the maps presented in Chapter 1 and throughout this book (and utilized by the World Stress Map project, see Zoback, 1992) come from wellbore failures and earthquake focal mechanisms. Hence, after introducing breakouts and drilling-induced tensile fractures in the first part of this chapter, we discuss the quality ranking criterion developed by Zoback and Zoback (1989, 1991) for mapping the intraplate stress field.

Next, I extend the discussion of tensile failures to discuss hydraulic fracturing and the determination of the least principal stress, S_3, from hydrofracs in reservoirs or extended leak-off tests at casing set points. As $S_3 \equiv S_{hmin}$ in normal and strike-slip faulting areas (the most common stress states around the world), establishing the magnitude of S_{hmin} is a critical component of determining the full stress tensor. When one principal stress is vertical, S_v is obtained by integration of density logs as discussed in Chapter 1. Pore pressure can be either measured directly or estimated using the techniques described in Chapter 2. With knowledge of the orientation of the horizontal principal stresses obtained from wellbore failures and the magnitude S_{hmin}, determination of the complete stress tensor requires only the magnitude of S_{Hmax} to be determined. In Chapters 7 and 8 we discuss determination of S_{Hmax} utilizing observations of wellbore failures with independently determined values of S_v, P_p and S_{hmin}. To set the stage for these discussions, we consider in this chapter both drilling-induced tensile wall fractures as well as compressive failures (breakouts) to include the influence of thermal stresses and excess mud weight on the formation of such fractures.

As a full understanding of compressive failure is also critically important for evaluation of wellbore stability (Chapter 10), we briefly consider at the end of this chapter a number of other processes that affect compressive failure around wellbores. These include the way in which the presence of weak bedding planes can induce anisotropic rock strength (previously introduced in Chapter 4) and briefly consider, theoretically, compressive failure when elastic–plastic constitutive laws (introduced in Chapter 3) are more appropriate for a given formation than the strength of materials approach used throughout most of this book. This is most applicable for the case of drilling through poorly cemented sands. Finally, we briefly broaden discussion of wellbore failure beyond geomechanics and consider briefly wellbore failure that is the result of

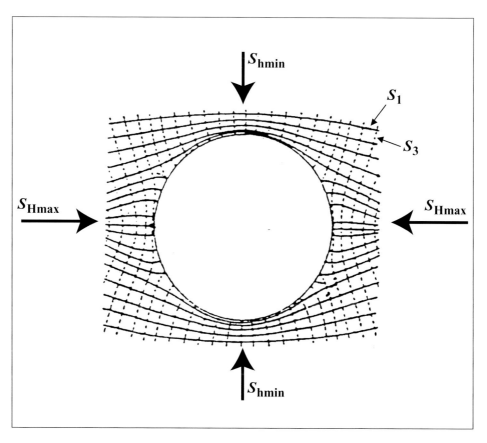

Figure 6.1. Principal stress trajectories around a cylindrical opening in a bi-axial stress field based on the Kirsch equations (Kirsch 1898). Note that as the wellbore wall is a free surface, the principal stress trajectories are parallel and perpendicular to it. Where the trajectories of maximum compressive stress converge, stresses are more compressive (at the azimuth of S_{hmin} in case of a vertical well). Where the trajectories diverge, the stresses are less compressive (at the azimuth of S_{Hmax}).

the chemical imbalance between drilling mud and the pore waters in shales that contain reactive clays.

Stress concentration around a cylindrical hole and wellbore failure

The stress concentration around a vertical well drilled parallel to the vertical principal stress, S_v, in an isotropic, elastic medium is described by the Kirsch equations (Kirsch 1898); see also Jaeger and Cook (1979). As illustrated in Figure 6.1 (taken from Kirsch's original paper), the creation of a cylindrical opening (like a wellbore) causes the stress trajectories to bend in such a way as to be parallel and perpendicular to the wellbore wall because it is a free surface which cannot sustain shear traction. Moreover, as the material removed is no longer available to support far-field stresses, there is a stress concentration around the well. This is illustrated by the *bunching up* of stress trajectories at the azimuth of S_{hmin}, which indicates strongly amplified compressive stress. In contrast, the *spreading out* of stress trajectories at the azimuth of S_{Hmax} indicates a decrease in compressive stress.

Mathematically, the effective stresses around a vertical wellbore of radius R are described in terms of a cylindrical coordinate system by the following:

$$\sigma_{rr} = \frac{1}{2}\left(S_{H\,max} + S_{h\,min} - 2P_p\right)\left(1 - \frac{R^2}{r^2}\right) + \frac{1}{2}\left(S_{H\,max} - S_{h\,min}\right)$$
$$\times \left(1 - \frac{4R^2}{r^2} + \frac{3R^4}{r^4}\right)\cos 2\theta + \frac{\Delta P R^2}{r^2} \tag{6.1}$$

$$\sigma_{\theta\theta} = \frac{1}{2}\left(S_{H\,max} + S_{h\,min} - 2P_p\right)\left(1 + \frac{R^2}{r^2}\right) - \frac{1}{2}\left(S_{H\,max} - S_{h\,min}\right)$$
$$\times \left(1 + \frac{3R^4}{r^4}\right)\cos 2\theta - \frac{\Delta P R^2}{r^2} - \sigma^{\Delta T} \tag{6.2}$$

$$\tau_{r\theta} = \frac{1}{2}(S_{H\,max} - S_{h\,min})\left(1 + \frac{2R^2}{r^2} - \frac{3R^4}{r^4}\right)\sin 2\theta \tag{6.3}$$

$$\sigma_{zz} = S_v - 2v(S_{H\,max} - S_{H\,min})\frac{r^2}{R^2}\cos 2\theta$$

where θ is measured from the azimuth of S_{Hmax}, r is radial distance from the center of the well, P_p is the formation pore pressure and ΔP is the difference between the mud weight, P_{mud}, and P_p. $\sigma^{\Delta T}$ represents thermal stresses arising from the difference between the mud temperature and formation temperature (ΔT). This will be ignored for the moment but is considered below. It can be shown that for any reasonable amount of elastic anisotropy, the stress concentration around a vertical well is not changed in any significant way (Lekhnitskii 1981). Hence, while anisotropic rock strength induced by weak bedding planes can have an important effect on wellbore failure (as described below), elastic anisotropy generally does not.

There are several important points about these equations that are illustrated in Figure 6.2 for the following parameters:

- $S_{Hmax} = 90$ MPa
- S_{Hmax} orientation is N90°E (east–west)
- $S_v = 88.2$ MPa (depth 3213m)
- $S_{hmin} = 51.5$ MPa
- $P_p = P_{mud} = 31.5$ MPa

First, the stress concentration varies strongly as a function of position around the wellbore and distance from the wellbore wall. Also, the stress concentration is symmetric with respect to the direction of the horizontal principal stresses. For an east–west direction of S_{Hmax}, Figure 6.2a shows that $\sigma_{\theta\theta}$ (the so-called effective *hoop* stress) is

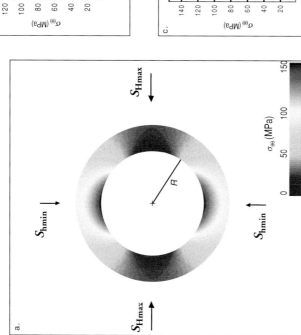

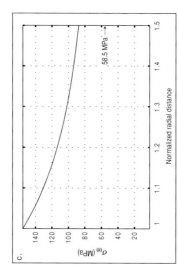

Figure 6.2. (a) Variation of effective hoop stress, $\sigma_{\theta\theta}$ around a vertical well of radius R subject to an east–west acting S_{Hmax}. Note that $\sigma_{\theta\theta}$ varies strongly with both position around the wellbore and distance from the wellbore wall. Values of stress and pore pressure used for the calculations are described in the text. (b) Variation of $\sigma_{\theta\theta}$ with normalized distance, r/R, from the wellbore wall at the point of maximum horizontal compression around the wellbore (i.e. at the azimuth of S_{hmin}). At the wellbore wall, $\sigma_{\theta\theta}$ is strongly amplified above the values of S_{Hmax} and S_{hmin} in accordance with equation (6.2). At $r/R = 1.5$, the hoop stress is approximately 30% greater than the effective far-field stress σ_{Hmax} that would be present at that position in the absence of the well. (c) Variation of $\sigma_{\theta\theta}$ with normalized distance, r/R, from the wellbore wall at the azimuth of S_{Hmax}, the point of minimum horizontal compression around the wellbore. At the wellbore wall, $\sigma_{\theta\theta}$ is close to zero. At $r/R = 1.5$, the hoop stress is slightly greater than the effective far-field stress σ_{hmin} that would be present at that position in the absence of the well. (*For colour version see plate section.*)

strongly compressive to the north and south, the azimuth of S_{hmin}, or 90° from the direction of S_{Hmax}. $\sigma_{\theta\theta}$ decreases rapidly with distance from the wellbore wall at the azimuth of S_{hmin} (Figure 6.2c) as given by equation (6.2). Note that at a radial distance equivalent to ~1.5 wellbore radii, the value of $\sigma_{\theta\theta}$ is about 50% greater than the far-field value of σ_{Hmax} (58.5 MPa), whereas it is almost three times this value at the wellbore wall.

In marked contrast, at the azimuth of S_{Hmax} the hoop stress is only slightly above zero because of the relatively large difference between S_{Hmax} and S_{hmin} (equation 6.2). Under such circumstances, the wellbore wall can go into tension which would lead to the formation of drilling-induced tensile wall fractures (Aadnoy 1990; Moos and Zoback 1990) because the tensile strength of rock is so low (Chapter 4). At the azimuth of S_{Hmax}, the hoop stress increases rapidly with distance from the wellbore wall. Note in Figure 6.2b that at $r = 1.5R$, the value of $\sigma_{\theta\theta}$ is slightly greater than the far-field value of σ_{hmin} which would be equivalent to $S_{hmin} - P_p = 20$ MPa ($\equiv 51.5 - 31.5$ MPa). Hence, drilling-induced tensile wall fractures are restricted to being extremely close (~several mm to cm) to the wellbore (Brudy and Zoback 1999), unless the pressure in the wellbore is sufficient to extend the fracture away from the wellbore as a hydrofrac (see below).

Note that the stress components described in equations (6.1)–(6.3) are independent of elastic moduli. For this reason, the manner in which stresses are concentrated does not vary from formation to formation. Moreover, the stress concentration around a wellbore is independent of R, the wellbore radius.

Because stresses are most highly concentrated at the wellbore wall, if either compressive or tensile failure is going to occur, it will initiate there. Figure 6.3a shows the variation of $\sigma_{\theta\theta}$, σ_{zz} and σ_{rr} at the wellbore wall for the same far field stresses used in Figure 6.2. Note the extremely large variations in $\sigma_{\theta\theta}$ with position around the well. σ_{zz} varies in a similar manner but the variations are much more subdued. The average value of σ_{zz} is the same as the far-field vertical effective stress of 56.7 MPa ($88.2 - 31.5$ MPa). In Figures 6.2a and 6.3a it is obvious that compressive failure of the wellbore wall is most likely to occur in the area of maximum compressive hoop stress (at the azimuth of S_{hmin}) if the stress concentration exceeds the rock strength (Bell and Gough 1979; Zoback, Moos et al. 1985). The zone of compressive failure around the well is shown in Figure 6.3c assuming a Mohr–Coulomb failure criterion and $C_0 = 45$ MPa, $\mu_i = 1.0$. The stress concentration exceeds the rock strength everywhere within the contour lines shown in Figure 6.3c on opposite sides of the hole. The breakouts have a finite width, w_{BO}, the span of failed rock around the wellbore wall on one side, and initial depth, both of which depend on rock strength for a given stress state (Zoback, Moos et al. 1985). The colors in Figure 6.3c indicate the value of rock strength required to prevent failure. Hence, hot colors means it takes high strength to prevent failure (because the stress concentration is high) whereas cold colors mean even a low-strength rock will not fail (because the stress concentration is low). The contour line describes the

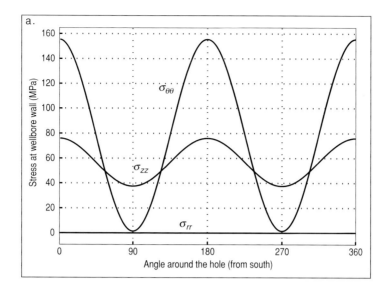

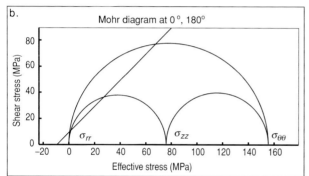

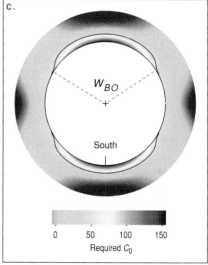

Figure 6.3. (a) Variation of effective principal stresses, $\sigma_{\theta\theta}$, σ_{rr} and σ_{zz} around a vertical wellbore as a function of azimuth. The far-field values of stress and pore pressure are the same as used for the calculations shown in Figure 6.2. As discussed in the text, the variation of $\sigma_{\theta\theta}$ around the wellbore is four times the difference between S_{Hmax} and S_{hmin} in the far field (equation 6.9). As the mud weight is assumed to equal the pore pressure $\sigma_{rr} = 0$. σ_{zz} varies around the well in the same manner as $\sigma_{\theta\theta}$ but without the extreme variation of values. (b) The three principal stresses at the wellbore wall at the point of maximum stress concentration ($\theta = 0$, 180°) shown as a three-dimensional Mohr diagram. Note that the strength of the rock is exceeded (a Mohr–Coulomb failure criterion is assumed, $C_0 = 45$ MPa, $\mu_i = 1.0$) such that the rock on the wellbore wall is expected to fail. (c) The zone of compressive failure around the wellbore wall for the assumed rock strength is indicated by the contour line. This is the expected zone of initial breakout formation with a width given by w_{BO}. Between the contour line and the wellbore wall, failure of even stronger rocks would have been expected (the scale indicates the magnitude of rock strength required to inhibit failure). Lower rock strength would result in a larger failure zone. (*For colour version see plate section.*)

boundary between the zones where the stress concentration exceeds the strength (as defined above) or does not.

To better visualize why breakouts and tensile fractures around a wellbore are such good indicators of far-field stress directions, let us first simplify equations (6.1)–(6.3) for the stresses acting right at the wellbore wall by substituting $r = R$. In this case, the effective hoop stress and radial stress at the wellbore wall are given by the following equation:

$$\sigma_{\theta\theta} = S_{\text{hmin}} + S_{\text{Hmax}} - 2(S_{\text{Hmax}} - S_{\text{hmin}}) \cos 2\theta - 2P_0 - \Delta P - \sigma^{\Delta T} \tag{6.4}$$

$$\sigma_{rr} = \Delta P \tag{6.5}$$

where ΔP is the difference between the wellbore pressure (mud weight, P_{m}) and the pore pressure. The effective stress acting parallel to the wellbore axis is:

$$\sigma_{zz} = S_{\text{v}} - 2\nu(S_{\text{Hmax}} - S_{\text{hmin}}) \cos 2\theta - P_0 - \sigma^{\Delta T} \tag{6.6}$$

where ν is Poisson's ratio. At the point of minimum compression around the wellbore (i.e. parallel to S_{hmin}) at $\theta = 0°$, $180°$, equation (6.4) reduces to

$$\sigma_{\theta\theta}^{\text{min}} = 3S_{\text{hmin}} - S_{\text{Hmax}} - 2P_0 - \Delta P - \sigma^{\Delta T} \tag{6.7}$$

whereas at the point of maximum stress concentration around the wellbore (i.e. parallel to S_{Hmax}) at $\theta = 90°$, $270°$,

$$\sigma_{\theta\theta}^{\text{max}} = 3S_{\text{Hmax}} - S_{\text{hmin}} - 2P_0 - \Delta P - \sigma^{\Delta T} \tag{6.8}$$

such that the difference between the two is

$$\sigma_{\theta\theta}^{\text{max}} - \sigma_{\theta\theta}^{\text{min}} = 4\,(S_{\text{Hmax}} - S_{\text{hmin}}) \tag{6.9}$$

which corresponds to the amplitude of the sinusoidal variation of hoop stress around the wellbore shown in Figure 6.3a and helps explain why observations of wellbore failures so effectively indicate far-field stress directions. Fundamentally, the variation of stress around the wellbore wall amplifies the far-field stress concentration by a factor of 4.

Introduction to breakouts

To understand the zone of compressive failure (breakouts) that results from the stress concentration around the wellbore (Figure 6.2), one simply has to consider the fact that like in a rock mechanics experiment, the rock surrounding the wellbore is subject to three principal stresses defined by equations (6.4)–(6.6). If these stresses exceed the rock strength, the rock will fail. The stress state at the wellbore wall at the azimuth of S_{hmin} (where the stress concentration is most compressive), is shown in Figure 6.3b using a three-dimensional Mohr diagram. This can then be compared to a failure law defining the strength of the rock. For this example, a Mohr–Coulomb failure law was used but, of course, any of the failure laws discussed in Chapter 4 could have been considered. In the general region of the maximum stress concentration around the well ($\theta = 0°$, $180°$), wherever the stress concentration exceeds the strength of the rock, failure is expected. Thus, the zone of compressive failure (initial breakout formation) within the contour line in Figure 6.3c indicates the region of initial breakout formation using the strength of materials concept first introduced in Chapter 3. The growth of breakouts after their initial formation is discussed later in this chapter.

The most reliable way to observe wellbore breakouts is through the use of ultrasonic image logs that were described in Chapter 5. As shown in Figure 6.4a, a standard *unwrapped* televiewer images breakouts as dark bands of low reflectance on opposite sides of the well. Interactive digital processing allows cross-sections of a well (such as that shown in Figure 6.4c) to be easily displayed (Barton, Tessler *et al.* 1991), which makes it straightforward to determine both the orientation and opening angle, w_{BO}, of the breakouts. Breakouts form symmetrically on both sides of the well, but during routine data analysis, the orientations of the breakouts are documented independently (*e.g.* Shamir and Zoback 1992). The two *out-of-focus* zones on opposites sides of the well in the electrical image shown in Figure 6.4b also correspond to breakouts. These result from poor contact between the wellbore wall and the pad upon which the electrode array is mounted. At any given depth, the azimuth of maximum horizontal stress is $90°$ from the mean of the azimuths of the breakouts on either side of the well. As illustrated below, comprehensive analysis of breakouts in wellbores can yield thousands of observations, thus enabling one to make profiles of stress orientation (and sometimes magnitude) along the length of a well.

It is easily seen in the equations above that if we raise mud weight, $\sigma_{\theta\theta}$ decreases (and σ_{rr} increases) at all positions around the wellbore. This is shown in Figure 6.5a for $\Delta P = 10$ MPa (compared to Figure 6.3a). As a point of reference, at a depth of 3213 m, this is equivalent to about a 10% increase in excess of hydrostatic pressure. Two phenomena are important to note. First, with respect to compressive failures, by increasing the mud weight, the zone of failure is much smaller in terms of both w_{BO} and breakout depth (the dashed lines indicate w_{BO} in Figure 6.3). This is shown in Figure 6.5b which was calculated with exactly the same stresses and rock strength as Figure 6.3c, except for the change in P_m. This is because as ΔP increases, $\sigma_{\theta\theta}$ decreases and σ_{rr} increases such that the size of the Mohr circle (Figure 6.3b) decreases markedly in the area of the wellbore wall subjected to most compressive stress. This demonstrates why increasing mud weight can be used to stabilize wellbores, a subject to be considered at length in Chapter 10.

Introduction to drilling-induced tensile fractures

The second point to note about wellbore failure is that as ΔP increases and $\sigma_{\theta\theta}$ decreases, the wellbore wall can locally go into tension at $\theta = 90°$, $270°$ and contribute to the occurrence of drilling-induced tensile fractures. This is illustrated in Figure 6.5a.

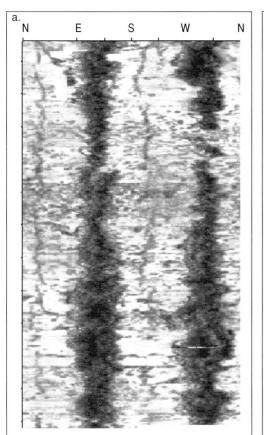

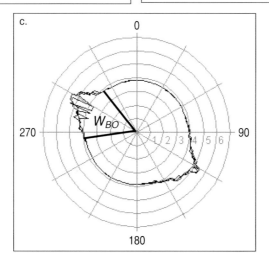

Figure 6.4. (a) Wellbore breakouts appear in an ultrasonic borehole televiewer image as dark bands on either side of a well because of the low-amplitude ultrasonic reflections off the wellbore wall. (b) Breakouts appear as out-of-focus areas in electrical image data because of the poor contact of the electrode arrays on the pads of the tool where breakouts are present. (c) A cross-sectional view of a well with breakouts can be easily made with televiewer data making determination of the azimuth of the breakouts and w_{BO} straightforward. Note the drilling-induced tensile fracture in the left image located 90° from the azimuth of the breakouts, just as expected from the Kirsch equations. From Zoback, Barton *et al.* (2003). *Reprinted with permission of Elsevier.* (*For colour version see plate section.*)

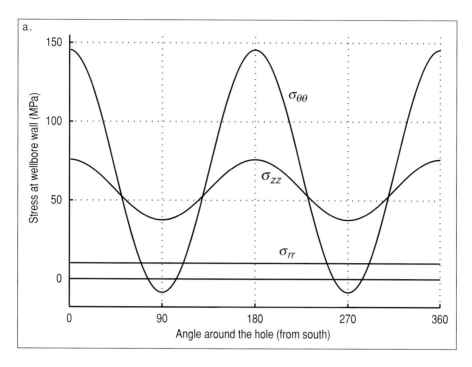

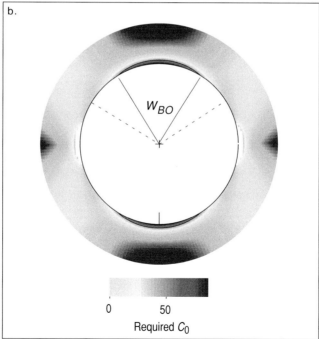

Figure 6.5. (a) Stress concentration at the wellbore wall and (b) zone of compressive failure around the wellbore (similar to Figure 6.3) when the mud weight has been raised 10 MPa above the mudweight. Figure 6.3b compares the width of breakouts for the two cases. Note that raising the mud weight decreases the size of the breakouts considerably. The area in white shows the region where tensile stresses exist at the wellbore wall. (*For colour version see plate section.*)

In Figure 6.5b, the area around the wellbore wall in which tensional stresses exist is at the azimuth of the maximum horizontal stress and is shown in white. As noted above, under normal circumstances, drilling induced tensile fractures are not expected to propagate more than a cm from the wellbore wall. Thus, the formation of drilling-induced tensile fractures will not lead to a hydraulic fracture propagating away from the wellbore (which could cause lost circulation) unless the mud weight exceeds the least principal stress. In the case of deviated wells, this is somewhat more complicated and is discussed in more detail in Chapter 8.

Because drilling-induced tensile fractures do not propagate any significant distance away from the wellbore wall (and thus have no appreciable effect on drilling), wellbore image logs are essentially the only way to know if drilling-induced tensile fractures are present in a well. This can be seen quite clearly in the two examples of electrical image logs in Figure 6.6 (from Zoback, Barton $et\ al.$ 2003). As predicted by the simple theory discussed above, the fractures form on opposite sides of the wellbore wall (at the azimuth of S_{Hmax}, $90°$ from the position of breakouts) and propagate along the axis of the wellbore. As discussed in Chapter 8, the occurrence of axial drilling-induced tensile fractures is evidence that one principal stress is parallel to the axis of the wellbore. Note that in the televiewer image shown in Figure 6.4, there are tensile fractures on opposite sides of the wellbore wall that are $90°$ from the midpoints of the wellbore breakouts. In other words, this well was failing simultaneously in compression and tension as it was being drilled. However, because the tensile fractures do not affect the drilling process, and because the breakouts were not excessively large (see Chapter 10) there were no problems with wellbore stability. In Figure 6.6c. the orientations of S_{Hmax} determined from breakouts and drilling-induced tensile fractures in a section of a well are shown (the orientation of breakouts were rotated $90°$ as they form at the azimuth of S_{hmin}). Note that the breakouts and tensile fractures form $180°$ apart, on opposite sides of the well and the breakouts and tensile fractures form $90°$ apart, exactly as predicted on the basis of the simple theory described above.

To illustrate how robust drilling-induced tensile fractures are as stress indicators, a stress map of the Visund field in the northern North Sea is shown in Figure 6.7 (after Wiprut and Zoback 2000). In the Visund field, an extremely uniform stress field is observed as a function of both depth and position in the oil field. Drilling-induced tensile fractures were observed in five vertical wells (A−E). The depth intervals logged are shown in white in the lower right corner of the figure and the intervals over which the tensile fractures were observed is shown by the black lines. The rose diagrams show the orientation and standard deviation of the drilling-induced tensile fractures observed in wells A–E as well as a compilation of the 1261 observations made in all of the wells. Note that numerous observations in each well indicate very uniform stress with depth (standard deviations of only $\sim 10°$). As these observations come from depths ranging between 2500 m and 5200 m and from wells separated by up to 20 km, a spatially uniform stress field is observed.

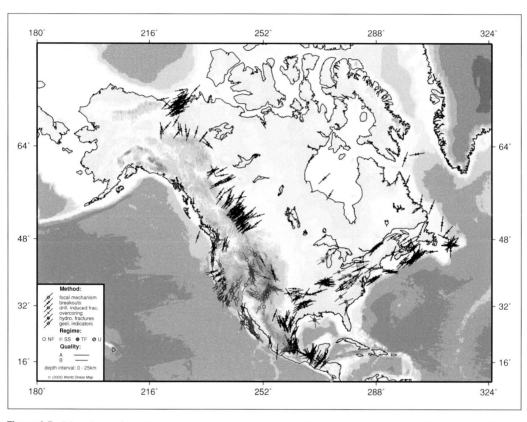

Figure 1.5. Directions of maximum horizontal stress (S_{Hmax}) in North America from the World Stress Map data base (http://www-wsm.physik.uni-karlsruhe.de/) superimposed on topography and bathymetry after Zoback and Zoback (1980, 1989, 1991). Only A and B quality data are shown. Data points characteristic of normal faulting are shown in red, strike-slip areas are shown in green, reverse faulting areas are shown in blue and indicators with unknown relative stress magnitudes are shown in black.

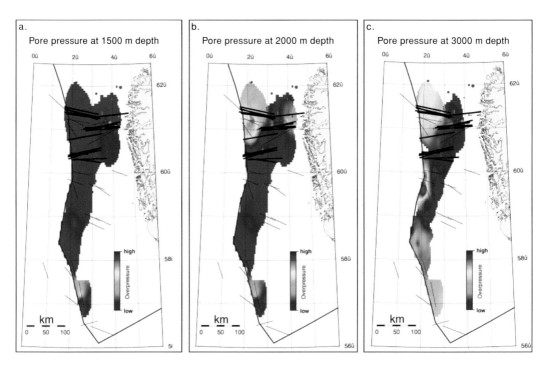

Figure 2.3. Spatial variations of pore pressure at various depths in the Norwegian sector of the northern North Sea (after Grollimund, Zoback *et al.* 2001). Note that at 1500 m depth, near hydrostatic values of pore pressure are observed. At greater depths, regions of elevated pore pressure are observed to develop in several areas. "Hard" overpressure (i.e. values near lithostatic) is observed in only a few restricted areas. Black lines indicate the direction of maximum horizontal compression determined from the orientation of drilling-induced tensile fractures and wellbore breakouts, as described in Chapter 6.

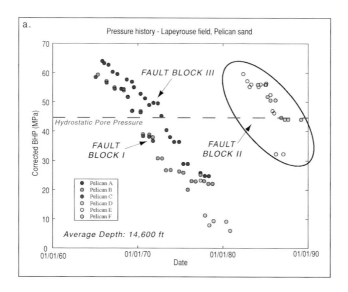

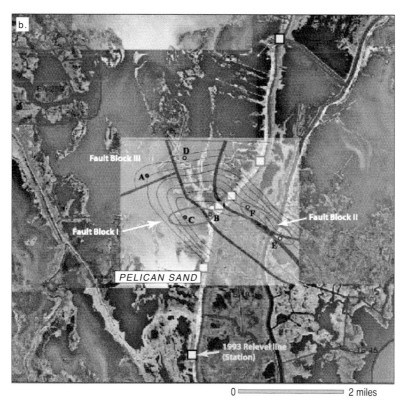

Figure 2.10. The Pelican sand of the Lapeyrouse field in southern Louisiana is highly compartmentalized. Note that in the early 1980s, while fault blocks I and III are highly depleted, fault block II is still at initial reservoir pressure (modified from Chan and Zoback 2006). Hence, the fault separating wells E and F from B and C is a sealing fault, separating compartments at pressures which differ by ∼55 MPa whereas the fault between wells B and C is not a sealing fault.

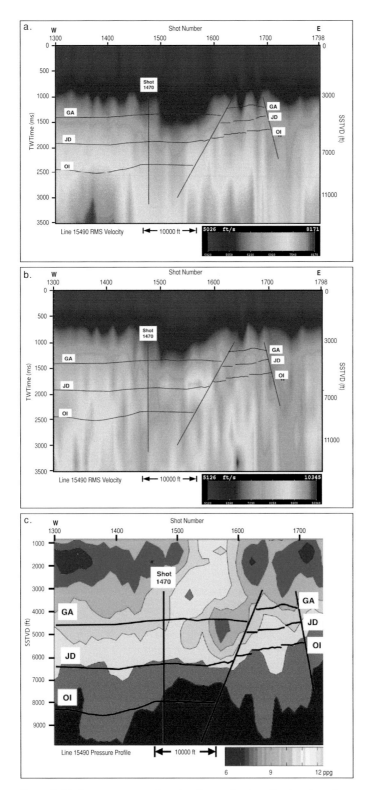

Figure 2.17. Pore pressure estimation from seismic reflection data along an E–W seismic line in the SEI-330 field (after Dugan and Flemings 1998). (a) Stacking, or RMS, velocities determined from normal moveout corrections. (b) Interval velocities determined from the RMS velocities. (c) Pore pressures inferred from interval velocity in the manner discussed in the text.

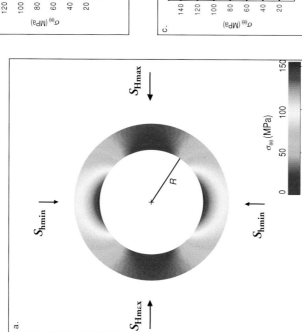

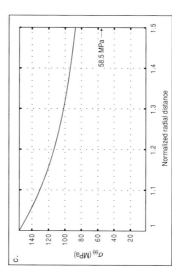

Figure 6.2. (a) Variation of effective hoop stress, $\sigma_{\theta\theta}$ around a vertical well of radius R subject to an east–west acting S_{Hmax}. Note that $\sigma_{\theta\theta}$ varies strongly with both position around the wellbore and distance from the wellbore wall. Values of stress and pore pressure used for the calculations are described in the text. (b) Variation of $\sigma_{\theta\theta}$ with normalized distance, r/R, from the wellbore wall at the point of maximum horizontal compression around the wellbore (i.e. at the azimuth of S_{hmin}). At the wellbore wall, $\sigma_{\theta\theta}$ is strongly amplified above the values of S_{Hmax} and S_{hmin} in accordance with equation (6.2). At $r/R = 1.5$, the hoop stress is approximately 30% greater than the effective far-field stress σ_{Hmax} that would be present at that position in the absence of the well. (c) Variation of $\sigma_{\theta\theta}$ with normalized distance, r/R, from the wellbore wall at the azimuth of S_{Hmax}, the point of minimum horizontal compression around the wellbore. At the wellbore wall, $\sigma_{\theta\theta}$ is close to zero. At $r/R = 1.5$, the hoop stress is slightly greater than the effective far-field stress σ_{hmin} that would be present at that position in the absence of the well.

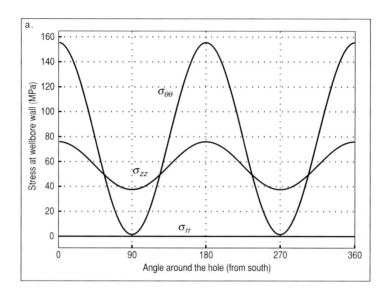

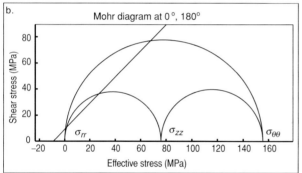

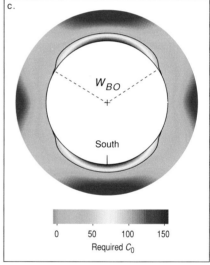

Figure 6.3. (a) Variation of effective principal stresses, $\sigma_{\theta\theta}$, σ_{rr} and σ_{zz} around a vertical wellbore as a function of azimuth. The far-field values of stress and pore pressure are the same as used for the calculations shown in Figure 6.2. As discussed in the text, the variation of $\sigma_{\theta\theta}$ around the wellbore is four times the difference between S_{Hmax} and S_{hmin} in the far field (equation 6.9). As the mud weight is assumed to equal the pore pressure $\sigma_{rr} = 0$. σ_{zz} varies around the well in the same manner as $\sigma_{\theta\theta}$ but without the extreme variation of values. (b) The three principal stresses at the wellbore wall at the point of maximum stress concentration ($\Theta = 0$, $180°$) shown as a three-dimensional Mohr diagram. Note that the strength of the rock is exceeded (a Mohr–Coulomb failure criterion is assumed, $C_0 = 45$ MPa, $\mu_i = 1.0$) such that the rock on the wellbore wall is expected to fail. (c) The zone of compressive failure around the wellbore wall for the assumed rock strength is indicated by the contour line. This is the expected zone of initial breakout formation with a width given by w_{BO}. Between the contour line and the wellbore wall, failure of even stronger rocks would have been expected (the scale indicates the magnitude of rock strength required to inhibit failure). Lower rock strength would result in a larger failure zone.

a.
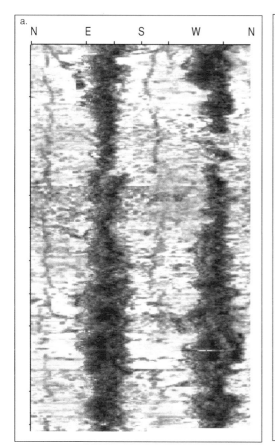

b.

c.
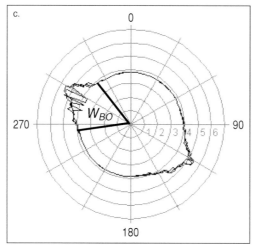

Figure 6.4. (a) Wellbore breakouts appear in an ultrasonic borehole televiewer image as dark bands on either side of a well because of the low-amplitude ultrasonic reflections off the wellbore wall. (b) Breakouts appear as out-of-focus areas in electrical image data because of the poor contact of the electrode arrays on the pads of the tool where breakouts are present. (c) A cross-sectional view of a well with breakouts can be easily made with televiewer data making determination of the azimuth of the breakouts and w_{BO} straightforward. Note the drilling-induced tensile fracture in the left image located $90°$ from the azimuth of the breakouts, just as expected from the Kirsch equations. From Zoback, Barton *et al.* (2003). *Reprinted with permission of Elsevier.*

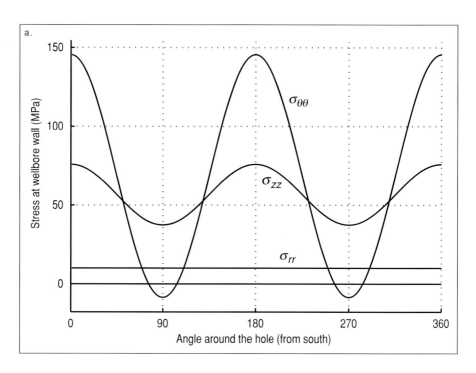

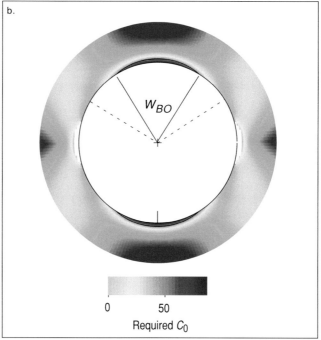

Figure 6.5. (a) Stress concentration at the wellbore wall and (b) zone of compressive failure around the wellbore (similar to Figure 6.3) when the mud weight has been raised 10 MPa above the mudweight. Figure 6.3b compares the width of breakouts for the two cases. Note that raising the mud weight decreases the size of the breakouts considerably. The area in white shows the region where tensile stresses exist at the wellbore wall.

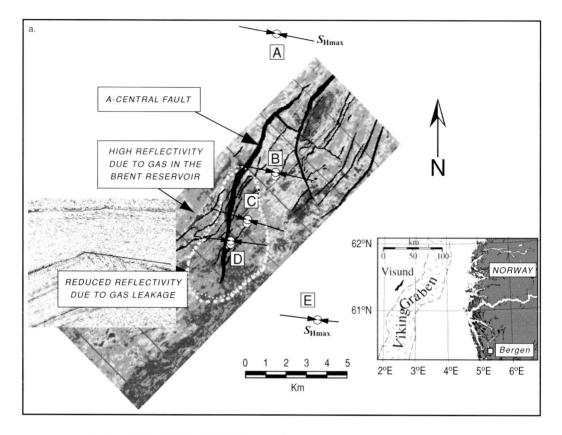

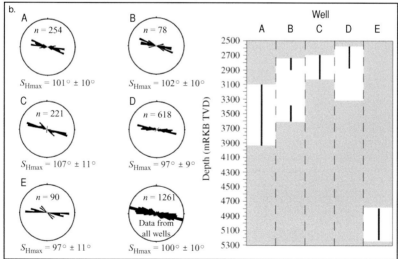

Figure 6.7. Drilling-induced tensile fractures in five wells in the Visund field in the northern North Sea indicate a remarkably uniform stress field both spatially and with depth (after Wiprut, Zoback *et al.* 2000). The rose diagrams illustrate how uniform the tensile fracture orientations are with depth in each well and the field as a whole. The length of each well logged with an electrical imaging device is shown in white in the diagram in the lower right. The drilling-induced tensile fractures are shown by the vertical black lines.

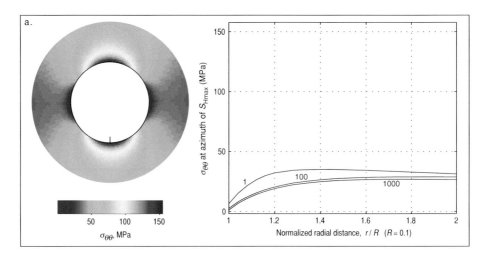

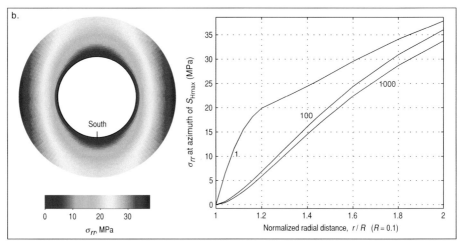

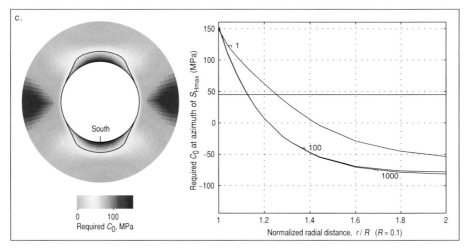

Figure 6.14. The effect of temperature on the state of stress around a wellbore for the same stress values used in Figures 6.2 and 6.3. (a) The thermally induced $\sigma_{\theta\theta}$ and the variation of $\sigma_{\theta\theta}$ with radial distance and time. (b) The thermally induced σ_{rr} and the variation of σ_{rr} with radial distance and time. (c) The effect of cooling on wellbore stability based on drilling with mud that is $10°$ cooler than the formation temperature. While the breakout is slightly smaller than that shown in 6.3c, it is probably not feasible to significantly improve wellbore stability through cooling.

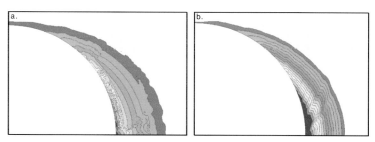

Figure 6.17. The area in which wellbore breakouts form around a cylindrical well can be modeled using a total plastic strain criterion rather than a stress criterion. These finite element calculations indicate the zone of expected breakouts assuming a critical strain level at which failure occurs (courtesy S. Willson). (a) Strain around a wellbore assuming a strain softening model of rock deformation (red indicates high strain). (b) Failure zone predicted using a strength of materials approach and Mohr–Coulomb failure criterion.

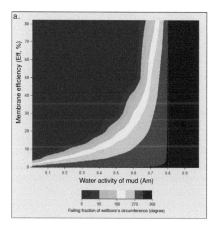

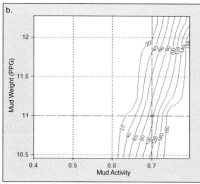

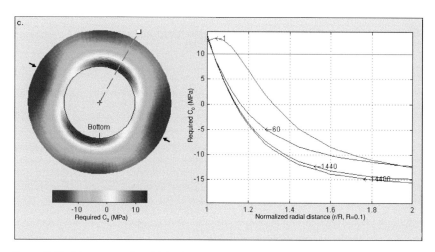

Figure 6.18. The manner in which chemical reactions between drilling mud and shale affect wellbore stability. (a) As the mud activity increases relative to the formation fluid, the failure zone becomes markedly larger. At moderate mud activity, increasing membrane efficiency increases wellbore stability. (b) In some cases, moderate increases in mud weight can offset the weakening effect due to mud/shale interaction. (c) When the mud activity is far below that of the formation, mud/shale interaction will result in strengthening of the wellbore wall with time as pore pressure decreases in the formation around wellbore. Possible, less beneficial, effects are discussed in the text.

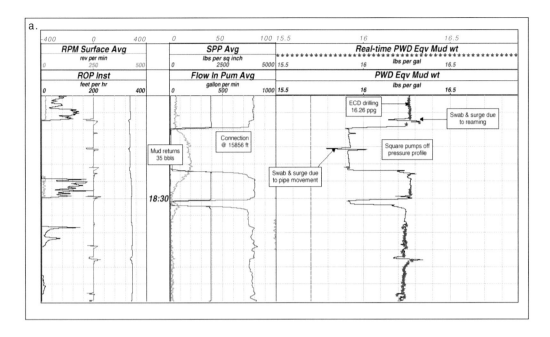

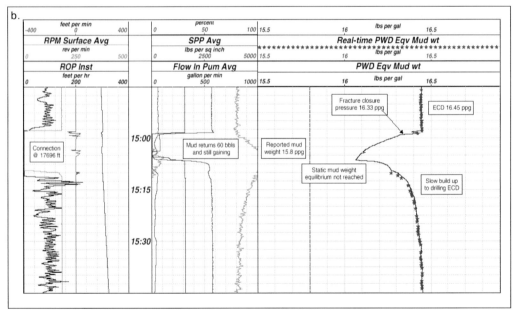

Figure 7.6. Pressure-while-drilling (PWD) records reconstructed from a well in the Gulf of Mexico. (a) A conventional pressure record indicating an abrupt decrease in pressure when pumping is stopped to make a connection while adding a new piece of pipe while drilling. (b) An example of wellbore ballooning where there is a gradual decrease in pressure when pumping is stopped and a corresponding gradual increase in pressure when pumping resumes (see text).

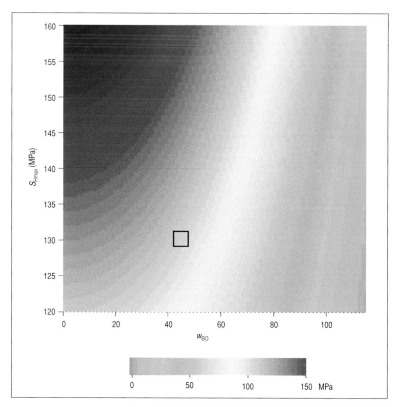

Figure 7.11. The dependence of S_{Hmax} on rock strength for given values of breakout width. The square indicates the analysis shown in Figure 7.10.

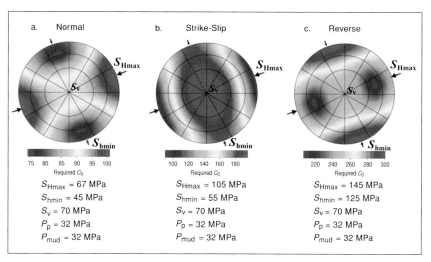

a. Normal	b. Strike-Slip	c. Reverse
S_{Hmax} = 67 MPa	S_{Hmax} = 105 MPa	S_{Hmax} = 145 MPa
S_{hmin} = 45 MPa	S_{hmin} = 55 MPa	S_{hmin} = 125 MPa
S_v = 70 MPa	S_v = 70 MPa	S_v = 70 MPa
P_p = 32 MPa	P_p = 32 MPa	P_p = 32 MPa
P_{mud} = 32 MPa	P_{mud} = 32 MPa	P_{mud} = 32 MPa

Figure 8.2. The tendency for the initiation of wellbore breakouts in wells of different orientation for normal, strike-slip and reverse faulting stress regimes. Similar to the figures in Peska and Zoback (1995). The magnitudes of the stresses, pore pressure and mud weight assumed for each case are shown. The color indicates the rock strength required to prevent failure, hence red indicates a relatively unstable well as it would take high rock strength to prevent failure whereas blue indicates the opposite. The strength scale is different for each figure as the stress magnitudes are progressively higher from normal to strike-slip to reverse faulting. Note that because these calculations represent the initiation of breakouts, they are not directly applicable to considerations of wellbore stability (see Chapter 10).

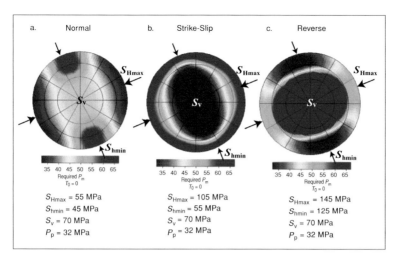

Figure 8.3. The tendency for the initiation of tensile fractures to form in wells of different orientation for normal, strike-slip and reverse faulting stress regimes. Similar to the figures in Peska and Zoback (1995). The magnitudes of the stresses, pore pressure and mud weight assumed for each case are shown. Note that the color indicates the mud pressure required to initiate tensile failure. Hence red indicates that tensile fractures are likely to form as little excess mud weight is required to initiate failure whereas blue indicates the opposite.

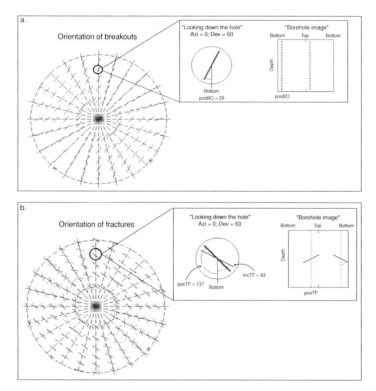

Figure 8.4. (a) The orientation of breakouts, if they were to form, in wells of different orientations. A looking-down-the-well convention is used as indicated in the inset. Similar to the figures in Peska and Zoback (1995). (b) The orientation of tensile fractures, if they were to form, in wells of different orientations is indicated by two angles that define the position of the tensile fracture around the wellbore's circumference as well as the orientation of the fracture trace with respect to the wellbore axis, as indicated in the inset. In both figures, a strike-slip faulting regime with S_{Hmax} acting in a NW–SE direction is assumed in the calculations.

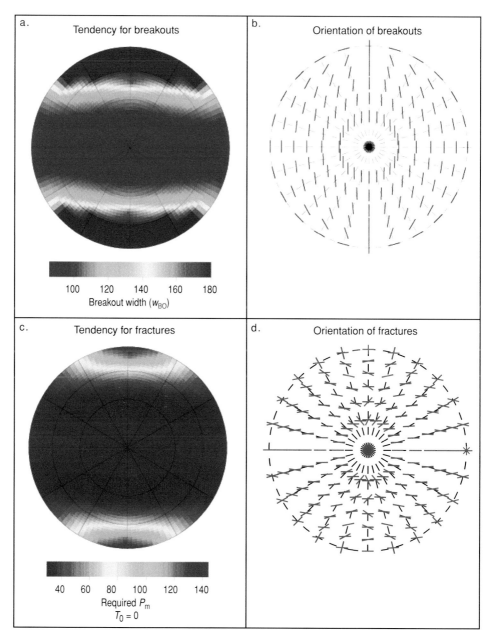

Figure 8.5. Lower hemisphere representations of the relative stability of wellbores of varied orientation with respect to the formation of wellbore breakouts (a, b) and drilling-induced tensile fractures (c, d). (a) The width of breakouts in red areas indicates the orientations of unstable wellbores as nearly the entire circumference of the well fails. (b) Orientation of wellbore breakouts (if they form) in a *looking-down-the-well* coordinate system. (c) The tendency of drilling-induced tensile fractures to form in terms of the magnitude of excess mud weight needed to initiate failure. (d) The orientation of induced tensile failures (colors are the same as in Figure 8.4b).

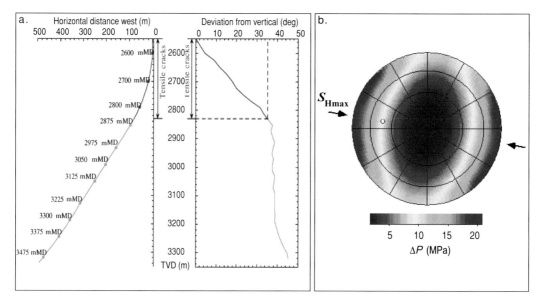

Figure 8.9. Drilling-induced tensile fractures were observed in the near-vertical portion of a well in the Visund field in the northern North Sea which abruptly ceased when the well deviated more than 35° (center). As shown in the figure on the right, this result is predicted by the stress state shown in Figure 8.10 (after Wiprut, Zoback *et al.* 2000). *Reprinted with permission of Elsevier.*

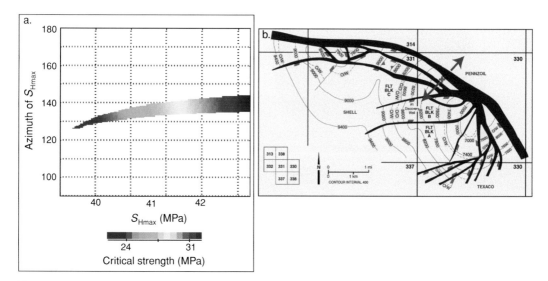

Figure 8.11. (a) The possible values of S_{Hmax} magnitude and orientation consistent with wellbore breakouts in a deviated well in the Gulf of Mexico (after Zoback and Peska 1995). © *2002 Society Petroleum Engineers.* (b) The stress orientation determined in this analysis indicates extension orthogonal to the strike of a major normal fault penetrated by the well. Original map after Holland, Leedy *et al.* (1990). *AAPG©1990 reprinted by permission of the AAPG whose permission is required for futher use.*

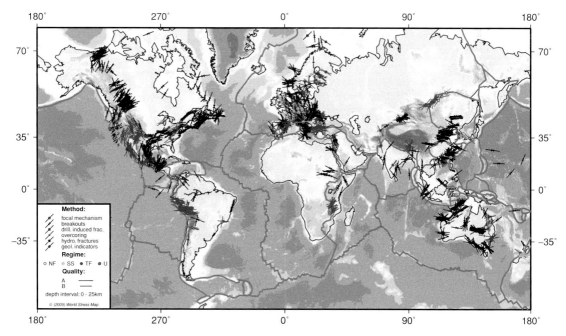

Figure 9.1. Directions of maximum horizontal stress from the World Stress Map data base. Colors are the same as in Figure 1.5. Again, only A and B quality data are shown.

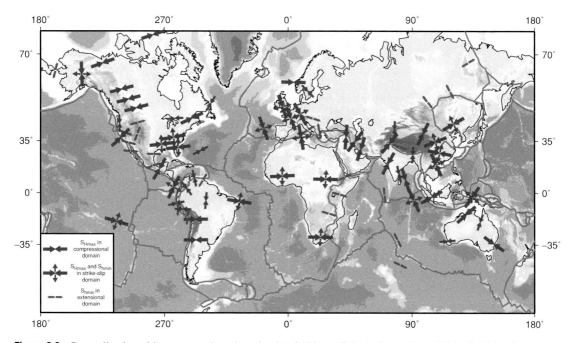

Figure 9.2. Generalized world stress map based on the data in Figure 9.1, similar to that of Zoback, Zoback *et al.* (1989). Inward pointing arrows indicate high compression as in reverse faulting regions. Paired inward and outward arrows indicate strike-slip faulting. Outward directed red arrows indicate areas of extension. Note that the plates are generally in compression and that areas of extension are limited to thermally uplifted areas.

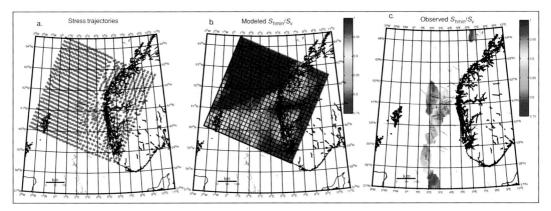

Figure 9.3. Comparison between observed directions of maximum horizontal stress in the northern North Sea and calculated stress directions incorporating the effects of lithospheric flexure due to removal of the Fennoscandian ice sheet (Grollimund and Zoback 2003). (a) The model accurately reproduces stress orientations in the northern North Sea. (b) The calculated stress magnitudes are shown as the ratio of the minimum horizontal stress to the overburden. (c) The magnitude of the minimum horizontal stress as determined from leak-off tests in wells is quite similar to that predicted by the model. *AAPG© 2003 reprinted by permission of the AAPG whose permission is required for futher use.*

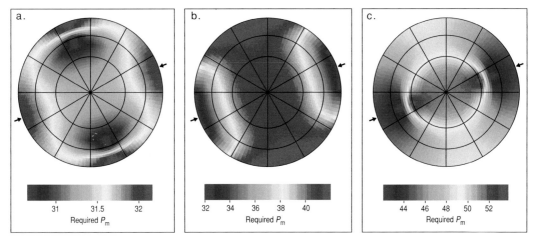

Figure 10.4. The effects of wellbore trajectory and stress state on wellbore stability. The parameters used in this figure are the same as those used for the calculations shown in Figure 8.2. The figure shows the mud pressure (in ppg) required to drill a stable well (maximum breakout width 30° for a relatively strong rock (UCS ~50 MPa) as a function of well orientation at a depth of 3 km for hydrostatic pore pressure: (a) normal faulting, (b) strike-slip faulting and (c) reverse faulting.

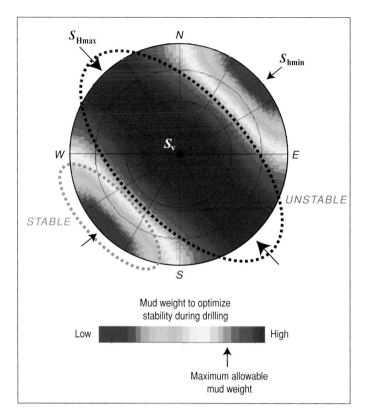

Figure 10.5. In this case study for an offshore Gulf of Mexico well, near vertical wells and those deviated to the northwest or southeast require unrealistically high mud weights (i.e. in excess of the frac gradient) to achieve an acceptable degree of wellbore stability. In contrast, wells that are highly deviated to the southwest or northeast are relatively stable.

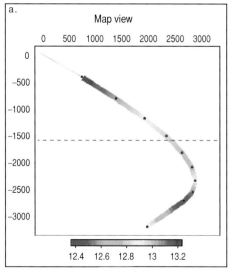

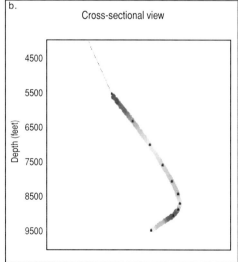

Figure 10.6. Two views of a well trajectory that takes advantage of the principles demonstrated in Figure 10.5. By turning the well to the southwest in the problematic area, wellbore stability could be achieved with a mud weight less than the fracture gradient. The color indicates the mud weight (in ppg) required to stabilize the well at a given depth.

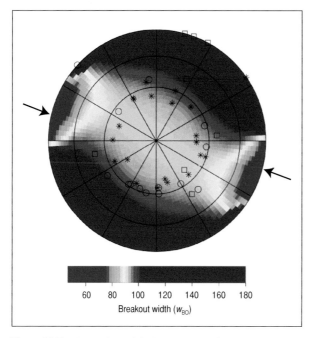

Figure 10.7. Illustration of the importance of drilling direction on success (after Zoback, Barton *et al*. 2003). The figure shows modeled breakout width in the shales in an oil field in South America at a depth of 2195 m TVD as a function of drilling direction assuming a mud weight of 10 ppg and $C_0 = 17.2$ MPa. The total circumference of the wellbore that fails is twice the breakout width. The symbols illustrate the number of days it took to drill the respective well, which is a measure of drilling problems associated with wellbore stability. The asterisks indicate wells that were not problematic (<20 days), the circles indicate wells that were somewhat problematic (20–30 days) and the squares indicate wells that were quite problematic (>30 days). The color scale ranges from an acceptable breakout width (70°, blue) in which less than half the wellbore circumference fails to an excessive amount failure corresponding to over half the circumference failing (breakout widths over 100°, dark red). *Reprinted with permission of Elsevier.*

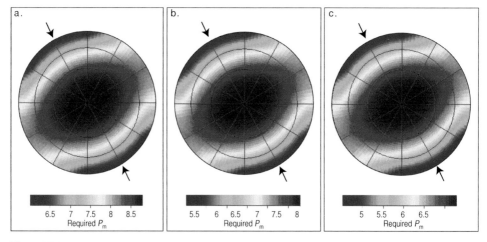

Figure 10.8. Wellbore stability as a function of mud weight. In each of these figures, all parameters are the same except rock strength. For a UCS of 7000 psi (a), a required mud weight of ~8.6 ppg (slightly overbalanced) is needed to achieve the desired degree of stability in near-vertical wells (the most unstable orientation). For a strength of 8000 psi (b), the desired degree of stability can be achieved with a mud weight of ~8 ppg (slightly underbalanced). If the strength is 9000 psi (c) a stable well could be drilled with a mud weight of ~7.3 ppg (appreciably underbalanced).

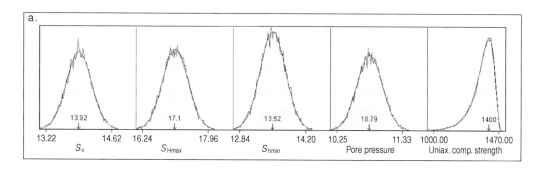

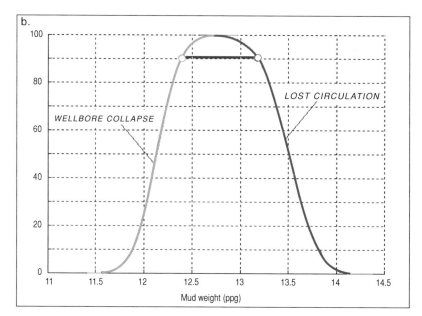

Figure 10.9. (a) Probability density functions (smooth, shaded curves) and the sampled values used in the QRA analysis (jagged lines) as defined by the minimum, most likely, and maximum values of the stresses, the pore pressure, and the rock strength. These quantify the uncertainties in the input parameters needed to compute the mud weight limits necessary to avoid wellbore instabilities. (b) Resulting minimum (quantified in terms of the likelihood of preventing breakouts wider than a defined collapse threshold) and maximum (to avoid lost circulation) bounds on mud weights at this depth. The horizontal bar spans the range of mud weights that ensure a greater than 90% likelihood of avoiding either outcome – resulting in a minimum mud weight of 12.4 ppg and a mud window of 0.75 ppg. After Moos, Peska *et al.* (2003). *Reprinted with permission of Elsevier.*

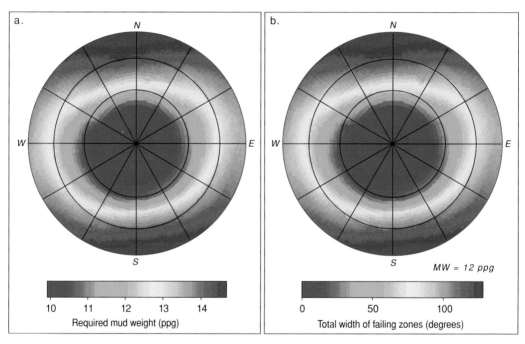

Figure 10.13. Drilling through sub-horizontal, weak bedding planes is only problematic in this case study when the wellbore deviation exceeds ~30°. Because there is little stress anisotropy, there are relatively minor differences in stability with azimuth. This can be seen in terms of the mud weight required to achieve an acceptable degree of failure (a) or the width of the failure zone at a mud weight of 12 ppg (b).

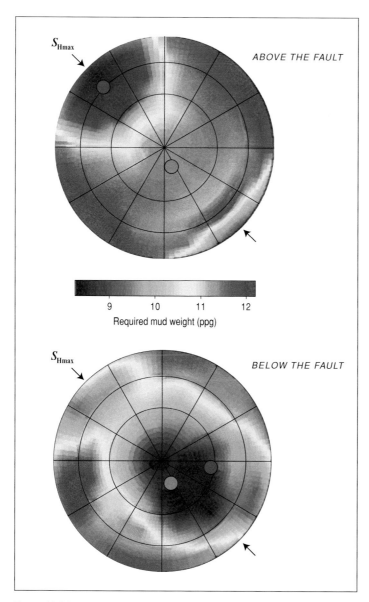

Figure 10.14. When bedding planes dip steeply, both the deviation and azimuth of wells have a strong effect on wellbore stability (similar to Willson, Last *et al*. 1999). (a) Wellbore stability diagram that shows the case above a fault at about 15,000 ft depth, where the bedding plane orientation (the red dot is the pole to the bedding planes) was such that drilling a near vertical well was quite problematic. Drilling orthogonal to the bedding planes (to offset the effect of bedding on strength) would require a steeply dipping well to the northwest. (b) Below the fault, the bedding orientation changes such that a near-vertical well is stable.

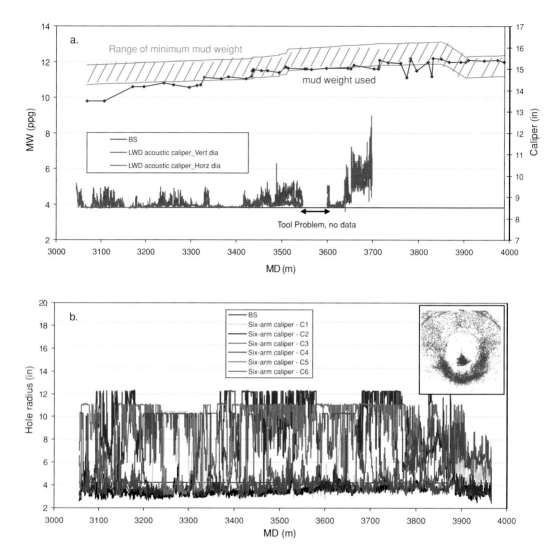

Figure 10.20. (a) Logging-while-drilling acoustic caliper data show relatively little borehole failure as the SAFOD borehole was being drilled in the vicinity of the San Andreas fault zone, confirming that the mud weights predicted by the wellbore stability analysis were essentially correct. A moderate degree of hole enlargement is seen in the deeper part of the interval logged with LWD (3630–3700 m). (b) Five weeks later, six-arm wireline calipers show deterioration of the borehole with time. In the inset, centralized six-arm caliper pads are plotted in a borehole coordinate system. The borehole diameter is greatly enlarged at the top (the blue points indicate the approximate center of the logging tool). From Paul and Zoback (2006).

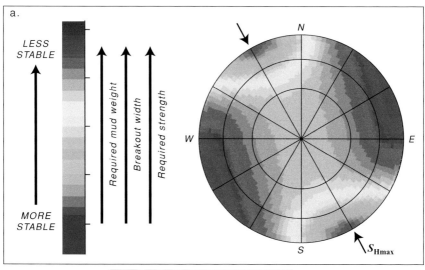

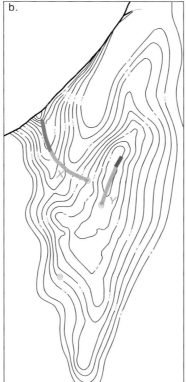

Figure 10.21. (a) Relative stability of multi-lateral wells drilled at various orientations in the Cook Inlet (modified from Moos, Zoback *et al.* 1999). Note that highly deviated wells drilled to the NW and SE are expected to be stable whereas those drilled to the NE and SW are not. (b) Following development of the analysis shown in (a) it was learned that well X (drilled to the NW) was drilled without difficulty whereas well Y (drilled to the NE) had severe problems with wellbore stability.
© *1999 Society Petroleum Engineers*

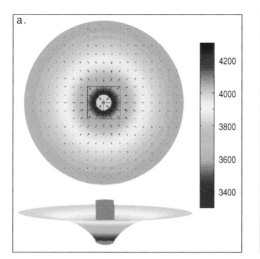
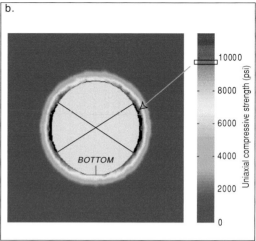
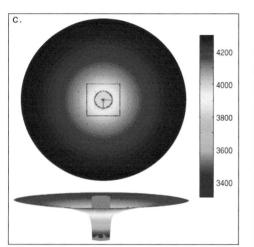
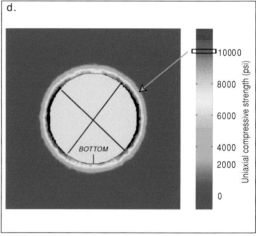

Figure 10.22. Pressure drawdown and predicted width of the failure zone for the Cook Inlet multilateral study shown in Figure 10.21 (modified from Moos, Zoback *et al.* 1999). (a) and (b) Calculations are for the case where a 500 psi drawdown is achieved slowly. Note that the zone of rock failure (breakout width) is limited to about 60°. (c) and (d) Calculations for a very rapid drawdown of 1000 psi. Note that the region of failure around the well is much more severe. © *1999 Society Petroleum Engineers*

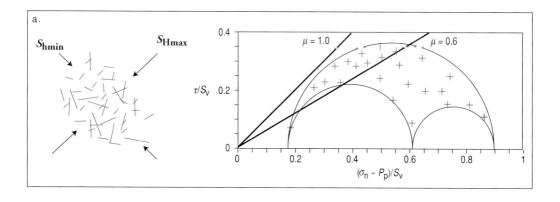

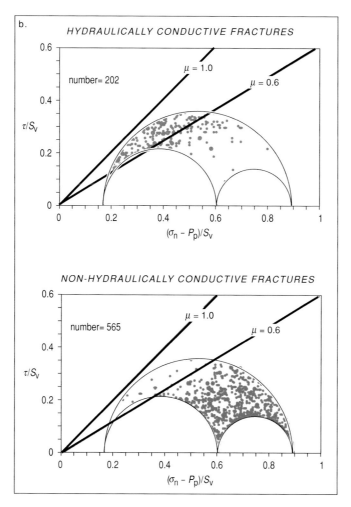

Figure 11.1. (a) A critically stressed crust contains many fractures, some of which are active in the current stress field (light line faults in cartoon on left and light + marks in the normalized Mohr diagram) and some of which are not (heavy line faults and heavy + marks). (b) In the context of the *critically-stressed-fault* hypothesis, hydraulically conductive faults are critically stressed faults (upper diagram) and faults that are not hydraulically conductive are not critically stressed. After Barton, Zoback *et al.* (1995).

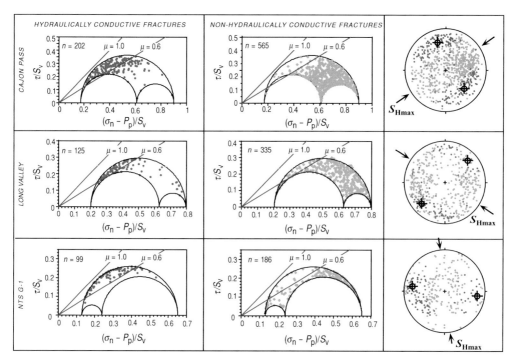

Figure 11.2. Normalized Mohr diagrams of the three wells that illustrate that most hydraulically conductive faults are critically stressed faults (left column) and faults that are not hydraulically conductive are not critically stressed (center column) along with stereonets that show the orientations of the respective fracture sets. The first row shows data from the Cajon Pass well (same Mohr diagrams as in Figure 11.1b), the second from the Long Valley Exploration Well and the third from well G-1 at the Nevada Test Site. After Barton, Zoback *et al.* (1995).

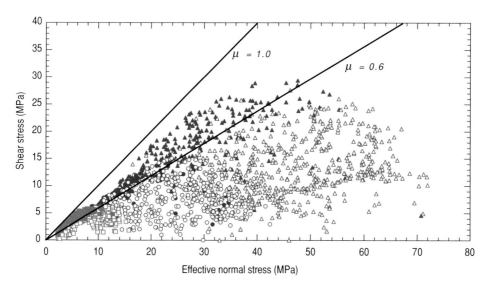

Figure 11.3. Shear and normal stresses on fractures identified with borehole imaging in Cajon Pass (triangles), Long Valley (circles), and Nevada Test Site (squares) boreholes. Filled symbols represent hydraulically conductive fractures and faults, and open symbols represent non-conductive fractures. From Townend and Zoback (2000) based on original data in Barton, Zoback *et al.* (1995).

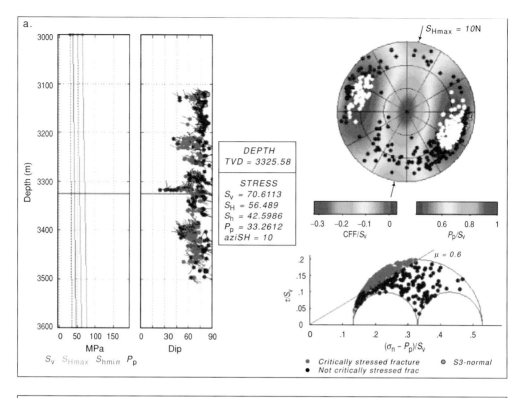

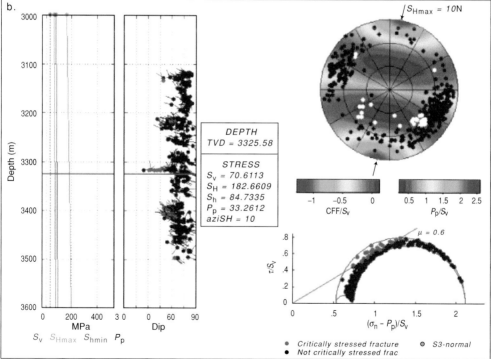

Figure 11.4. Illustration of the relationship between critically stressed fault orientations and absolute stress magnitudes (left column). In both cases, the direction of S_{Hmax} is N10°E. (a) For a normal faulting stress state, a large fraction of the fault population is critically stressed as they are well-oriented for slip in a normal faulting stress field (i.e. they strike NNE–SSW and dip relatively steeply). (b) In a reverse faulting stress state, very few of the faults are critically stressed because of the steep dip of the faults.

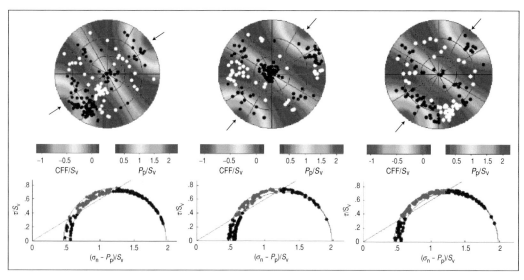

Figure 11.5. Faults identified in image logs from wells A, B, C in the Monterey formation of western California shown previously in Figure 5.8. As in Figure 11.4, the color of the stereonets indicates the tendency for fault slip to occur for a fault of given orientation, in terms of either the CFF or pore pressure needed to induce fault slip. Critically stressed faults are shown in white on the stereonet and red in the Mohr diagram. Note that although each well is considered to be in the same reverse/strike-slip stress state, the distribution of critically stressed faults in each well is quite different because of the distribution of faults that happen to be present in the three respective wells.

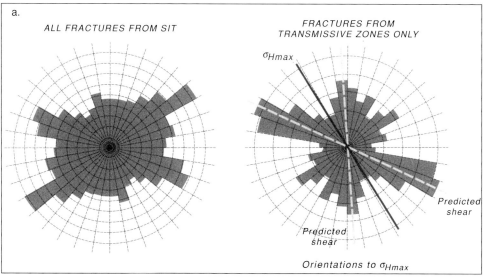

a.

ALL FRACTURES FROM SIT

FRACTURES FROM
TRANSMISSIVE ZONES ONLY

σ_{Hmax}

Predicted
shear

Predicted
shear

Orientations to σ_{Hmax}

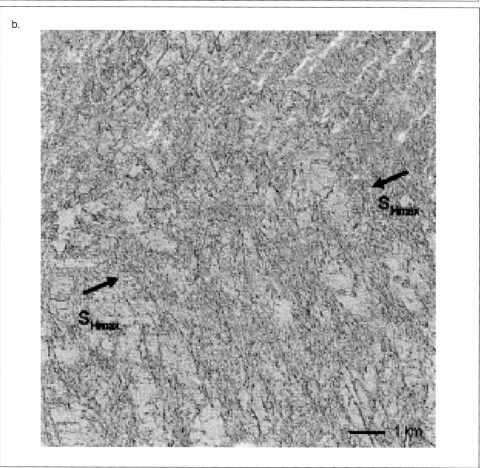

b.

S_{Hmax}

S_{Hmax}

1 km

Figure 11.7. (a) Rose diagrams showing the strike direction of all faults in a test well (Engelder and Leftwich) and only the strike direction of faults shown to be permeable in numerous packer tests (right). The orientation of permeable faults is consistent with the critically-stressed-fault hypothesis for strike-slip faulting (from Rogers 2002). (b) An attribute analysis of a depth slice from a 3D seismic survey in the Mediterranean Sea (from Ligtenberg 2005). Faults that are fluid migration pathways (yellow) are at the appropriate orientation to S_{Hmax} for a strike-slip faulting regime.

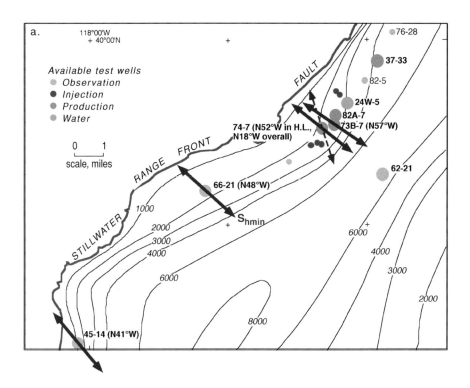

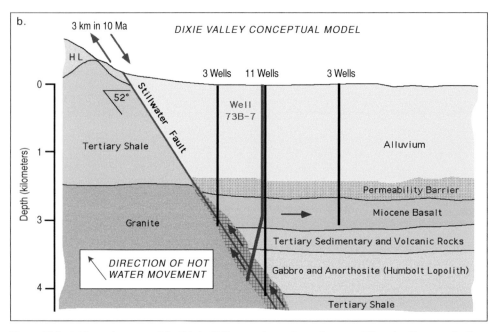

Figure 11.8. (a) Location map of the Dixie Valley geothermal area in central Nevada. Contours indicate depth to basement. Arrows indicate the direction of S_{hmin} from observations of wellbore breakouts and drilling-induced tensile fractures in the wells indicated. (b) Schematic cross-section of the Dixie Valley system showing that hot water comes into the fault zone reservoir at ~4 km depth. Prior to exploitation of geothermal energy (hot water that flashes to steam in producing wells), fluid flowed out of the fault zone and into fractured basalts beneath the valley. After Hickman, Barton *et al.* (1997).

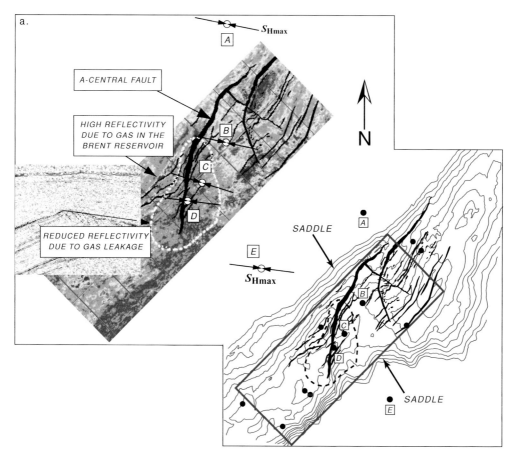

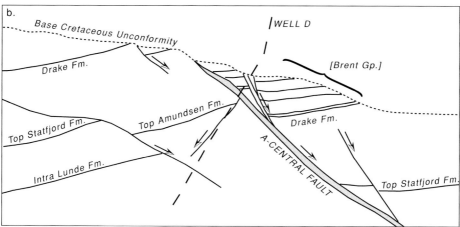

Figure 11.13. (a) Seismic reflectivity map of the top of the Brent formation (Engelder and Leftwich 1997) and a structure contour map (right). The dashed lines indicate the region of apparent gas leakage from along the southern part of the A-central fault. The stress orientations in Visund wells were shown previously in Figure 6.7. The inset is a portion of a seismic section showing an apparent gas chimney in the overburden above the leakage point. (b) A generalized geologic cross-section showing the manner in which well D penetrates a splay of the A-central fault. Note that hydrocarbons in the Brent formation between the A-central fault and the splay fault would have trapped the hydrocarbons in a footwall reservoir (after Wiprut and Zoback 2002). *Reprinted with permission of Elsevier.*

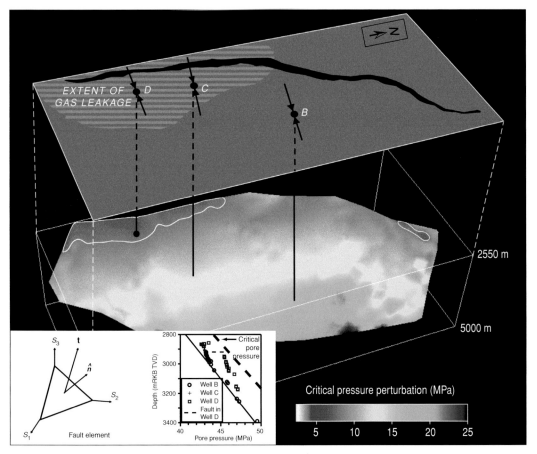

Figure 11.14. Perspective view of the A-central fault and area of fault leakage shown in Figure 11.13a. The stress orientations associated with wells B, C and D were previously shown in Figure 6.7. The color on the fault plane (dipping toward the viewer) indicates the critical pore pressure perturbation (or pressure above ambient pore pressure at which fault slip is likely to occur). The inset illustrates how stress is resolved on individual segments of the fault and the fact that the calculated critical pore pressure (dashed line) is within one MPa of measured pressure in well D on the foot wall side of the fault. After Wiprut and Zoback (2000).

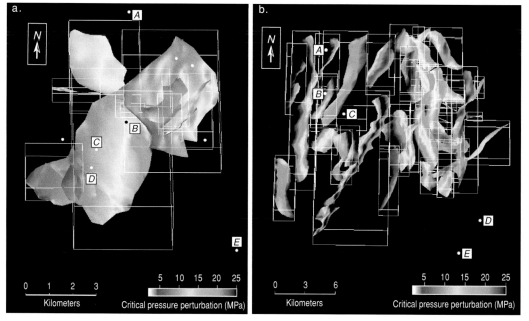

Figure 11.15. (a) Leakage potential map for the faults of the Visund field (similar to that shown for the A-central fault, the large fault in the center dipping to the east) in Figure 11.14. Note that most of the faults in this field have high leakage potential because they strike almost normal to S_{Hmax} (see Figure 6.7) and have shallow dips making them relatively easy to reactivate as reverse faults. The letters refer to well locations. It should be noted that because of the perspective view, the scale is only approximate. (b) A leakage potential map for a field that is relatively near Visund with a very similar stress state. Because these faults have steeper dips, S_{Hmax} tends to resolve high normal stress on these faults making them unlikely to be reactivated. After Wiprut and Zoback (2002). *Reprinted with permission of Elsevier.*

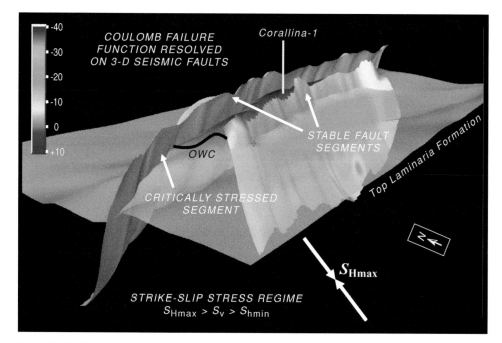

Figure 11.16. Leakage potential map for the Corallina field in the Timor Sea. In this strike-slip faulting regime, there is significant leakage potential (which correlates with the oil–water contact) where the fault has the appropriate strike to the direction of S_{Hmax}. After Castillo, Bishop *et al.* (2000).

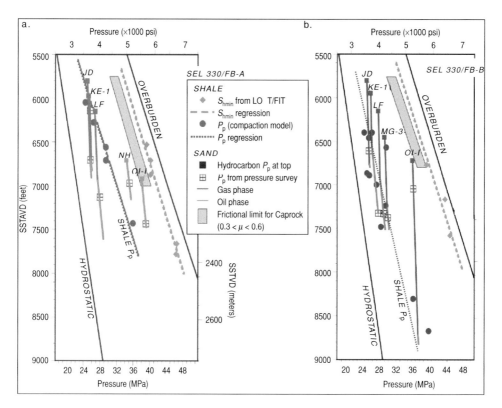

Figure 11.20. Pressure in various reservoirs in South Eugene 330 Field in (a) fault block A and (b) fault block B. The map of the OI sand in Figure 2.7 identifies the location of fault blocks A and B. The geologic cross-section shown in Figure 2.6a identifies the various reservoirs. The square-with-cross symbol indicates the measurement point with the pressures extrapolated to greater and lesser depth from knowledge of the hydrocarbon column heights and fluid densities. Note that only the pressure at the top of the OI sand columns are near the dynamic limit for inducing slip on reservoir bounding faults. After Finkbeiner, Zoback *et al.* (2001). *AAPG© 1994 reprinted by permission of the AAPG whose permission is required for futher use.*

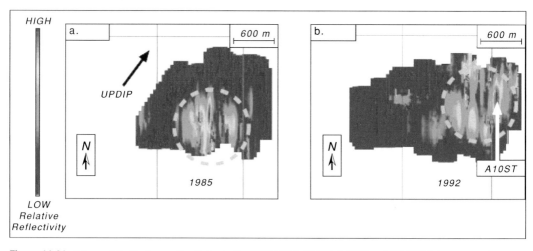

Figure 11.21. Maps of fault plane reflectivity in the SEI-330 field in (a) 1985 and (b) 1992 that appear to show a pocket of hydrocarbons moving updip along the fault plane. After Haney, Snieder *et al.* (2005). The location of the fault along which this *burp* of hydrocarbons is moving (and position of well A10ST along the fault) can be deduced by comparing the maps in Figures 8.11b and 2.7.

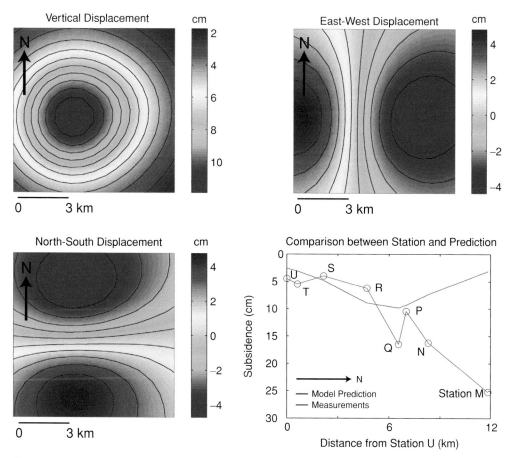

Figure 12.19. Cumulative subsidence and horizontal displacement calculated from superposition of many disk-shaped reservoirs in the Lapeyrouse field using the solution (Geertsma 1973) and DARS (with a viscoplastic rheology) for calculating the total reservoir compaction (after Chan and Zoback 2006). The predicted surface displacements are measured in cm. The predicted subsidence (lower right) is comparable to the measured elevation change from the leveling survey (red line) in the middle of the field but underpredicts the apparent subsidence at the north end of the field near benchmark M.

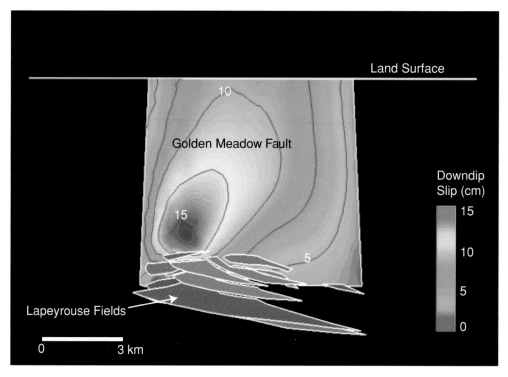

Figure 12.20. Perspective view (looking north) of the calculated slip on the Golden Meadow Fault using Poly3D (Thompson 1993) and a reasonable representation of the actual shape of each of the reservoirs. Note that the highest amount of slip on the fault is just above the shallowest reservoir (after Chan and Zoback 2006).

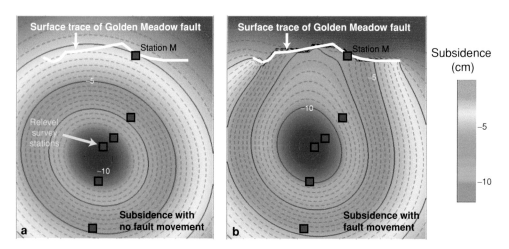

Figure 12.21. Estimated surface subsidence assuming (a) no slip on the Golden Meadow fault and (b) the slip on the fault shown in Figure 12.19 utilizing Poly3D. The shape of the subsidence bowl is altered when the fault is allowed to move freely but the predicted subsidence is still much less than the apparent elevation change at benchmark M (after Chan and Zoback 2006).

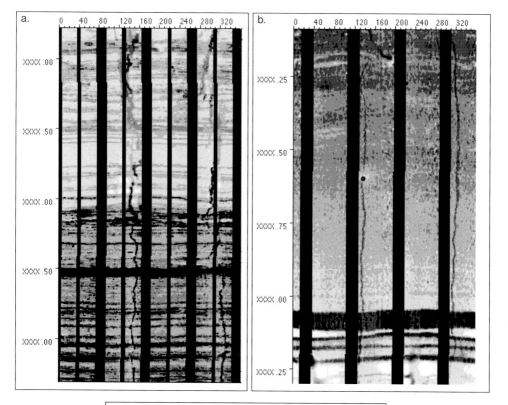

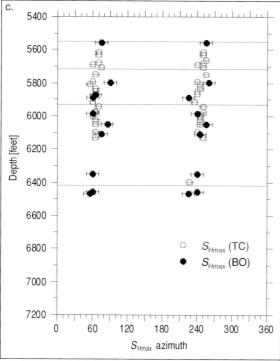

Figure 6.6. (a) and (b) Electrical image logs showing drilling-induced tensile fractures that are on opposite sides of the wellbore (in the direction of maximum horizontal stress) and parallel to the axes of these two vertical wells (indicating that S_v is a principal stress). (c) Example of a well in which the same stress orientation is indicated by breakouts and tensile fractures. From Zoback, Barton *et al.* (2003). *Reprinted with permission of Elsevier.*

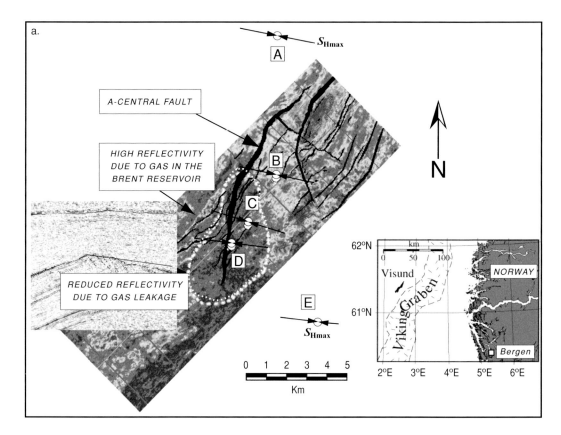

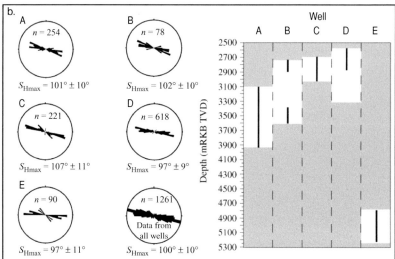

Figure 6.7. Drilling-induced tensile fractures in five wells in the Visund field in the northern North Sea indicate a remarkably uniform stress field both spatially and with depth (after Wiprut, Zoback *et al*. 2000). The rose diagrams illustrate how uniform the tensile fracture orientations are with depth in each well and the field as a whole. The length of each well logged with an electrical imaging device is shown in white in the diagram in the lower right. The drilling-induced tensile fractures are shown by the vertical black lines. (*For colour version see plate section.*)

Numerous studies have shown that if carefully analyzed, breakout orientations are also remarkably consistent in a given well and in a given oil-field (Bell and Babcock 1986; Klein and Barr 1986; Mount and Suppe 1987; Plumb and Cox 1987), and yield reliable measures of stress orientation in many parts of the world. The basis and criteria associated with creating integrated maps of contemporary tectonic stress are those developed by (Zoback and Zoback 1980, 1989, 1991) and subsequently used in the World Stress Map Project (Zoback 1992). From the perspective of regional stress studies, the depth range of breakouts, typically 1–4 km, provide an important bridge between the depths at which most crustal earthquakes occur (3–15 km) and the depths of most *in situ* stress measurements (<2 km) and geologic observations at the surface. Such data are an integral part of the World Stress Map data base.

Figure 6.8a is a stress map of the Timor Sea (Castillo, Bishop *et al.* 2000), constructed by compiling abundant breakout and tensile fracture orientations at depth in the numerous wells shown. In each well the variation of the maximum horizontal stress direction determined from the breakouts and tensile fractures is less than 10°. Note that in each subregion, the stress field is remarkably uniform. Although the average stress orientation in the area seems clearly to correspond to the convergence direction between Australia and Indonesia, the origin of the variations among the subregions in this tectonically active area is not known.

Determination of breakout orientation from caliper logs

Traditionally, most of the data used to determine the orientation of breakouts in wells come from magnetically oriented four-arm caliper data which are part of the dipmeter logging tool that is commonly used in the petroleum industry. Despite the relatively low sensitivity of this technique, with sufficient care it is possible to use four-arm caliper data to reliably determine breakout orientations. This has been clearly demonstrated in western California (Figure 6.8b, from Townend and Zoback 2004) where S_{Hmax} orientations obtained from analysis of breakouts with four-arm caliper data (inward pointed arrows) yield consistent stress orientations that correlate well with earthquake focal plane mechanisms (data points with circle in center) and young geologic indicators of deformation (the trends of fold axes and active reverse faults).

While the analysis of wellbore breakouts with four-arm caliper data appears to be quite straightforward, it is important not to misinterpret key seats (grooves in the side of the well caused by the rubbing of pipe) or washouts (enlargements of the entire wellbore circumference) as stress-induced breakouts. In fact, relatively little information comes from a standard dipmeter log (Figure 6.9a) – the diameter of the well as measured by the orthogonal pairs of caliper arms (termed the C1–3 and C2–4 pairs as the arms are numbered sequentially "looking down the hole"), the pad 1 azimuth, the deviation of the well and the hole azimuth. The actual resistivity data measured on each pad are not used in the breakout analysis.

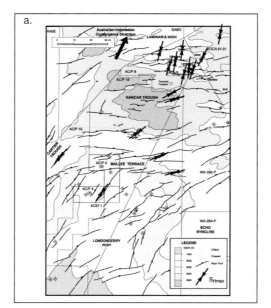

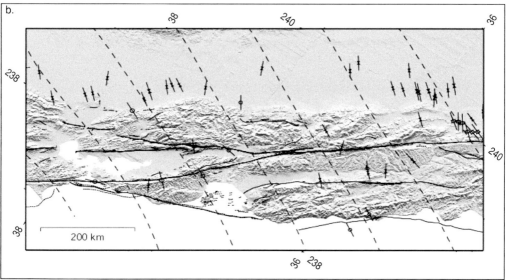

Figure 6.8. (a) Regional stress map of the Timor Sea area (courtesy Woodside Energy) and (b) the coast ranges of central California (after Townend and Zoback 2004). *Reprinted with permission from International Geology Review V. H. Winton and Son, Inc.* The Timor Sea stress map was constructed using observations of drilling-induced tensile fractures and wellbore breakouts in near-vertical wells. The coast ranges stress map was constructed principally on the basis of using breakouts (inward pointed arrows) and earthquake focal plane mechanisms (straight lines with open circle).

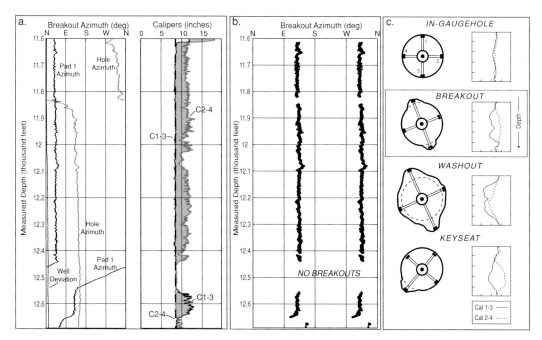

Figure 6.9. (a) Illustration of data derived from *uncomputed* dipmeter logs that shows the azimuth of the well, deviation of the well from vertical, the azimuth of a reference arm as determined from a magnetometer in the tool (pad 1 azimuth) and the diameters of the well as determined from the 1–3 and 2–4 caliper pairs. (b) By utilizing strict quality control criteria, it is possible to properly identify the orientation of stress-induced wellbore breakouts. Note that at the depths near 12,500 feet where both sets of caliper arms are the same and equal to the bit size (8.5 inches), it is clear that no breakouts (or key seats) are present and the tool rotates as it comes up the hole. Where breakouts (or keyseats) are present, the tool does not rotate and one pair of caliper arms measures bit size and the other pair indicates an enlarged wellbore diameter. Note that near the bottom of the interval, the C2–4 calipers are in gauge and C1–3 are enlarged, whereas at shallower depth the opposite is true. As the tool rotated 90°, the breakout orientations are the same. (Figure courtesy D. Wiprut.) (c) Examples of how variations of hole shapes derived from caliper data can be used to identify stress-induced breakouts and distinguish them from washouts and keyseats (after Plumb and Hickman 1985).

Breakouts are wellbore enlargements caused by stress-induced failure of a well occurring 180° apart. In vertical wells breakouts occur at the azimuth of minimum horizontal stress (S_{hmin}) and generally (when analyzed properly) show a remarkably consistent orientation within a given well or field. One exception to this is localized perturbations of the stress field due to slip on active faults discussed in Chapter 11. Another is perturbations associated with salt bodies, as discussed in Chapter 1.

Plumb and Hickman (1985) offered straightforward criteria for the analysis of caliper data and suggested the following definitions to help properly interpret four-arm caliper logs (Figure 6.9): One pair of caliper arms measures the size of the drill bit whereas the

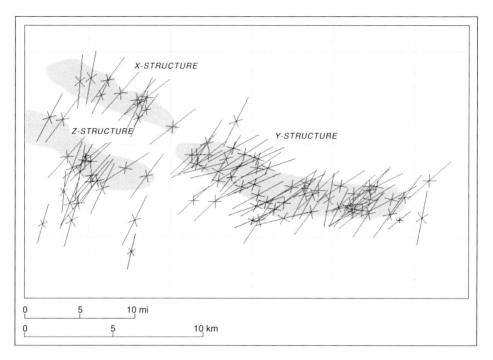

Figure 6.10. A detailed stress map of an oil field in an area of active faulting as determined with four-arm caliper data. The length of the arrow corresponds to the quality of the data as explained in the text. Note that while the stress field shows many local variations due to the processes associated with active faulting and folding, these variations are straightforwardly mappable (courtesy D. Castillo).

orthogonal pair measures a larger diameter. The logging tool does not rotate with depth in the breakouts. *Washouts* represent essentially complete failure of the wellbore such that both sets of arms of the four-arm caliper are larger than the diameter of the drill bit. *Keyseats* are an asymmetrical notching of the well caused by mechanical wear of the borehole at its top or bottom (or on the side of the well if there is a rapid turn in the well trajectory). It is also useful to use a wellbore deviation criterion to help distinguish keyseats from breakouts. Even in near-vertical wells, great care needs to be taken when considering enlargements of the wellbore that are essentially parallel (within 10–15°) to the well deviation direction as indicators of wellbore breakouts.

Figure 6.10 is a detailed stress map of an oil field in central California constructed through analysis of breakouts with four-arm caliper data. The lengths of the individual wells correspond to the quality ranking (A, B or C) described in the next section. There are distinct variations in the direction of maximum horizontal stress from one part of the field to another due to the fact that this is an area of active reverse and strike-slip faulting (and folding). Nonetheless, the consistency of the stress measurements in each well (and between wells) is quite remarkable.

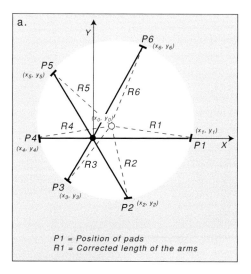

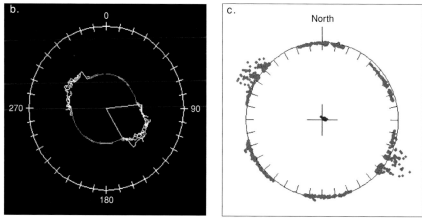

Figure 6.11. (a) A schematic image showing how an off-centered six-arm caliper tool in a circular well results in caliper data that are extremely hard to interpret (after Jarosinski 1998). (b) Analysis of interval of a well logged with an ultrasonic imager that is accurately reproduced by analysis of breakouts with six-arm data (c). *Reprinted with permission of Elsevier.*

There are commercially available dipmeters which are six-arm tools that turn out to be particularly difficult to analyze. The principal reason for this is that there are six different radii that are measured (Figure 6.11a), each of which will be different even in a well with a circular cross-section if the tool is slightly off-center. A special algorithm developed by Jarosinski (1998) for analysis of 6-arm data is particularly useful for identification of breakouts. Note that the breakout identified in the ultrasonic televiewer data in Figure 6.11b is accurately reproduced through analysis of the six-arm caliper data as shown in Figure 6.11c. Figure 6.12 compares the analysis of data from breakouts in a well via ultrasonic data (a) six-arm caliper analysis (b) and out-of-focus

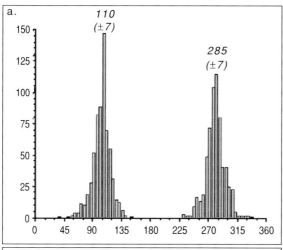

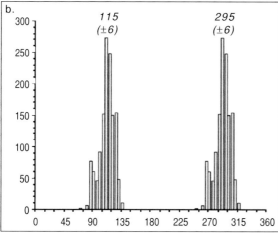

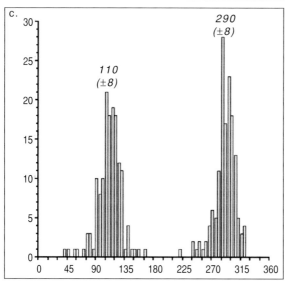

zones in electrical image data (c). Note that the results of the three sets of data are essentially identical.

Quality ranking system for stress indicators

Throughout this book we focus on stress determination techniques that yield reliable estimates of stress orientation and relative magnitude at depth *and* are applicable to the types of geomechanical problems being addressed in subsequent chapters. Hence, there are four types of stress indicators of most interest; well-constrained earthquake focal mechanisms, stress-induced wellbore breakouts, drilling-induced tensile fractures and open-hole hydraulic fracturing stress measurements. As each of these stress indicators has already been discussed at some length, they are discussed below only in terms of the quality table (Table 6.1, modified after Zoback and Zoback 1991; Zoback 1992). As noted previously, the correlation among stress orientations determined from different stress indicators is quite good. Although surface observations of fault slip and volcanic vent alignments yield valuable information about the stress field, in general, such information is not available in the regions of interest.

It is worth making a few comments about the basic logic behind this table. While the rankings are clearly subjective, there are three factors that affect data quality. The greater the depth interval over which wellbore observations are made, the more reliable the data are likely to be. As discussed at some length in Chapter 11, there may be distinct, localized variations of stress due to processes such as fault slip, but, in general, the greater the depth range over which observations are made, the more reliable the observations are likely to be. Also, it is clear that the larger the number of observations, and the smaller the standard deviation of the observations, the more reliable they are likely to be. Each of these criteria are thus used in the quality rankings. A-quality data is of higher quality (and thus more reliable) than B-quality, etc. A, B or C quality are all considered to be of sufficient quality to warrant *putting on a map* and interpreting with some confidence. Commonly, arrows of successively shorter length (such as in Figure 6.10) are used to indicate A, B and C quality, respectively. In marked contrast, D-quality data are thought to be so unreliable that they should not appear on maps and should not be used with confidence to assess the stress field.

As discussed in Chapter 1, the advantages of utilizing well-constrained earthquake focal plane mechanisms to map the stress field are fairly obvious – earthquakes record

Figure 6.12. Comparison of breakouts observed with ultrasonic borehole televiewer data (a), six-arm caliper data (b) and electrical imaging data (c) yield essentially identical breakout orientations. Identification of out-of-focus zones is the least robust of the methods used. Note that there are many fewer observations with such data (right) but nonetheless, their orientation is the same.

stress-induced deformation at mid-crustal depths and sample relatively large volumes of rock, and focal plane mechanisms provide data on both the orientation and relative magnitude of the *in situ* stress field. The principal disadvantage of utilizing earthquake focal plane mechanisms as stress indicators is that focal plane mechanisms record deformation and not stress. Thus, the *P*- and *T*-axes are, by definition, the bisectors of the dilatational and compressional quadrants, respectively, and are not the maximum and minimum principal stress directions (as is often assummed) but are the compressional and extensional strain directions for the two possible faults. Most crustal earthquakes appear to occur on pre-existing faults (rather than resulting from new fault breaks), the slip vector is a function both of the orientation of the fault and the orientation and relative magnitude of the principal stresses. The *P*- and *T*-axes of the focal plane mechanism do not correlate directly with principal stress directions. Because the coefficient of friction of many rocks is often in the range of 0.6–0.8 (Byerlee 1978), it has been suggested that the expected angle between the fault plane and the direction of maximum principal stress should be plotted 30° from the fault plane (Raleigh, Healy *et al.* 1972). However, this requires knowledge of the actual fault plane, which is frequently not the case. If the auxiliary plane of the focal mechanism was mistakenly selected as the fault plane (Chapter 5), the presumed stress orientation would be off by ~45°.

To optimize the use of focal plane mechanism data for determining stress orientations it is necessary to consider multiple events in a given region and use either the average *P*-axis direction as the maximum horizontal stress direction or to formally invert a group of focal plane mechanisms to determine the orientation and relative magnitude of the principal stress tensor (see, for example, Angelier 1979; Gephart and Forsyth 1984; Michael 1987).

The assignment of A, B, C and D qualities to wellbore breakout data is based on the frequency, overall length and consistency of occurrence of breakouts in a given well. The standard deviations associated with the respective qualities shown in Table 6.1 were determined from our empirical experience working with breakout data in a number of boreholes. High standard deviations ($> \sim 25°$) represent either very scattered, or bi-modal, distributions due to a variety of factors – if the rock is strong compared to the stress concentration, breakouts will be poorly developed. On the other hand, the rock might be very weak and failure so pervasive that it is difficult to distinguish between stress-induced breakouts and more pervasive washouts. This is why utilizing the criterion for distinguishing breakouts from key seats and washouts (as discussed later in this chapter) is so important.

To calculate the mean breakout direction, θ_{m}, and the standard deviation, sd, of a set of breakouts on a given side of a well, we utilize Fisher statistics and let θ_i ($i = 1, \ldots, N$) denote the observed breakout directions in the range 0–360°. First, we define

$$l_i = \cos \theta_i \ \text{ and } \ m_i = \sin \theta_i \tag{6.10}$$

Table 6.1. *Quality ranking system*

	A	B	C	D
Earthquake focal mechanisms	Average P-axis or formal inversion of four or more single-event solutions in close geographic proximity(at least one event $M \geq 4.0$, other events $M \geq 3.0$)	Well-constrained single-event solution ($M \geq 4.5$) or average of two well-constrained single-event solutions ($M \geq 3.5$) determined from first motions and other methods (e.g. moment tensor wave-form modeling, or inversion)	Single-event solution (constrained by first motions only, often based on author'squality assignment) ($M \geq 2.5$). Average of several well-constrained composites ($M \geq 2.0$)	Single composite solution. Poorly constrained single-event solution. Single-event solution for $M < 2.5$ event
Wellbore breakouts	Ten or more distinct breakout zones in a single well with sd $\leq 12°$ and/or combined length >300 m. Average of breakouts in two or more wells in close geographic proximity with combined length >300 m and sd $\leq 12°$	At least six distinct breakout zones in a single well with sd $\leq 20°$ and/or combined length > 100 m	At least four distinct breakouts with sd $< 25°$ and/or combined length > 30 m	Less than four consistently oriented breakout or >30 m combined length in a single well. Breakouts in a single well with sd $\geq 25°$
Drilling-induced tensile fractures	Ten or more distinct tensile fractures in a single well with sd $\leq 12°$ and encompassing a vertical depth of 300 m, or more	At least six distinct tensile fractures in a single well with sd $\leq 20°$ and encompassing a combined length > 100 m	At least four distinct tensile fractures with sd $< 25°$ and encompassing a combined length > 30 m	Less than four consistently oriented tensile fractures with <30 m combined length in a single well. Tensile fracture orientations in a single well with sd $\geq 25°$
Hydraulic fractures	Four or more hydrostatic orientations in a single well with sd $\leq 12°$ depth >300 m. Average of hydrofrac orientations for two or more wells in close geographic proximity, sd $\leq 12°$	Three or more hydrofrac orientations in a single well with sd $< 20°$. Hydrofrac orientations in a single well with $20° < $ sd $< 25°$	Hydrofac orientations in a single well with $20° < $ sd $< 25°$. Distinct hydrofrac orientation change with depth, deepest measurements assumed valid. One or two hydrofrac orientations in a single well	Single hydrofrac measurements at <100 m depth

and then calculate

$$l = \frac{\sum\limits_{i=1}^{n} l_i}{R} \quad \text{and} \quad m = \frac{\sum\limits_{i=1}^{n} m_i}{R} \tag{6.11}$$

$$R^2 = \sum_{i=1}^{n} l_i + \sum_{i-1}^{n} m_i \tag{6.12}$$

The mean breakout direction is given by

$$\theta_m = \tan^{-1}\left(\frac{m}{l}\right) \tag{6.13}$$

We define

$$k = \frac{N-1}{N-R} \tag{6.14}$$

such that the standard deviation is given by

$$\text{sd} = \frac{81°}{\sqrt{k}} \tag{6.15}$$

The rationale for the A, B, C and D quality assignments for drilling-induced tensile fractures is similar to that for breakouts as discussed above. When 10 or more consistently oriented tensile fractures are seen over a 300 m depth interval, such observations are given an A quality. B and C quality involve fewer tensile fracture observations, greater variation of orientation and a smaller depth interval. Again a standard deviation greater than 25° is interpreted as an indication of unreliable data (quality D) such that it should not be presented on maps.

As discussed in Chapter 7, in an *open-hole* hydraulic fracturing stress measurement, an isolated section of a well is pressurized until a tensile fracture is induced at the point of least compressive stress around the well. Under ideal circumstances, impression packers (oriented with respect to magnetic north) can be used to determine the azimuth of the induced hydrofrac. However, this is both a time-consuming and difficult process and rarely yields reliable results in oil and gas wells.

More on drilling-induced tensile fractures

In this section we return to the subject of drilling-induced tensile fractures to make a few additional points. First, drilling-induced tensile fractures occur in vertical wells only when there is a significant difference between the two horizontal stresses. In fact, it is straightforward to show that the conditions for the occurrence of drilling-induced tensile fractures around a vertical wellbore in the absence of excess mud weight or wellbore cooling are essentially identical to the values of S_{hmin} and S_{Hmax} associated with a strike-slip faulting regime in frictional equilibrium.

Tensile fractures and strike-slip faulting

Consider the state of stress in a strike-slip faulting environment previously discussed in Chapter 4. By rewriting equation (4.46) for the case of frictional equilibrium and $\mu = 0.6$, we have

$$\frac{S_{Hmax} - P_p}{S_{hmin} - P_P} = (\sqrt{\mu^2 + 1} + \mu)^2 = 3.1 \tag{6.16}$$

which can be simplified to

$$S_{hmax} = 3.1 S_{Hmin} - 2.1 P_p$$

and, for reasons that will soon be evident, rewritten as

$$S_{Hmax} = 3 S_{hmin} - 2 P_p + 0.1 (S_{hmin} - P_p) \tag{6.17}$$

If we now revisit equation (6.7) that describes the formation of tensile fracture in the wall of a vertical wellbore and assume that the cooling stress, $\sigma^{\Delta T}$, excess mud weight, ΔP, and tensile strength are negligible, a tensile fracture will form at the wellbore wall when

$$\sigma_{\theta\theta}^{min} = 3 S_{hmin} - S_{Hmax} - 2 P_p = 0 \tag{6.18}$$

or

$$S_{Hmax} = 3 S_{hmin} - 2 P_p \tag{6.19}$$

It is obvious that because the last term in equation (6.17) $(0.1(S_{hmin} - P_p))$ is extremely small, equations (6.17) and (6.19) are nearly equal in magnitude. In other words, for any combination of S_{hmin} and S_{Hmax} which results in frictional equilibrium in the crust for a strike-slip domain (and $\mu = 0.6$), the wellbore wall will go into tension at the azimuth of S_{Hmax}, even without excessive wellbore pressure or cooling of the wellbore wall.

   This can be illustrated graphically using the type of plot shown in Figure 4.28 that defines possible stress magnitudes at any given depth based on the frictional strength of the crust. In Figure 6.13a, we illustrate the fact that for $\Delta T = \Delta P = 0$, the line on the periphery of the polygon (indicating the magnitude of S_{Hmax} as a function of S_{hmin} for strike-slip faults that are active) is almost exactly the same as the line representing equation (6.14) for zero rock strength as shown.

Thermal effects

It was noted above that additional stresses are applied to the rock at the borehole wall if the wellbore fluid is at a significantly different temperature than the rock. These stresses can be compressive or tensile depending on whether the temperature of the fluid is higher or lower, respectively, than the ambient *in situ* temperature. When the

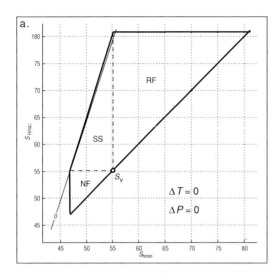

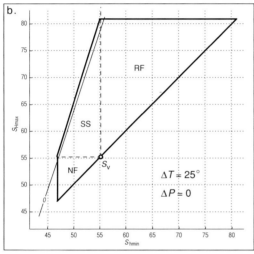

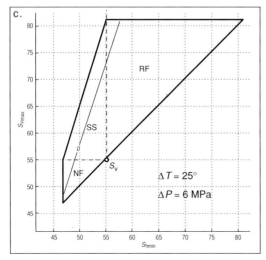

mud is cooler than the formation (the usual case at the bit) thermal stresses make the stress concentration around a well more tensile at all azimuths in the same manner as increasing mud pressure.

The effect of temperature is time-dependent, in the sense that the longer the rock is in contact with the wellbore fluid the further away from the hole the temperature perturbation will propagate. To simplify this problem, one can assume that the material is impermeable, and relatively simple integral equations can be written for the magnitudes of $\sigma_{\theta\theta}$ and σ_{rr} as a function of radial position r and time t (Stephens and Voight 1982). Although the exact solution for the temperature distribution near a constant-temperature wellbore is a series expansion (Ritchie and Sakakura 1956), solutions which approximate the temperature using the first two terms of the expansion give sufficiently accurate results close to the hole, where the stresses are given by:

$$\sigma_{\theta\theta} = \left[\frac{\alpha_t E \,\Delta T}{1 - v}\right]\left[\left(\frac{1}{2\rho} - \frac{1}{2} - \ln\rho\right)I_0^{-1} - \left(\frac{1}{2} + \frac{1}{2\rho}\right)\right] \tag{6.20}$$

$$\sigma_{rr} = \left[\frac{\alpha_t E \,\Delta T}{1 - v}\right]\left[\left(-\frac{1}{2\rho} + \frac{1}{2} - \ln\rho\right)I_0^{-1} - \left(\frac{1}{2} - \frac{1}{2\rho}\right)\right] \tag{6.21}$$

$$I_0^{-1} = \frac{1}{2\pi\,\mathrm{i}} \int_{-\infty}^{0+} \frac{e^{[4\tau_z/\sigma_2]^z}}{z\ln z}\,dz$$

Once steady state has been reached, the change in the hoop stress is given by

$$\sigma_{\theta\theta}^{\Delta T} = \frac{\alpha_t E \,\Delta T}{1 - v} \tag{6.22}$$

where α_t is the linear coefficient of thermal expansion and E is the static Young's modulus. Figure 3.14 shows the coefficient of thermal expansion for different rock types. As shown, α_t is a strong function of the silica content because the coefficient of thermal expansion of quartz is an order of magnitude higher than other common rock forming minerals.

Figure 6.13b incorporates the effect of wellbore cooling of 25 °C on the formation of drilling-induced tensile fractures described by equation (6.17). As seen through

Figure 6.13. Polygons showing the possible values of S_{hmin} and S_{Hmax} at a given depth and pore pressure that are constructed in the manner of Figure 4.28. A coefficient of friction of 0.6 for faults in the crust is assumed. In addition, the equation describing the magnitude S_{Hmax}, as a function of S_{hmin} that is required to cause drilling-induced tensile fractures in a vertical well. (a) No cooling stress and no excess pore pressure are considered. (b) When the mud is 25° cooler than the formation, drilling-induced tensile fractures can be induced at a slightly lower value of S_{Hmax} for a given value of S_{hmin} because the thermal stress slightly decreases the $\sigma_{\theta\theta}$. (c) When there is 6 MPa of excess mud weight, tensile wall fractures occur at still lower values of S_{Hmax}.

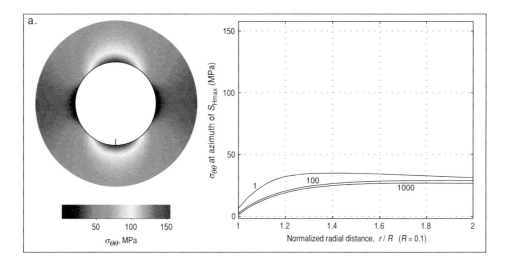

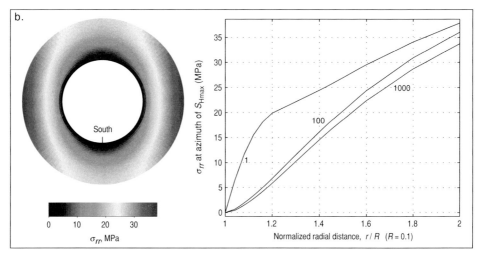

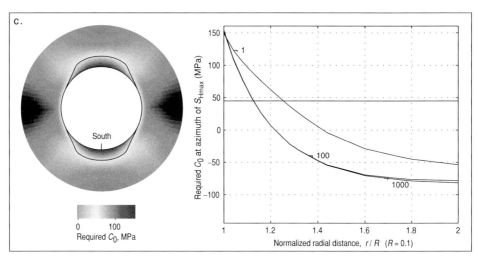

comparison with Figure 6.13a, the result of moderate cooling makes it slightly easier for drilling induced tensile fractures to be induced. That is, for a given value of S_{hmin}, tensile fractures can occur at a slightly lower value of S_{Hmax}. Of course, in geothermal wells, where very significant cooling occurs, this effect can be much greater – so much, in fact, that enlargements due to pervasive tensile failure have sometimes been mistaken for breakouts in geothermal wells (D. Moos, personal communication). For the drilling-induced tensile fractures in well D in the Visund field of the northern North Sea (Figure 6.7), wellbore cooling of $\sim$30 °C at a depth of $\sim$2750 m resulted in $\sigma_{\theta\theta}^{\Delta T} = 1.7$ MPa based on $\alpha = 2.4 \times 10^{-6} \,°C^{-1}$ (corresponding to a rock composed of 30% quartz), $E = 1.9 \times 10^4$ MPa (from the measured P-wave velocity) and $\nu = 0.2$ (based on the P- to S-wave velocity ratio) (Wiprut, Zoback *et al.* 2000).

As noted above, mud weights above the pore pressure encourage the formation of drilling induced tensile fractures. Figure 6.13c shows how 25 °C of wellbore cooling and 6 MPa of excess mud weight affect the formation of tensile fractures. Note that modest increases in mud weight are much more influential on the formation of tensile fractures than modest amounts of wellbore cooling. This will be important in Chapter 7 when we attempt to use observations of drilling-induced tensile fractures for estimating the magnitude of S_{Hmax}. Nonetheless, Pepin, Gonzalez *et al.* (2004) note that in a deepwater Gulf of Mexico well, cooling seems to have decreased the *frac gradient* leading to lost circulation.

Because equation (6.17) is a simplification, it is important to consider how accurately it predicts the change in hoop stress at the wellbore wall. Comparison with the exact analytic thermoporoelastic solution (Li, Cui *et al.* 1998) demonstrates that within a relatively short period of time (10 hours or less) the difference between the two solutions is quite small (a few MPa). Figure 6.14 illustrates the effects of cooling on both the hoop stress (Figure 6.14a) and radial stress (Figure 6.14b) as a function of time and distance from the wellbore wall using the exact analytic solutions. The analytical calculations are done at the azimuth of S_{Hmax}, where tensile fractures are expected to form. The stress conditions used in these calculations are the same as those used in Figures 6.2 and 6.3. Note that both the hoop stress and radial stress become slightly less compressive with time, but are close to steady state after 100 min (the 1000 min calculations are nearly the same). Because $\sigma_{rr} = 0$ (when $\Delta P = 0$) is a boundary condition, its value at the wellbore wall doesn't change with time.

Figure 6.14. The effect of temperature on the state of stress around a wellbore for the same stress values used in Figures 6.2 and 6.3. (a) The thermally induced $\sigma_{\theta\theta}$ and the variation of $\sigma_{\theta\theta}$ with radial distance and time. (b) The thermally induced σ_{rr} and the variation of σ_{rr} with radial distance and time. (c) The effect of cooling on wellbore stability based on drilling with mud that is 10° cooler than the formation temperature. While the breakout is slightly smaller than that shown in 6.3c, it is probably not feasible to significantly improve wellbore stability through cooling. (*For colour version see plate section.*)

More on wellbore breakouts

We discuss here several aspects of breakout formation that will be important when we use breakout observations to estimate stress magnitudes (Chapters 7 and 8) and examine excessive breakout formation associated with wellbore instabilities (Chapter 10).

As discussed above, breakouts form in the area around a wellbore where the stress concentration exceeds the rock strength. As first pointed out by Zoback, Moos *et al.* (1985), once a breakout forms, the stress concentration around the wellbore is such that breakouts will tend to deepen. This was illustrated theoretically as shown on the left side of Figure 6.15a (Zoback, Moos *et al.* 1985). Subsequent work on the manner in which breakout growth would eventually stabilize confirmed this result (Zheng, Kemeny *et al.* 1989), as did laboratory studies of breakout formation by Haimson and Herrick (1989) who presented photographs of breakouts formed in laboratory experiments (Figure 6.15a,b) and also found an excellent correlation between measured breakout widths and the theoretically predicted ones (right side of Figure 6.15b) using a relatively simple failure theory presented by Zoback, Moos *et al.* (1985). While this will not be true for extremely weak formations such as uncemented sands, it appears to be the case for cemented rocks of at least moderate strength. As discussed in Chapter 10, the fact that after initial formation, breakouts deepen (until reaching a stable shape) but do not widen allows us to establish a relatively simple criterion for assessing wellbore stability. As long as drilling conditions result in breakouts that do not have excessive width, wells can be drilled successfully.

In general, it is quite difficult to predict the evolution of the failure zone around a well once a breakout has formed. Zheng, Kemeny *et al.* (1989) attempted to model this analytically, but the rather pointed breakout shapes they predicted are not generally seen when viewing actual breakouts in cross-section. This appears to be because as the rock begins to fail, strain energy is absorbed through inelastic deformation, thus allowing the breakout shape to stabilize with a relatively flat-bottomed shape.

In the same manner that the stresses induced by cool drilling mud in the wellbore affect the formation of tensile fractures, thermally induced stresses also affect the formation of breakouts. The decrease in circumferential stress at the wellbore wall will decrease the tendency for breakouts to occur although the effect is relatively small. Note that for the case in which the hoop stress decreases by several MPa (as in the Visund example cited above), the temperature change would only decrease the maximum hoop stress shown in Figures 6.2 and 6.3 by about 2%. Such a decrease in hoop stress would not be as effective in increasing wellbore stability because there would be no comparable increase in σ_{rr}, as mentioned above. Thus, the initial area of wellbore failure when 10 °C of cooling occurs (as shown in Figure 6.14c) is only slightly smaller than that when there is no cooling (Figure 6.3). However, with time, cooling changes both $\sigma_{\theta\theta}$ and σ_{rr} in such a way as to lessen the tendency for rock failure away from the wellbore

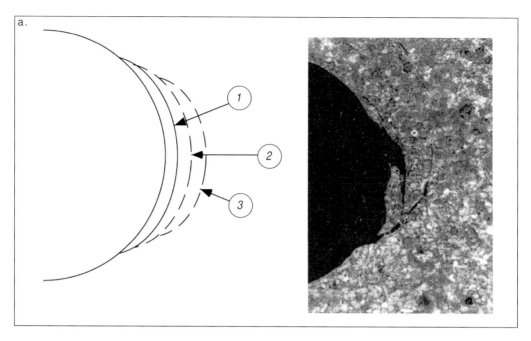

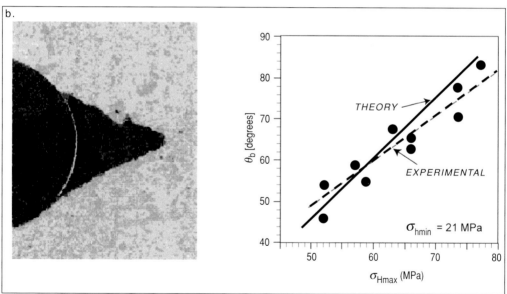

Figure 6.15. After the formation of wellbore breakouts, they are expected to increase in depth, but not width. This is as shown theoretically in (a) after Zoback, Moos *et al.* (1985) and confirmed by laboratory studies (Haimson and Herrick 1989). It can be seen photographically that breakouts in laboratory experiments deepen but do not widen after formation. A shown in (b), measured breakout widths compare very well with those predicted by the simple thoery presented in Zoback, Moos *et al.* (1985) which form the basic for the breakout shapes illustrated in Figures 6.2 and 6.3.

with time (Figure 6.14c). Thus, wellbore cooling does tend to stabilize breakout growth with time, but will not affect breakout width observations used for estimation of S_{Hmax} in Chapter 7 nor will it be a viable technique for minimizing problems of wellbore stability (Chapter 10).

There are other complexities affecting the formation of breakouts that will arise when we consider wellbore stability in Chapter 10:

- the effect of rock strength anisotropy that results from the presence of weak bedding planes in shale;
- the possibility that the *strength of materials* approach utilized here oversimplifies the breakout formation process;
- the relation between mud chemistry, rock strength and wellbore stability;
- other modes of breakout formation; and
- penetration of mud into fractured rock surrounding a wellbore.

Rock strength anisotropy

The theory of compressive failure of rocks with weak bedding planes was discussed in Chapter 4. An example of how slip on weak bedding planes affects breakout formation is illustrated in Figure 6.16. The breakouts seen in the *unwrapped* ultrasonic televiewer data in Figure 6.16a are unusual in that there are four dark zones around the circumference of the well (indicating low reflectivity of the acoustic pulse from the borehole wall) rather than two as seen in Figure 6.4. The cross-sectional view (Figure 6.16b) indicates that the breakouts on each side of the well are each *double-lobed*, as originally hypothesized by Vernik and Zoback (1990) when considering the formation of breakouts in the KTB hole when there were steeply dipping foliation planes cutting across the hole.

Modeling the formation of breakouts when weak bedding or foliation planes cut across a wellbore at a high angle is shown in Figure 6.16c. In such cases, breakouts form due to two processes: when the stress concentration exceeds the intact rock strength and when the stress concentration activates slip on the weak planes thus enlarging the failure zone. The reason that the double lobes appear is related to the fact that the stress trajectories bend around the wellbore as shown in Figure 6.1 such that there are zones where the orientation of S_1 becomes optimal for inducing slip on the bedding plane. As shown in Figure 6.16c, this occurs at the edges of places around the wellbore where *normal* breakouts would form (*i.e.* those forming as a result of the concentrated stresses exceeding the intact rock strength). As is the case with breakout formation in homogeneous and isotropic rock, increases in mud weight tend to stabilize wellbores and reduce the size of breakouts (Figure 6.16d). This problem was also investigated by Germanovich, Galybin *et al.* (1996). The issue of breakout formation in such cases will be important in Chapter 10 in two contexts: in areas where there are steep bedding planes due to the tilting of overburden units and when highly deviated wells are drilled

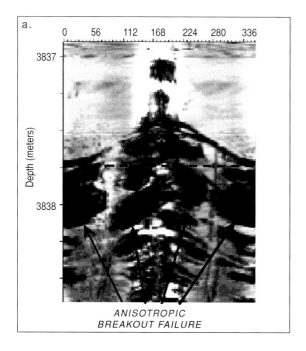

a.

ANISOTROPIC
BREAKOUT FAILURE

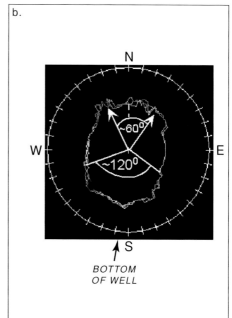

b.

BOTTOM
OF WELL

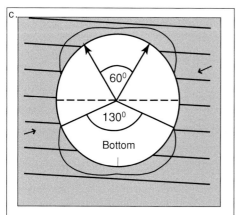

c.

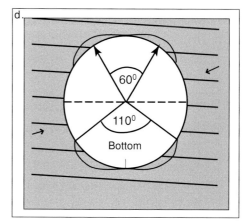

d.

Figure 6.16. (a) Ultrasonic televiewer image of breakouts influenced by rock strength anisotropy associated with the presence of weak bedding planes cutting across a wellbore at a high angle. Note that there are four vertical bands of low reflectivity rather than two as shown in Figure 6.4. (b) Cross-sectional view of a breakout influenced by the presence of weak bedding planes shows a distinctive four-lobed shape. (c) This can be modeled by slip on bedding planes as the stress trajectories bend around the well. (d) When mud weight is increased, the size of the breakouts decreases.

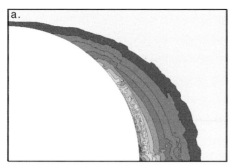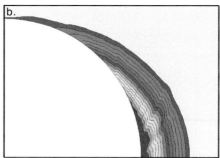

Figure 6.17. The area in which wellbore breakouts form around a cylindrical well can be modeled using a total plastic strain criterion rather than a stress criterion. These finite element calculations indicate the zone of expected breakouts assuming a critical strain level at which failure occurs (courtesy S. Willson). (a) Strain around a wellbore assuming a strain softening model of rock deformation (red indicates high strain). (b) Failure zone predicted using a strength of materials approach and Mohr–Coulomb failure criterion. (*For colour version see plate section.*)

through zones of near-horizontal bedding. In the first case, wellbore stability, even when drilling near-vertical wells, must take into account the presence of weak bedding planes (Willson, Last *et al.* 1999). In the second, small changes in wellbore deviation and azimuth can have a significant effect on wellbore stability depending on whether slip on weak bedding planes is activated by the stress concentration around the well. As shown in two case studies presented in Chapter 10, when weak bedding planes are present, their presence needs to be incorporated into wellbore stability calculations.

With respect to predicting breakout widths using the strength of materials approach adopted here, it is important to note that a number of relatively comprehensive theories have been developed to evaluate the formation of breakouts. For example, Vardulakis, *et al.* (1988) investigated breakout formation in terms of bifurcation theory and Germanovich and Dyskin (2000) investigated breakout formation in terms of micro-crack growth utilizing fracture mechanics theory. There is no doubt that such theories may eventually lead to a more complete and useful understanding of breakout formation than the relatively simple theory discussed here. Nonetheless, we shall see in future chapters that even relatively simple theories of rock failure can be quite effective in predicting wellbore failure with sufficient accuracy to be quite useful for both stress estimation (Chapter 7) and prediction of wellbore stability (Chapter 10).

One important approach for predicting the zone of failure around a well is to utilize an elastic–plastic failure criterion and predict the zone of failure around a well in terms of a total plastic strain failure criterion. In practice, such calculations are performed using a numerical analysis technique such as the finite element method. An example of such calculations is shown in Figure 6.17 (courtesy S. Willson). The colors indicate the plastic strain calculated with both a strain softening model (Figure 6.17a) and a standard Mohr–Coulomb model (Figure 6.17b). Note that the shapes of the breakouts are quite similar in the two cases and, in fact, quite similar to the shape of breakouts predicted using the elastic stress concentration predicted by the Kirsch equations (Figures 6.3 and 6.5). The similarity of these calculations might bring into question the value of using a more complicated theory and analysis method to assess wellbore failure. This said, it is important to recognize that the true value of this methodology is in addressing problems such as sand production and wellbore failure when drilling in extremely weak and/or plastic formations. In such cases, it would be inappropriate to use an elastic analysis and the strength of materials approach in assessing wellbore failure.

The principal drawback of utilizing numerical methods with a total plastic strain failure criterion (such as that illustrated in Figure 6.17a) is that the strain at which failure occurs needs to be determined empirically, although it is argued that estimates of the strain at failure can be guided by thick-walled hollow cylinder tests on core samples. Such tests involve axially loading a rock cylinder with an axial hole until sand is noted and assuming the strain at which sanding is noted will be approximately the same *in situ* as in the lab. Nonetheless, there are 50 years of laboratory tests characterizing failure as a function of stress in moderately strong sedimentary rocks (*i.e.* rocks with cemented grains) and there is relatively little knowledge of how to express failure as a function of strain for applications such as wellbore stability during drilling.

Chemical effects

Chemical interactions between drilling mud and clay-rich (shaley) rocks can affect rock strength and local pore pressure and thus exacerbate wellbore failure. While the utilization of oil-based muds can mitigate such problems, it can be very expensive to implement or precluded by regulatory restrictions. In the context of the theory described by Mody and Hale (1993), there are three important factors that affect wellbore stability when chemical effects may be important: First is the relative salinity of the drilling mud in relation to the formation pore fluid. This is expressed as the water activity A_m, which is inversely proportional to salinity. If the activity of the mud is greater than that of the formation fluid (A_w), osmosis will cause the formation pore pressure to increase and the wellbore to be more unstable. Simply put, osmotic pressures cause the movement of the less saline fluid toward the more saline fluid. Second, the change in pore pressure is limited by the membrane efficiency, which describes how easily ions can pass from the drilling mud into the formation. The concept of osmotic pressure differentials impacting wellbore stability is most easily understood when using water-based drilling muds. As oil has perfect membrane efficiency, it prevents ion exchange and utilization of oil-based mud is usually considered as a means to obviate this effect. However, Rojas, Clark *et al.* (2006) have shown that if oil-based mud incorporates emulsified water, it is still important to optimize the salinity of the drilling fluid to minimize wellbore instability. Finally, the ion exchange capacity of the shale is important as the replacement of cations such as Mg^{++} by Ca^{++} and Na^+ by K^+ weakens the shale. Simply stated, if $A_m < A_p$, *virtual* excess mud pressure is introduced that would tend to stabilize the wellbore whereas if $A_m > A_p$, a *virtual* underbalance results which would be destablilizing (Mody and Hale 1993).

In Figure 6.18a, the portion of the wellbore wall that goes into failure is shown as a function of membrane efficiency and water activity of the mud. As discussed at some length in Chapter 10, a convenient rule-of-thumb is that if less than half the wellbore wall goes into failure, wellbore stability will not be particularly problematic. As can be seen, for very high A_m (low mud salinity), the wellbore is very unstable, regardless of membrane efficiency. At intermediate values of A_m, increasing membrane efficiency can dramatically improve wellbore stability. Another way of saying this is that where $A_m < A_p$, restricting ion transfer enhances wellbore stability. It should be pointed out that a possible consequence of excessive mud salinity is desiccation and fracturing of the shale in the borehole wall. This has the potential to mechanically weaken the shale although the importance of this effect is controversial.

Figure 6.18b illustrates the fact that because the effect of chemical interactions between drilling mud and shaley formations effectively weakens the rock, one can sometimes use mud weight to offset the effect of weakening (although this effect may diminish over time due to chemo-poroelastic processes). For example, for $A_m = 0.7$, a mud weight of 11 ppg results in breakout widths of $100°$. As this results in more than half the wellbore circumference failing, it would result in a relatively unstable wellbore. Raising the mud weight to 11.5 ppg reduces breakout widths to about $60°$, thus resulting in a much more stable wellbore. A similar result could have been achieved with 11 ppg mud by lowering A_m to 0.67. Figure 6.18c shows what happens to the zone of wellbore failure when $A_m = 0.5$ and $A_w = 0.88$ for a membrane efficiency of 0.1. When the mud is much more saline than the formation fluid, the wellbore actually becomes more stable with time as indicated by the strength of the rock required to avoid failure as a function of distance from the wellbore wall and time. This is because the saline mud actually causes pore pressure to decrease in the wellbore wall as a consequence of induced fluid flow from the formation into the wellbore.

Comprehensive discussion of chemical effects on wellbore stability is beyond the scope of this book. However, it needs to be remembered that the time dependence of ion exchange is not considered in the calculations shown (*i.e.* ion exchange is only considered in the context of membrane efficiency). Another issue affecting ion exchange is the physical size of the ions involved such that the rate of ion exchange is slower for large ions (such as K^+ and Ca^{++}) than small ions (such as Na^+ and Mg^{++}) (Van Oort, Hale *et al.* 1995).

Multiple modes of breakout formation

In the discussion of breakouts so far in this chapter, I have focused on the case when $\sigma_{11} = \sigma_{\theta\theta}$, $\sigma_{zz} = \sigma_{22}$ and $\sigma_{33} = \sigma_{rr}$ but this is not the only case of compressive

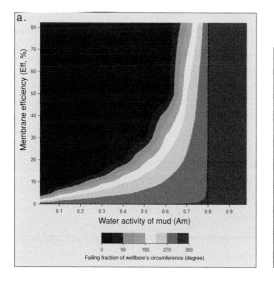

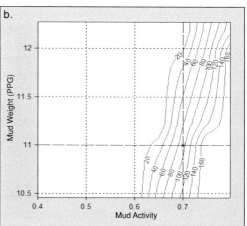

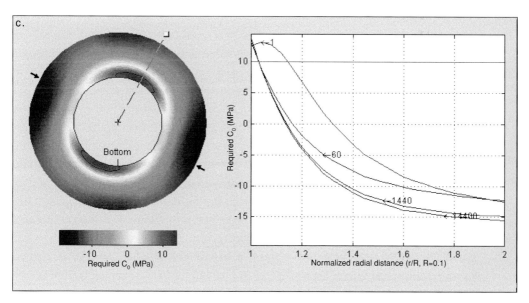

Figure 6.18. The manner in which chemical reactions between drilling mud and shale affect wellbore stability. (a) As the mud activity increases relative to the formation fluid, the failure zone becomes markedly larger. At moderate mud activity, increasing membrane efficiency increases wellbore stability. (b) In some cases, moderate increases in mud weight can offset the weakening effect due to mud/shale interaction. (c) When the mud activity is far below that of the formation, mud/shale interaction will result in strengthening of the wellbore wall with time as pore pressure decreases in the formation around wellbore. Possible, less beneficial, effects are discussed in the text. (*For colour version see plate section.*)

Table 6.2. *Multiple modes of compressive wellbore failure*

Mode	σ_1	σ_2	σ_3	Comment
B	$\sigma_{\theta\theta}$	σ_{zz}	σ_{rr}	Conventional breakout
X	σ_{zz}	σ_{rr}	$\sigma_{\theta\theta}$	Forms on opposite side of well as a conventional breakout but the failed rock will not fall into the wellbore as $\sigma_{rr} \equiv \sigma_2$
Z	σ_{zz}	$\sigma_{\theta\theta}$	σ_{rr}	Results in failure all the way around the wellbore
X2	$\sigma_{\theta\theta}$	σ_{rr}	σ_{zz}	Requires high mud weights. Failed rock will not fall into the wellbore as $\sigma_{rr} \equiv \sigma_2$
R1	σ_{rr}	σ_{zz}	$\sigma_{\theta\theta}$	Requires unreasonably high mud weights
R2	σ_{rr}	$\sigma_{\theta\theta}$	σ_{zz}	Requires unreasonably high mud weights

wellbore failure. Guenot (1989) pointed out that there are a variety of types of breakouts depending on the relative magnitudes of the three principal stresses $\sigma_{\theta\theta}$, σ_{rr} and σ_v. This subject has also been discussed by Bratton, Bornemann *et al.* (1999) who illustrate the various modes of wellbore failure.

Table 6.2 summarizes six possible modes of compressive wellbore failure depending upon whether $\sigma_{\theta\theta}$, σ_{rr} or σ_{zz} corresponds to σ_1, σ_2 or σ_3. Conventional breakouts (referred to as mode B in the table), correspond to failure being driven by the magnitudes of $\sigma_{\theta\theta}$, and σ_{rr}, that correspond to σ_1 and σ_3, respectively. As discussed above, because σ_{rr} does not vary as you go around the wellbore, failure occurs in the region where it has its maximum value (in a vertical well, this is in the region of the borehole wall near the azimuth of the minimum horizontal principal stress, Figure 6.2). Because the intermediate principal stress, σ_2 ($\sigma_2 = \sigma_{zz}$ in this case), is in the shear failure plane when the stress concentration exceeds rock strength, the fractured rock spalls into the wellbore. With modes X and X2, σ_2 corresponds to σ_{rr} such that when failure occurs, the fracture planes will form perpendicular to the wellbore wall with a slip direction parallel to the wellbore wall. Hence, significant zones of failure do not occur because the failed rock does not spall into the wellbore. Also, the modes of failure referred to as X2, R1 and R2 in Table 6.2 all require very high mud weights as σ_{rr} ($\sigma_{rr} \equiv \Delta P \equiv P_m - P_p$) corresponds to either σ_1 and σ_2. Moreover, the very high mud weight associated with these modes of failure are greater than those that would cause lost circulation due to inadvertent hydraulic fracturing of the formation. Finally, mode Z failures can occur, but would not be confused with breakouts because failure usually would be expected to occur all the way around the well because σ_{rr} (that corresponds to σ_3) does not vary around the well and if σ_{zz} (in this mode corresponding to σ_1) is large enough to cause failure when it exceeds both $\sigma_{\theta\theta}$ and σ_{rr} (near the azimuth of S_{Hmax}, Figure 6.2), it will have even larger values at other azimuths (compare equations 6.4 and 6.6). Hence, such failures would result in washouts (failure all around the wellbore) but not breakouts occurring at an azimuth that is $90°$ from the direction of conventional breakouts. The consistency of hundreds to thousands of breakout orientations in thousands of wells world-wide with independent stress observations (as illustrated in Chapters 1 and 9) demonstrate that conventional breakout formation (mode B) is the dominant mode of compressive wellbore failure. When wellbore stability is addressed in Chapter 10, the Z mode of failure is included in the failure analysis.

Penetration of mud into fractured rock surrounding a wellbore

As a final note on the factors affecting the formation of wellbore breakouts, it is important to emphasize the fact that the ability for mud weight to stabilize a wellbore is based on the precondition that the wellbore wall is impermeable. Hence, σ_{rr} is equal to the difference between the pressure in the well and that in the formation (equation 6.5) and mud weight in excess of pore pressure tend to decrease the zone of failure around a wellbore. This was discussed conceptually in Figure 4.5 and illustrated quantitatively by the difference of the width of the failure zones in Figures 6.3 and 6.5. When drilling through fractured rock, penetration of the mud pressure into rock over time could potentially result in time-dependent wellbore instability due to two processes. First, to the degree that pressure tends to equalize around the wellbore and ΔP decreases, σ_{rr} will decrease and $\sigma_{\theta\theta}$ will increase leading to increased instability of the wellbore wall. Paul and Zoback (2006) discuss this process in the context of time-dependent wellbore failure of the SAFOD scientific research borehole that was drilled through the San Andreas fault in 2005. This is discussed at further length in Chapter 10.

A second process that can lead to enhanced wellbore failure if there is an increase in pressure surrounding a wellbore due to penetration of the drilling mud, results from a temporary underbalance of wellbore pressure. This can occur when the pressure within the wellbore drops suddenly, either when the mud pumps are shut off or when the drillpipe and bottom-hole assembly are tripped out of the well. As noted above, underbalanced conditions tend to exacerbate wellbore failure. To avoid problems associated with fluid penetration into the formation surrounding a wellbore one needs to consider utilizing various additives in the drilling mud (in effect, lost circulation materials) to stop penetration of the drilling mud into the region surrounding the wellbore in fractured formations.

7 Determination of S_3 from mini-fracs and extended leak-off tests and constraining the magnitude of S_{Hmax} from wellbore failures in vertical wells

As mentioned at the outset of this book, arriving at practical solutions to many problems in geomechanics requires knowledge of the magnitude and orientation of all three principal stresses. This is well illustrated by the range of geomechanical topics and case studies presented in Chapters 10–12. The first subject discussed in this chapter is the magnitude of the least principal stress, S_3, as obtained by hydraulic fracturing, specifically mini-frac (or micro-frac) tests done specifically for the purpose of measuring stress. As discussed at length below, because hydraulic fracturing frequently occurs during leak-off tests (LOT's) and especially extended leak-off tests (XLOT's), these tests also can be used to determine S_3. In normal and strike-slip faulting environments, S_3 is equivalent to S_{hmin}. In reverse faulting environments, S_3 is equivalent to S_v. Methods for determination of S_{hmin} from Poisson's ratio (obtained from P- and S-wave sonic logs) are based on questionable physical and geologic assumptions. These methods will be discussed briefly in Chapter 9. Suffice it to say at this point that direct measurement of the least principal stress through *some form* of hydraulic fracturing is the only reliable method known that is practical to use in wells and boreholes at any appreciable depth.

One can determine the magnitude of the least principal stress from a *micro-frac*, a very small-scale hydraulic fracture induced only to measure stress at a particular depth, usually at a specific depth through perforations in cemented casing. One could also determine the least principal stress from a *mini-frac*, or the relatively small-scale frac made at the beginning of a larger hydrofrac operation intended to stimulate production in a low-permeability formation. It can also be determined at the beginning of *frac-pack* operations, where a hydrofrac is made through some sort of gravel-pack screen set in a well in weak sands, principally to spread out the depletion cone around the well to reduce the likelihood of sand production. One can also determine the least principal stress from a *leak-off* test: after the casing has been cemented in place at a given depth and the well is drilled a short distance (usually 10–20 ft) the open section of the well is pressurized to the point that a hydraulic fracture is created, and the magnitude of the least principal stress can be determined. When leak-off tests are carried out fully (as described below), they are referred to as extended leak-off tests.

When significant mud losses are noted during drilling, it can denote the accidental hydraulic fracturing of a well, requiring that the mud weight be reduced to a value less than the least principal stress, or *frac-gradient*. Finally, wellbore ballooning noted during logging-while-drilling (LWD) operations indicate that the wellbore pressure is very close to the least principal stress (see Chapter 10).

In many problems encountered in geomechanics, knowledge of the magnitude of the maximum horizontal principal stress at depth, S_{Hmax}, is especially important. For example, an accurate determination of S_{Hmax} is usually very important in problems related to wellbore stability such as the determination of optimal mud weights, well trajectories, casing set points, etc. (Chapter 10). As explained in Chapter 6, in the area of maximum stress concentration where breakouts form in vertical wells, the hoop stress results from a value of S_{Hmax} that is amplified by a factor of 3 at the wellbore wall. Hence, an accurate estimate of S_{Hmax} is often a critically important element of a wellbore stability analysis. The same thing is true when trying to assess the likelihood of shear failure on pre-existing faults (Chapter 11). As discussed in detail in Chapter 5, determination of shear and normal stress on an arbitrarily oriented fault requires knowledge of all three principal stresses.

Despite the importance of the determination of S_{Hmax} in geomechanics, it has long been recognized that this is the most difficult component of the stress tensor to accurately estimate, particularly as it cannot be measured directly. Because making stress measurements at great depth offers a unique set of challenges, we review in this chapter techniques developed that have proven to be especially efficacious for determination of S_{Hmax} in relatively deep wells. These techniques were reviewed by Zoback, Barton *et al.* (2003). The type of *integrated stress measurement strategy* utilized here was first employed to estimate the magnitude of the three principal stresses in the Cajon Pass scientific research borehole (Zoback and Healy 1992) and KTB scientific drilling project in Germany (Zoback, Apel *et al.* 1993; Brudy, Zoback *et al.* 1997). Hydraulic fracturing was used to estimate the least principal stress, S_{hmin}, to 6 km depth. Knowing this, observations of drilling-induced tensile fractures and/or the width of wellbore breakouts (w_{BO}) were used to constrain the magnitude of S_{Hmax}. While these stress techniques support the concept that brittle crust is in a state of frictional failure equilibrium (Chapters 1 and 9) in the context of laboratory friction measurements and Coulomb faulting theory (Chapter 4), the viability of these techniques for application to a variety of practical problems encountered in geomechanics has been confirmed through numerous case studies world wide. A sampling of these types of studies is presented in subsequent chapters.

The widespread use of wellbore imaging devices has been an important development that has made possible the application of the techniques for estimating S_{Hmax}, described below. As illustrated in Chapter 6, ultrasonic borehole televiewers (Zemanek, Glenn *et al.* 1970) and electrical imaging devices (Ekstrom, Dahan *et al.* 1987) yield detailed information about wellbore failure that is critically important in assessing stress orientation and magnitude at depth. In fact, we illustrate below that when drilling-induced tensile fractures are present, it is possible to make inferences about rock strength *in situ* from the presence, or absence, of wellbore breakouts. Analysis of data obtained from multiple wells (and different stratigraphic levels in each) allows a fairly comprehensive model of the stress field to be developed. While such models are only accurate within certain limits (obviously, the more information used to derive a stress model at depth, the better the model is likely to be), the way in which uncertainties in the stress estimates affect wellbore stability calculations can be addressed using rigorous, probabilistic methods (Ottesen, Zheng *et al.* 1999; Moos, Peska *et al.* 2003). This will be illustrated through case studies applied to wellbore stability in Chapter 10. We conclude this chapter by discussing estimation of the magnitude of S_{Hmax} by modeling breakout rotations associated with slip on faults (Shamir and Zoback 1992; Barton and Zoback 1994).

Hydraulic fracturing to determine S_3

In this section we consider two fundamental aspects of hydraulic fracture initiation and propagation that were addressed in a classic paper by Hubbert and Willis (1957) – the way in which the stress concentration around a well affects the initiation of hydraulic fractures at the wellbore wall and the manner in which the orientation of the minimum principal stress away from the well controls the orientation of a hydraulic fracture as it propagates. In Chapter 12, we briefly consider the use of hydraulic fracturing for stimulating production from depleted low-permeability reservoirs. The problem of inadvertent hydraulic fracturing of wells and problems associated with lost circulation during drilling is discussed in Chapter 8.

Hubbert and Willis (1957) presented a compelling physical argument that hydraulic fractures in the earth will always propagate perpendicular to the orientation of the least principal stress, S_3. Because the work done to open a Mode I fracture a given amount is proportional to the product of the stress acting perpendicular to the fracture plane times the amount of opening (*i.e.* work is equal to force times distance), hydraulic fractures will always propagate perpendicular to the least principal stress because it is the least energy configuration. They confirmed this with simple sand-box laboratory tests (Figure 7.1) and pointed out that igneous dike propagation is also controlled by the orientation of the least principal stress. This fundamental point is the basis for using hydraulic fracturing to measure the magnitude of the least principal stress as discussed below. In strike-slip and normal faulting environments where $S_3 \equiv S_{hmin}$, hydraulic fracture (and dike) propagation will be in a vertical plane perpendicular to S_{hmin} (and parallel to S_{Hmax}). In reverse faulting environments where $S_3 \equiv S_v$, hydraulic fracture propagation will be in a horizontal plane. At the time that the Hubbert and Willis (1957) paper was written, their arguments put to rest a great deal of argument

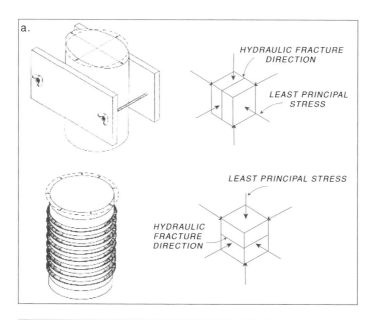

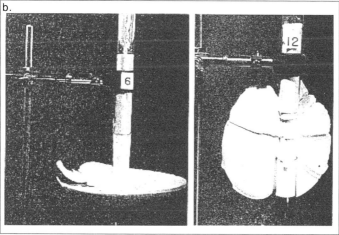

Figure 7.1. Schematic diagram of the laboratory sand-box experiments that illustrated that hydraulic fractures will propagate perpendicular to the orientation of the least principal stress. The photographs illustrate hydrofracs made with plaster of Paris as a *frac fluid* in a stressed container of unconsolidated sand. From Hubbert and Willis (1957). © *1957 Society Petroleum Engineers*

and debate over whether hydraulic fractures in oil wells and gas were propagating in vertical or horizontal planes and whether they were following pre-existing fractures and faults. Dike studies and hydrofrac mine-back experiments (Warren and Smith 1985) have shown that while pre-existing fractures and faults have some influence on fracture propagation, the overall trajectory of fracture propagation is controlled by the orientation of the least principal stress.

The other issue addressed by Hubbert and Willis (1957) is the manner of hydraulic initiation at the wellbore wall. They were the first to note that a tensile wall fracture will be induced when equation (6.7) equals $-T_0$, the tensile strength of the rock. Because $T_0 \sim 0$, a tensile fracture will form at the wellbore wall when the hoop stress goes into tension, as in the formation of a drilling-induced tensile fracture. As mentioned in Chapter 6, what distinguishes a drilling-induced tensile fracture from a hydraulic fracture is the fact that during hydraulic fracturing, the fluid pressure in the wellbore is above the magnitude of the least principal stress so that the fracture will propagate away from the wellbore. In some cases, the wellbore pressure required to initiate a tensile fracture is greater than the least principal stress so that the pressure drops after fracture initiation. In other cases, the fracture initiation pressure is significantly lower than the least principal stress such that the wellbore pressure slowly climbs to the value of the least principal stress after a tensile fracture initiates at the wellbore wall (see Hickman and Zoback 1983). This point should now be obvious in the context of the formation of drilling-induced tensile fractures discussed in Chapter 6. It is obvious that if the interval being hydraulically fractured already has drilling-induced tensile fractures present, no additional pressurization is needed to initiate them.

A schematic pressure–time history illustrating an XLOT or mini-frac is shown in Figure 7.2 (modified after Gaarenstroom, Tromp *et al.* 1993). In the schematic example shown in Figure 7.2, the pumping rate into the well is constant. Thus, the pressure should increase linearly with time as the volume of the wellbore is fixed. At the pressure where there is a distinct departure from a linear increase of wellbore pressure with time (referred to as the LOP, the leak-off point) a hydraulic fracture must have formed. The reason for this is that there cannot be a noticeable decrease in the rate of wellbore pressurization unless there is a significant increase in the volume of the system into which the injection is occurring. In other words, the pressure in the wellbore must be sufficient to propagate the fracture far enough from the wellbore to increase system volume enough to affect the rate of wellbore pressurization. Thus, there must be a hydraulic fracture propagating away from the wellbore, perpendicular to the least principal stress in the near-wellbore region, once there is a noticeable change in the pressurization rate. Thus, a clear LOP (a distinct break-in-slope) is approximately equal to the least principal stress (as shown in Figure 7.2) although the wellbore pressure may also reflect some near-wellbore resistance to fracture propagation. If the hydrofrac is being made through perforations in a cased and cemented wellbore (as is the case in mini- or micro-fracs), the tortuosity of the perforation/fracture system may cause the pressure to increase in the wellbore above the least principal stress. The same is true if the injection rate is high or if a relatively high viscosity fluid is used.

It should be noted that Figure 7.2 represents pressure at the surface during a mini-frac or LOT (note that the pressure is zero at the beginning of the test). To determine the magnitude of the least principal stress at the depth of the test, it is necessary to add

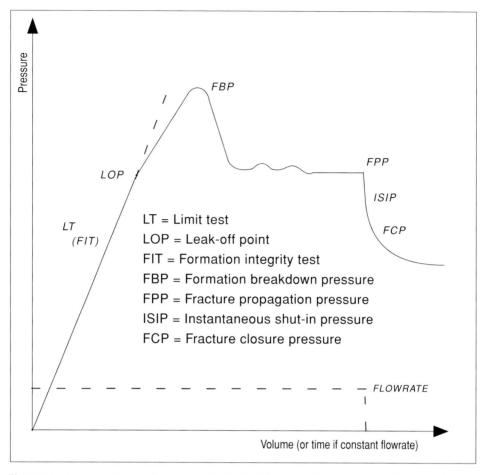

Figure 7.2. A schematic mini-frac or extended leak-off test showing pressure as a function of volume, or equivalently time (if the flow rate is constant). Modified after Gaarenstroom, Tromp *et al.* (1993). The significance of the various points indicated on the pressure record is discussed in the text.

the pressure in the wellbore due to the column of wellbore fluid. In fact, it is always preferable to measure pressure downhole during such tests.

If the LOP is not reached, a limit test, or formation integrity test (LT, or FIT), is said to have been conducted. Such tests merely indicate that at the maximum pressure achieved, the fluid pressure did not propagate away from the wellbore wall, either because the maximum wellbore pressure did not exceed the least principal stress or was not sufficient to initiate a fracture of the wellbore wall in the case of an open-hole test. The peak pressure reached during a LOT or mini-frac is termed the formation breakdown pressure (FBP) and represents the pressure at which unstable fracture propagation away from a wellbore occurs (fluid flows into the fracture faster from the wellbore than the pump supplies it; hence the pressure drops). The difference between the LOP and FBP is a complex function of the conditions immediately surrounding the well (especially when a frac is being initiated through perforations). If pumping continues at a constant rate, the pumping pressure will drop after the FBP to a relatively constant value called the fracture propagation pressure (FPP). This is the pressure associated with propagating the fracture away from the well. In the absence of appreciable near-wellbore resistance mentioned above (i.e. if the flow rate and fluid viscosity are low enough), the FPP is very close to the least principal stress (*e.g.* Hickman and Zoback 1983). Hence, the FPP and LOP values should be similar. It should be emphasized that a distinct FBP need not be present in a reliable mini-frac or XLOT. This correspondence between the LOP and FPP is the reason why, in typical oil-field practice, leak-off tests are taken only to the LOP, rather than performing a complete, extended leak-off test.

An even better measure of the least principal stress is obtained from the instantaneous shut-in pressure (ISIP) which is measured after abruptly stopping flow into the well, because any pressure associated with friction due to viscous pressure losses disappears (Haimson and Fairhurst 1967). In carefully conducted tests, constant (and low) flow rates of $\sim$200 liter/min (1 BBL/min), are maintained and low viscosity fluid (such as water or thin oil) is used and pressure is continuously measured. In such tests, the LOP, FPP, and ISIP have approximately the same values and can provide redundant and reliable information about the magnitude of S_3. If a viscous frac fluid is used, or a frac fluid with suspended propant, FPP will increase due to large friction losses. In such cases the fracture closure pressure (FCP) is a better measure of the least principal stress than the FPP or ISIP. In such, tests, the FCP can be determined by plotting pressure as a function of $\sqrt{\text{time}}$ and detecting a change in linearity of the pressure decay (Nolte and Economides 1989). However, if used inappropriately, fracture closure pressures can underestimate the least principal stress and care must be taken to assure that this is not the case.

Figure 7.3 illustrates two pressurization cycles of a mini-frac test conducted in an oil well in Southeast Asia. Note that the flow rate is approximately constant at a rate of $\sim$0.5 BBL/min during the first cycle (in which 10 BBLS was injected before shut-in), and was held quite constant during the second (in which 15 BBLS was injected before shut-in). It is not clear if a constant FPP was achieved before shut-in on the first pressurization cycle, but it is quite clear that it was on the second. Pressures after shut-in are shown for the two tests. The ISIPs were determined from the deviation in the rate of rapid pressure decrease to a more gradual decay on the linear plots of pressure as a function of time. The FCP's were determined from the deviation from linearity in the $\sqrt{\text{time}}$ plots that are shown. As shown, these two pressures vary by only a few tens of psi. Once the hydrostatic head is added to the measured values, the variation between these tests results in a variance of estimates of S_{hmin} that is less than 1% of its value.

Figure 7.4 shows a compilation of pore pressure and LOT data from the Visund field in the northern North Sea (Wiprut, Zoback *et al.* 2000). Pore pressure is hydrostatic

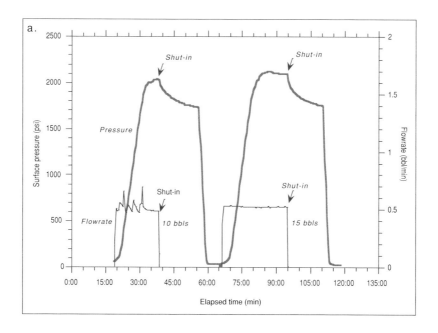

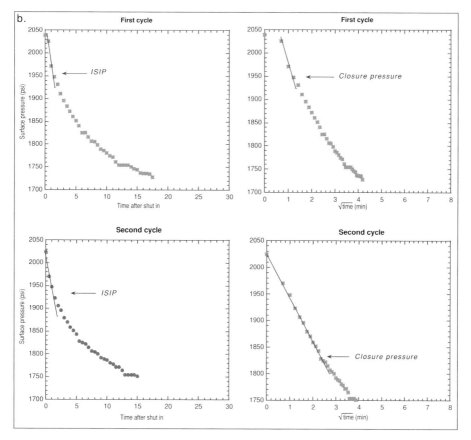

Figure 7.3. Pressure and flow data from two cycles of a mini-frac test in the Timor Sea. In such a test there is very little difference between the shut-in pressure and the closure pressures on each of the two cycles. In fact, the pumping pressure is only about 100 psi higher than the shut-in pressure. As the records shown indicate surface pressure, the variance of downhole pressure (after adding the hydrostatic pressure from the surface to the depth of measurement) varies by less than 2%.

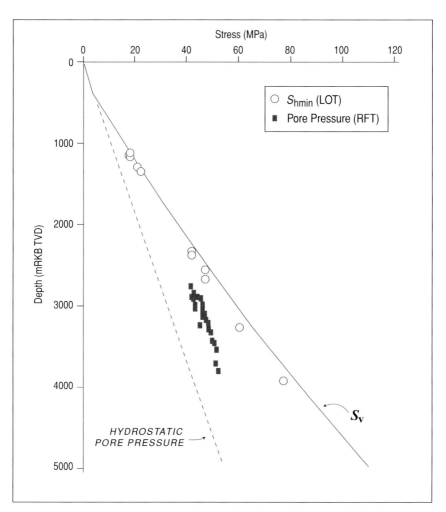

Figure 7.4. Least principal stress as a function of depth determined from extended leak off tests in the Visund field (after Wiprut, Zoback *et al*. 2000). The least principal stress is slightly below the overburden stress (determined from integration of density logs). The pore pressure is somewhat above hydrostatic (shown by the dashed line for reference). *Reprinted with permission of Elsevier.*

to about 1500 m depth (not shown) but at $\sim$3000 m it is approximately 75% of the vertical stress. There are three important features to note about the least principal stress values at depth. First, the measurements are repeatable and indicate a consistent trend throughout the field. Second, the measurements clearly indicate a compressional stress state because even at relatively shallow depth (where pore pressure is hydrostatic), the magnitude of the least principal stress is extremely close to the vertical stress. We show below that the magnitude of S_{Hmax} is greater than S_v such that a strike-slip faulting regime exists in this region. However, because S_{hmin} is extremely close to S_v if the magnitude of S_v was slightly over-estimated (due to uncertainties in density), or if S_3 is slightly higher than the values shown (if the measurements were not carefully made), it might be the case that S_3 would appear to be equal to S_v such that a reverse-faulting regime would be indicated. If one were concerned with the propagation of hydraulic fractures (to stimulate production in low-permeability reservoirs, for example), this point is quite important. As noted above, if $S_3 \equiv S_{hmin}$, vertical hydrofracs would be initiated at the wellbore wall. However, if $S_3 \equiv S_v$, vertical fractures would be expected to form at the wellbore wall (when the increase in wellbore pressure causes $\sigma_{\theta\theta}$ to go into tension). However, the hydraulic fracture will rotate into a horizontal plane (perpendicular to S_v) as the fracture propagates away from the wellbore (Baumgärtner and Zoback 1989). After fracture propagation away from the wellbore, the FPP or ISIP can be used to determine S_3. In cases where $S_3 \sim S_v$ it is particularly important to carefully integrate density logs to determine S_v and to determine if S_3 corresponds to S_v or S_{hmin} with confidence. In fact, in the Visund field, considerable effort was taken to estimate rock density at extremely shallow depth to derive the curve shown. Had this not been the case, it would have been extremely difficult to determine whether or not the least principal stress was less than, or equal to, the vertical stress.

Another way to measure the least principal stress is to conduct step-rate tests. In such tests injection into the well is performed at a number of fixed flow rates as illustrated in Figure 7.5. It is easy to detect the pressure at which a hydrofrac opens; injection can take place at increasingly higher flow rates with only minimal increases in wellbore pressure. Prior to the hydrofrac opening, there is a strong increase in pressure with flow rate as expected for a system dominated by diffusion into the formation and/or a closed hydraulic fracture.

A number of methods for the analysis of shut-in pressure data for determination of the least principal stress have been proposed over the years. A discussion of various techniques was reviewed by Zoback and Haimson (1982), Baumgärtner and Zoback (1989), Rummel and Hansen (1989), Hayashi and Haimson (1991) and Guo, Morgenstern et al. (1993). An alternative way to measure the least principal stress is to pressurize the wellbore in steps and measure the rate of pressure decrease after pressurization is stopped. The logic behind such tests is that once a fracture has formed, the rate of pressure decrease with time will be faster. If a sufficient number of closely spaced pressure steps is used, the magnitude of the least principal stress can be determined with corresponding accuracy. Similarly, there are other techniques referred to as pump in/flow back methods (see Raaen and Brudy 2001) that yield reliable results. Hayashi and Haimson (1991) discuss the interpretation of shut-in data from a theoretical perspective and present a technique they argue is most optimal for determination of the least principal stress.

Unfortunately, many LOT's are conducted using extremely poor field procedures. When trying to analyze such tests, two questions must be kept in mind to know whether the test can be used to obtain a measure of the least principal stress. First, is there an indication that the LOP was reached? If so, the LOP can be considered an approximate

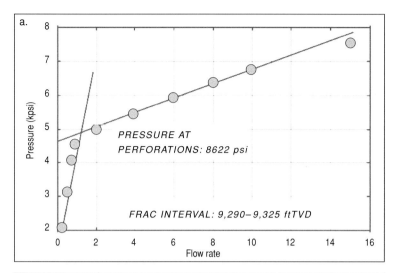

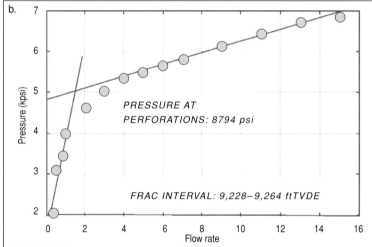

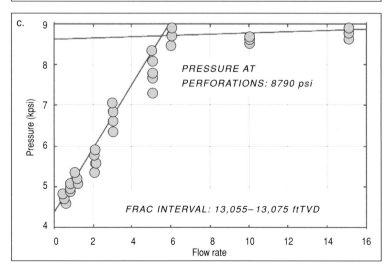

measure of the least principal stress. If not, then the test must be considered a FIT and the maximum pressure achieved cannot be used to estimate the least principal stress. Second, was a stable FPP achieved? If so, the fracture clearly propagated away from the well and the shut-in pressure is likely a good measure of S_3. While these two questions are straightforwardly answered when there is a good record of the test, it is sometimes necessary to rely on a single reported value, not knowing whether it refers to a reasonable estimate of the least principal stress. In fact, in some cases the pressure–time record is approximated by a few distinct data points only, obtained by reading pressure on a fluctuating gauge and estimating flow rate by counting pump strokes. In such cases, determination of accurate LOT values is essentially impossible. Values of LOT's that are markedly lower than the expected trend for a given area should also be treated with extreme caution as these tests may simply indicate a poor-quality cement job rather than an anomalously low value of the least principal stress.

Pressure-while-drilling (PWD) is a measurement-while-drilling (MWD) or logging-while-drilling (LWD) sensor that continually measures annular pressures during the drilling process and has the potential for providing information about the magnitude of the least principal stress, especially in difficult drilling situations (Ward and Clark 1998). This measurement is generally taken some 5–10 m behind the bit and allows for accurate downhole determination of mud weight, equivalent circulation density (ECD), swab and surge pressures. Pressure values are transmitted to the surface in real time during drilling and recorded downhole in memory that can be read when the bottom hole assembly is brought to the surface or transmitted to the surface once circulation is resumed. Figure 7.6 shows a number of drilling-related parameters as a function of time (including PWD in the right column) that are measured during drilling operations. Note the step-like nature of the pressure drop that occurred when drilling was stopped to connect a section of drill pipe. When drilling resumed the pressure abruptly increased. This step in pressure allows us to define the difference between the ECD (the equivalent circulating density which corresponds to the bottom-hole pressure during drilling) to the static mud weight. In this case the viscous resistance to mud circulation during drilling results in a difference between the ECD and static mud weight of 0.33 ppg.

There are three ways that PWD data can be used to better constrain the minimum horizontal stress: improving the accuracy of leak-off test (LOT) measurements, identifying lost circulation incidents and identifying *ballooning* incidents. With respect to LOT's, it is important to recognize that such tests are normally recorded at the surface

Figure 7.5. Determination of the least principal stress from step-rate tests conducted in a well in Alaska. This test is similar to a conventional mini-frac or extended leak-off test except that the injection rate is varied in incremental steps. The least principal stress (i.e. the pressure at which the hydrofrac opens) is indicated by the distinct change in slope.

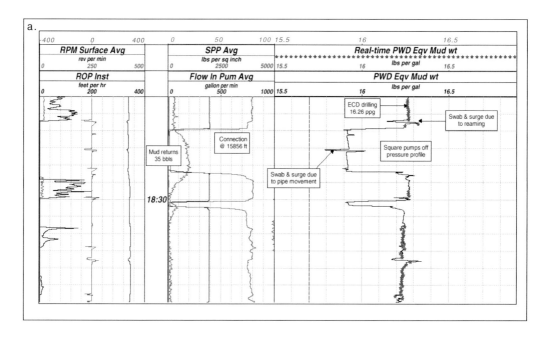

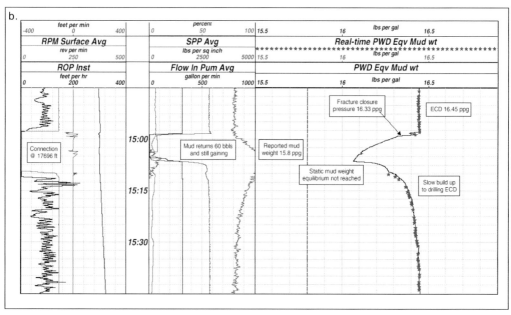

Figure 7.6. Pressure-while-drilling (PWD) records reconstructed from a well in the Gulf of Mexico. (a) A conventional pressure record indicating an abrupt decrease in pressure when pumping is stopped to make a connection while adding a new piece of pipe while drilling. (b) An example of wellbore ballooning where there is a gradual decrease in pressure when pumping is stopped and a corresponding gradual increase in pressure when pumping resumes (see text). (*For colour version see plate section.*)

and the pressure downhole is determined by adding a pressure corresponding to the static mud column to the surface. PWD records pressure downhole directly and a number of comparisons have shown that there can be a significant differences between downhole pressures calculated from surface measurements and actual downhole LOT measurements (Ward and Beique 1999). This difference could be caused by suspended solids, pressure and temperature effects on mud density, or measurement error. There is an additional error during the pumping and shut-in phases that could be due to the mud gels, mud compressibility or pressure loss in the surface lines.

PWD also accurately measures the pressures imposed on the formation during a lost circulation incident (Ward and Beique 1999). There is often some uncertainty about exactly where the losses are happening in a long open-hole section so the PWD measurement may need to be referenced to the appropriate depth. Sometimes repeated resistivity logs can help identify the depths at which the losses occur. Similar to what happens in a LOT, losses of drilling mud will occur at pressures slightly higher than the least principal horizontal stress. The accurate determination of such pressures with PWD data yield reliable estimates of the least principal stress because the fracture must be propagating into the far field away from the wellbore in order for circulation to be lost.

Finally, ballooning, sometimes called loss/gain or wellbore breathing, is now generally thought to be caused by the opening and closing of near wellbore fractures (Ward and Clark 1998). This phenomenon is especially likely to occur when pore pressure is significantly above hydrostatic and drilling is occurring with an ECD close to the least principal stress. In such cases, small mud losses can occur during drilling that, when the pumps are turned off, bleed back into the wellbore. In Figure 7.6a, we observe that the pressure drop from the ECD to the static mud weight is quite abrupt when the pump stops and then increases abruptly when drilling resumes. Note the markedly different behavior in Figure 7.6b – when the pump is shut off for a connection, the pressure slowly decays, then slowly builds up when the pump is turned back on. This behavior is reminiscent of a balloon because it implies the storage of drilling fluid upon pressurization and the return of this fluid into the wellbore when the pumps are shut off. Thus, the PWD signature during ballooning has a distinctive curved pressure profile (Figure 7.6b) as closing fractures bleed fluid back into the wellbore and fractures are refilled as circulation is resumed. The ECD at which ballooning occurs can be used as a lower bound for the magnitude of the least principal stress (if S_3 was lower, lost circulation would have occurred). In fact, it has been argued by Ward and Clark (1998) that unless the ECD was close to S_3, ballooning cannot occur. Modeling by Ito, Zoback *et al.* (2001) indicates that the most likely reason ballooning occurs is that en echelon tensile fractures form in the wall of a deviated well (see Chapter 8) that store fluid at the pressure corresponding to the ECD during drilling. When the pump is shut off and the pressure drops to the static mud weight, the mud comes out of the fractures and back into the wellbore.

Can hydraulic fracturing be used to estimate the magnitude of S_{Hmax}?

Following the work of Hubbert and Willis (1957), Haimson and Fairhurst (1970) proposed *open-hole* hydraulic fracturing in vertical wellbores as a technique for determination of the orientation and magnitude of S_{Hmax}. While a number of minor modifications have been made to the techniques they proposed over the years, suffice it to say that many successful stress measurements have been made around the world using the basic technique they proposed over 30 years ago. Amadei and Stephansson (1997) offer a fairly comprehensive review of the experimental techniques and analytical procedures associated with using the hydraulic fracturing technique for *in situ* stress measurements.

We briefly review here the *classical* use of hydraulic fracturing for determination of the magnitude of S_{Hmax}. However, because these techniques are best suited for relatively shallow holes where both stress and temperatures are low (generally about 2 km, or less) and relatively strong rocks (so that breakouts are not present), this technique has very limited application in the petroleum industry. In fact, we will conclude this section with a brief summary of the reasons why classical hydraulic fracturing is not particularly useful for determining the magnitude of S_{Hmax} in the oil and gas (or geothermal) industries.

Following the discussions in Chapter 6, at the point of minimum compression around the wellbore (*i.e.* at $\theta = 0$, parallel to S_{Hmax}), a hydraulic fracture will be induced when

$$\sigma_{\theta\theta}^{min} = -T_0 = 3S_{hmin} - S_{Hmax} - 2P_p - \Delta P - \sigma^{\Delta T} \tag{7.1}$$

Ignoring $\sigma^{\Delta T}$, a tensile fracture will form at the wellbore wall when

$$P_b = 3S_{hmin} - S_{Hmax} - P_p + T_0 \tag{7.2}$$

where P_b is called the breakdown pressure, similar to the FBP (formation breakdown pressure) referred to in Figure 7.2. Assuming that S_{hmin} is measured from the pumping pressure (FPP) or shut-in pressure (ISIP), and that P_0 and T_0 have either been measured or estimated,

$$S_{Hmax} = 3S_{hmin} - P_b - P_p + T_0 \tag{7.3}$$

Bredehoeft, Wolf *et al.* (1976) pointed out that to avoid the problem of determining T_0, a secondary pressurization cycle can be used (*i.e* after a hydraulic fracture is initiated at the wellbore wall), and this reduces to

$$S_{Hmax} = 3S_{hmin} - P_b(T = 0) - P_p \tag{7.4}$$

where $P_b(T = 0)$ indicates the breakdown pressure after an initial hydrofrac has been created at the wellbore wall.

Numerous papers have been written that both use this basic technique for measurement of S_{Hmax} or propose modifications to result in improvements. Two compilations of papers related to hydrofrac stress measurements summarize much of the relevant experience (Zoback and Haimson 1983; Haimson 1989). Most of the successful application of this technique has been related to determination of S_{Hmax} for scientific purposes or for application to problems in civil engineering, but as mentioned above, at relatively shallow depth.

As alluded to above, it is important to recognize the reasons why hydraulic fracturing is not a viable method for determination of S_{Hmax} in relatively deep and/or hot wells. First, most wells are cased at the time of hydraulic fracturing and hydrofrac initiation in the perforations made through the casing and cement and into the formation. In this case, hydraulic fracturing tests still yield accurate estimates of the least principal stress (because once a fracture propagates away from the wellbore, it propagates perpendicular to the least principal stress), but the equations above are no longer relevant because the stress concentration around the well does not govern fracture initiation. Second, during a leak-off test, the open hole section below the casing may not be well described by the Kirsch equations. In other words, in classical hydraulic fracturing, we assume the well is circular and there are no pre-existing fractures or faults present. When doing *classical* hydraulic fracturing in open hole, it is common to use wellbore imaging devices to assure this. In the case of leak-off tests, this is never done and there can be serious washouts at the bottom of the casing (or other irregularities) that might affect fracture initiation. The most important reason that hydraulic fracturing cannot be used to determine S_{Hmax} in oil and gas (or geothermal) wells is that it is essentially impossible to detect fracture initiation at the wellbore wall during pressurization. In other words, equation (7.2) assumes the P_b is the pressure at which fracture initiation occurs. In point of fact, depending on the stress state, the breakdown pressure may not be the fracture initiation pressure (as originally discussed by Zoback and Pollard 1978; Hickman and Zoback 1983). Regardless of the stress state, however, in oil and gas wells it is straightforward to show that it is essentially impossible to detect the pressure at which the fracture initiates at the wellbore wall. To see this, consider the volume of fluid associated with conducting a hydrofrac, V_s, consisting of the well, pump and surface tubing, etc. which has a system compressibility, β_s, given by

$$\beta_s = \frac{1}{V_s}\frac{\Delta V_s}{\Delta P} \tag{7.5}$$

which can be rewritten as

$$\Delta P = \frac{1}{\beta_s V_s}\Delta V_s$$

If we divide both sides by Δt, we obtain

$$\frac{\Delta P}{\Delta t} = \frac{1}{\beta_s V_s}\frac{\Delta V_s}{\Delta t} \tag{7.6}$$

which expresses the rate at which the pressurization rate $\Delta P/\Delta t$, as a function of the rate at which the system volume, changes with time, $\frac{1}{V_s}\frac{\Delta V_s}{\Delta t}$. Thus, for a constant pumping rate, the change in pressurization rate (the parameter being observed during the hydrofrac) would be proportional to the change in system volume. However, when a tensile fracture opens at the wellbore wall, the change of V_s is negligible because V_s is so large. Thus, P_b represents unstable fracture propagation into the far-field (fluid is flowing into the fracture faster than the pump is supplying it) but fracture initiation could have occurred at any pressure. Because equation (7.4) assumes that hydrofracs initiate at the breakdown pressure, if the actual initiation pressure cannot be observed due to the large system volume, it is obvious that hydraulic fracturing pressure data cannot be used to determine S_{Hmax} in most circumstances.

Wellbore failure and the determination of S_{Hmax}

As mentioned above the type of *integrated stress measurement strategy* utilized here and summarized by Zoback, Barton *et al.* (2003) was first employed to estimate the magnitude of the three principal stresses in the Cajon Pass (Zoback and Healy 1992) and KTB scientific drilling projects (Zoback, Apel *et al.* 1993; Brudy, Zoback *et al.* 1997). Figure 7.7 presents a summary of the stress results for the KTB Project. Hydraulic fracturing was used to estimate the least principal stress, S_{hmin}, to 6 km depth, as well as the magnitude of S_{Hmax} to a depth of $\sim$3 km using a modification of the conventional hydraulic fracturing method described above (Baumgärtner, Rummel *et al.* 1990). The magnitude of S_{hmin} determined from hydraulic fracturing and estimates of rock strength from laboratory measurements along with observations of wellbore breakouts were used to constrain the magnitude of S_{Hmax} between depths of 1.7 and 4 km (the open and filled triangles indicate lower and upper bound estimates). Observations of drilling-induced tensile fractures between 3 and 4 km allowed us to independently estimate the lower and upper bound of S_{Hmax} (+'s and ×'s, respectively), again using the magnitude of S_{hmin} determined from hydraulic fracturing. Note how well the estimates of S_{Hmax} from the three techniques compare between $\sim$1.7 and 4 km. At greater depth, it was necessary to combine the observations of tensile fractures and breakouts (the wellbore was failing simultaneously in compression and tension in the manner illustrated in the left panel of Figure 6.4) to constrain the magnitude of S_{Hmax}. Because of the large uncertainty in temperature at which the tensile fractures formed, there is a correspondingly large uncertainty in the magnitude of S_{Hmax} at great depth (see Brudy, Zoback *et al.* 1997). Modeling of a breakout rotation at 5.4 km depth using the technique described at the end of this chapter provided an independent estimate of the magnitude of S_{Hmax} consistent with the combined analysis (Barton and Zoback 1994).

As discussed in Chapter 6, breakouts form in the area around a wellbore where the stress concentration exceeds the rock strength and once a breakout forms, the stress concentration around the wellbore is such that breakouts will tend to deepen. Because breakout width is expected to remain stable even as breakout growth occurs after initiation, Barton, Zoback *et al.* (1988) proposed a methodology for determination

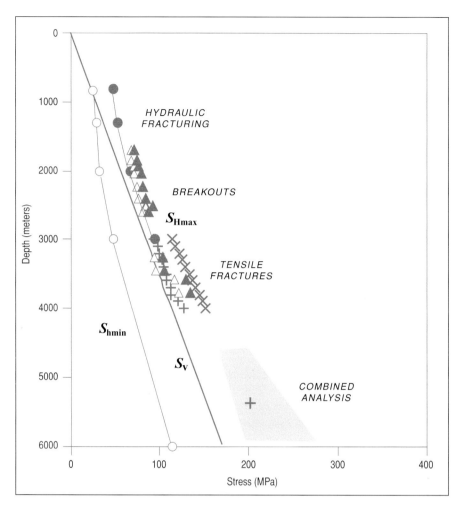

Figure 7.7. Summary of stress measurements made in the KTB scientific research borehole in Germany obtained using the integrated stress measurement strategy described in the text (Zoback, Apel *et al.* 1993).

of S_{Hmax} when the rock strength is known utilizing observations of breakout width. Because the stress concentration at the edge of a breakout is in equilibrium with the rock strength, they derived the following:

$$S_{Hmax} = \frac{(C_0 + 2P_p + \Delta P + \sigma^{\Delta T}) - S_{hmin}(1 + 2\cos 2\theta_b)}{1 - 2\cos 2\theta_b} \qquad (7.7)$$

where $2\theta_b \equiv \pi - w_{bo}$.

As illustrated in Chapter 6, ultrasonic televiewer data allow for the accurate determination of breakout width at any depth in a well. Figure 7.8 shows a compilation of breakout width and azimuth data from a geothermal research well in Fenton Hill, New

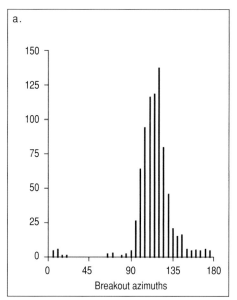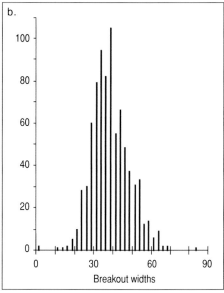

Figure 7.8. Breakout orientation and width observations in the Fenton Hill geothermal research well (Barton, Zoback *et al.* 1988).

Mexico (after Barton, Zoback *et al.* 1988). Over 900 observations of breakout azimuth (and over 600 measurements of breakout width) were made over a 262 m interval at about 3.3 km depth. The mean breakout azimuth is $119 \pm 9°$ and the mean breakout width is $38°$.

Utilizing the least principal stress measurements made in the Fenton Hill well at 3 and 4.5 km depth via hydraulic fracturing and estimates of rock strength between 124 and 176 MPa based on laboratory measurements in equation (7.7) yields a value of S_{Hmax} approximately equal to S_v (Figure 7.9). From a geologic perspective, the strike-slip/normal faulting stress state implied by these measurements (Chapter 4) is consistent with numerous normal faulting and strike-slip faulting earthquakes that occurred as the result of large-scale fluid injection associated with hydraulic fracturing to stimulate geothermal energy production. Moreover, for the value of S_v at 3.3 km depth, the least principal stress has exactly the value predicted for normal faulting from frictional faulting theory for a coefficient of friction of 0.6 and hydrostatic pore pressure (equation 4.45 and Figure 4.25a). In addition, the difference between S_{hmin} and S_{Hmax} is exactly as predicted for strike-slip faulting in terms of equation (4.46) such that both strike slip faults and normal faults are expected to be active. Graphically, this stress state corresponds to the point on the periphery of the stress polygons shown in Figure 4.28 where the normal faulting and strike-slip faulting field meet ($S_v = S_{Hmax} > S_{hmin}$), corresponding to the stress states shown in Figure 5.1b and c.

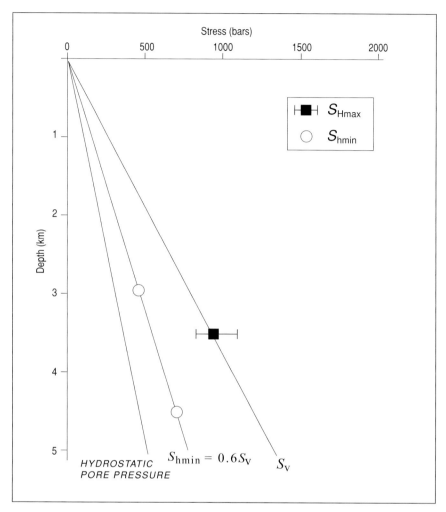

Figure 7.9. Estimate of S_{Hmax} from breakout widths (error bar) using the least principal stress measurements obtained from hydraulic fracturing (squares). Note that as the value of S_{hmin} is consistent with a normal faulting stress state and $S_{Hmax} \approx S_v$, a normal/strike slip stress regime is indicated, consistent with the focal mechanisms of injection-induced earthquakes. From Barton, Zoback *et al.* (1988).

Another way to implement equation (7.7) for determination of S_{Hmax} is in the context of the diagrams shown in Figure 6.13, but using the occurrence of breakouts of a given observed width at a particular depth, as well as the occurrence of tensile fractures. The example shown in Figure 7.10 is for a deep oil well in Australia that was drilled into very strong rock (unconfined compressive strength 138 ± 14 MPa). Both wellbore breakouts (average width 45°) and drilling induced tensile fractures were present in this well. The light diagonal line is analogous to those shown in Figure 6.13 (the value

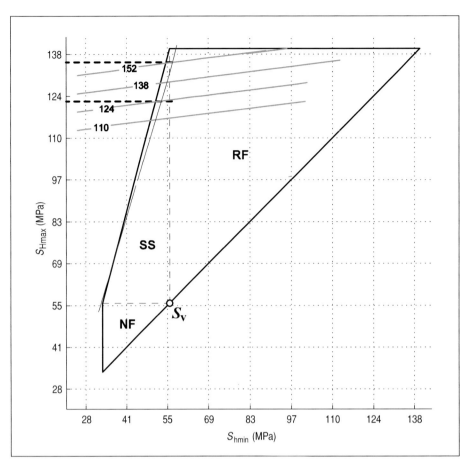

Figure 7.10. Polygon showing possible stress states at depth for a case study in Australia as described in the text. The sub-horizontal lines labeled 110, 124, 138 and 152 MPa indicate the magnitude of S_{Hmax} (as a function of S_{hmin}) that is required to cause breakouts with a width of 45° for the respective rock strengths. As in Figure 6.13, the light diagonal line indicates the stress values associated with the initiation of tensile fractures in the wellbore wall.

of S_{Hmax} required to cause drilling-induced tensile fractures). The darker, subhorizontal lines correspond to the value of S_{Hmax} required to cause breakouts with a width of 45° for rocks of the different strengths indicated. A modified Lade failure criterion was utilized. As the magnitude of S_{hmin} is approximately the vertical stress, S_{Hmax} is approximately 130 MPa. If S_{Hmax} had been lower, the breakout widths would have been smaller. Similarly, if S_{Hmax} had been appreciably larger, the breakout width would have been greater. We will revisit Figure 7.10 momentarily to discuss the significance of the drilling-induced tensile fractures.

While breakouts are common in many wells and it is straightforward to estimate breakout width, in order to utilize this technique, knowledge of pore pressure, the

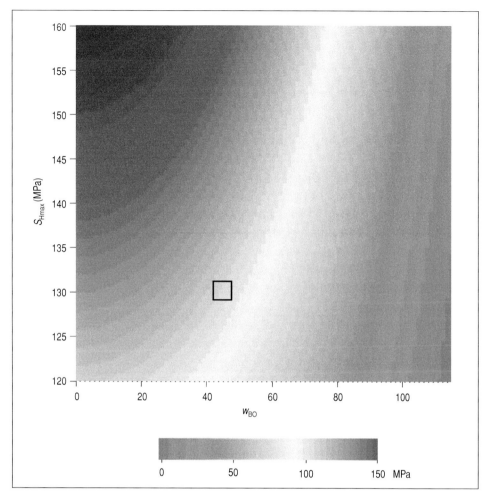

Figure 7.11. The dependence of S_{Hmax} on rock strength for given values of breakout width. The square indicates the analysis shown in Figure 7.10. (*For colour version see plate section.*)

vertical stress, the least principal stress as well as a reasonable estimate of rock strength are also needed. Figure 7.11 shows the sensitivity of S_{Hmax} determined from breakout width as a function of rock strength. The square box shows the result from Figure 7.10; for a breakout width of $45°$ and a rock strength of about 138 MPa, S_{Hmax} is estimated to be about 130 MPa. Had the rock strength been mistakenly assumed to be about 150 MPa, a value of S_{Hmax} of about 140 MPa would have been indicated. Had the strength been mistakenly assumed to be as low as 100 MPa, a value of S_{Hmax} of about 125 MPa would have been inferred.

In the case studies presented in Chapters 9–12 (as well as many others), the technique of estimating S_{Hmax} from breakout width has proved to provide reasonably useful estimates of the value of S_{Hmax}. As a historical note, Hottman, Smith *et al.* (1979)

used variations of the occurrence of breakouts (as indicated by wellbore spalling) with changes in mud weight to make an estimate of the maximum horizontal stress, after first constraining the other parameters associated with wellbore failure as described above. While no detailed observations of the shape of breakouts were available to them, the approach they used was fundamentally similar to that described above.

Drilling-induced tensile fractures and the magnitude of S_{Hmax}

As discussed in Chapter 6, drilling-induced tensile fractures occur in vertical wells whenever there is a significant difference between the two horizontal stresses. From equation (6.8), it can easily be shown that the condition for tensile fracture formation in the wellbore wall in a vertical well leads to

$$S_{Hmax} = 3S_{hmin} - 2P_p - \Delta P - T_0 - \sigma^{\Delta T} \qquad (7.8)$$

As mentioned in Chapter 6, it is straightforward to show that the conditions for the occurrence of drilling-induced tensile fractures (the light diagonal line in Figure 7.10) around a vertical wellbore are essentially the same as the values of S_{hmin} and S_{Hmax} associated with a strike-slip faulting regime in frictional equilibrium (the upper left periphery of the polygon). Following the logic used in the discussion of wellbore breakouts, the value of S_{Hmax} required to explain the occurrence of drilling induced tensile fractures in the well considered in Figure 7.10 requires a value of S_{Hmax} to be approximately 130 MPa. A lower value of S_{Hmax} would not have been sufficient for the tensile fractures to form. A higher value is not reasonable as it would imply a value of S_{Hmax} that exceeds the frictional strength of the earth's crust. Thus, the observations of breakout width and occurrence of tensile fractures in this well yield the same values of S_{Hmax} (about 130 MPa) and indicate a strike-slip stress state in frictional equilibrium. Note that we assumed $T_0 = 0$ in this analysis. As noted in Chapter 4, T_0 is always quite small (a few MPa, at most), so drilling-induced tensile fractures can initiate at small flaws on the wellbore wall and the influence of T_0 on the computed value of S_{Hmax} (equation 7.8) is quite small (Figure 6.13). In Figure 7.5, the impact of a finite T_0 of 2 MPa, for example, would be to shift the light-diagonal line down by that amount such that uncertainty in tensile strength has a comparable (but relatively very small) effect on the estimated value of S_{Hmax}.

Mud weights during drilling (*i.e.* that which correspond to the Equivalent Circulating Density, or ECD) above the pore pressure also encourage the formation of drilling-induced tensile fractures. Because of this, it is necessary to assure that the occurrence of drilling-induced tensile fractures is not significantly influenced by increases in mud pressure associated with such drilling activities as running in the hole (potentially resulting in a piston-induced increase in mud pressure at the bottom of the well), surges in mud pressure associated with wash and reaming operations or *pack-off* events

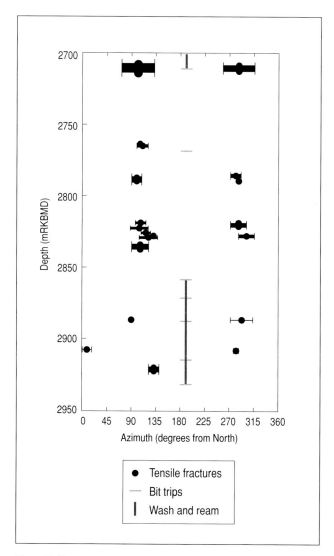

Figure 7.12. Depths at which drilling-induced tensile fractures were observed in one of the Visund wells studied by Wiprut, Zoback *et al.* (2000), along with indications of depths where there were wash-and-ream operations or bit trips that may have caused sudden increases in wellbore pressure. *Reprinted with permission of Elsevier.*

when excessive formation failure blocks the annulus around the bottom-hole assembly during drilling. To assure that this was not the case, Wiprut, Zoback *et al.* (2000) carefully noted depths at which such operations occurred (Figure 7.12) and restricted analysis of the drilling-induced tensile fractures to depths at which such activities did not occur.

The results of the S_{Hmax} analysis based on drilling-induced tensile fractures in the Visund field are shown in Figure 7.13 (after Wiprut, Zoback *et al.* 2000) which shows

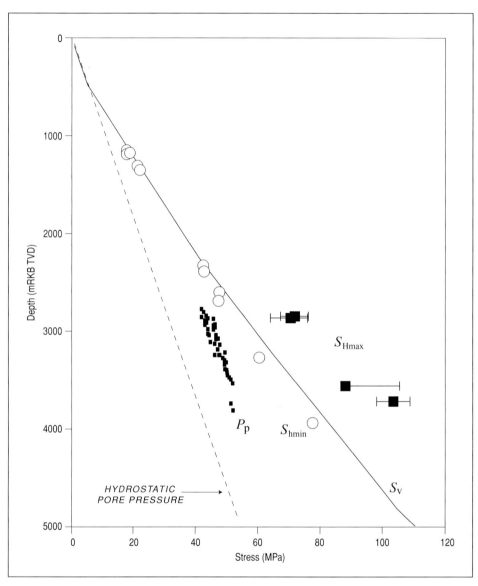

Figure 7.13. Similar to Figure 7.4 with estimates of S_{Hmax} (dots with error bars) determined from analysis of drilling-induced tensile fractures in the Visund field of the northern North Sea (after Wiprut, Zoback *et al.* 2000). *Reprinted with permission of Elsevier.*

S_{Hmax} is greater than the overburden. As shown, there is a significant amount of uncertainty in the estimates of S_{Hmax}. Brudy, Zoback *et al.* (1997) pointed out that the value of S_{Hmax} required to induce drilling-induced tensile fractures (after correcting for excess mud weight and cooling) must be considered as a lower-bound estimate. This is because the drilling-induced tensile fractures might have occurred even if there had been no excess mud weight or cooling of the wellbore wall. This represents an upper bound value of S_{Hmax}. The uncertainty in the values of S_{Hmax} (such as those shown in Figure 7.13) can be taken into account when the values are used for consideration of wellbore stability or fault reactivity as described in subsequent chapters.

A point not mentioned above is that for the case study shown in Figure 7.10, it was important to have observations of tensile fractures in the well in order to constrain the magnitude of both S_{hmin} and S_{Hmax}. This is because the least principal stress was so close to the magnitude of the vertical stress that it was not clear if S_3 corresponded to S_v or S_{hmin}. If the former, it would have indicated a thrust faulting regime and the magnitude of S_{hmin} would have been unknown. However, because there are drilling-induced tensile fractures in the well, the stress state had to be a strike-slip/reverse faulting regime with a stress state corresponding to the upper left corner of the stress polygon. If S_{hmin} had been appreciably larger than S_v, drilling-induced tensile fractures would not have formed.

Estimating rock strength from breakouts when tensile fractures are present

There is an interesting extension of the discussions above when both breakouts and drilling-induced tensile fractures occur in a well. As noted above, if there is no information on rock strength available, one cannot use observations of breakout width to constrain S_{Hmax} because breakout width is a function of both S_{Hmax} and rock strength. When both breakouts and drilling induced tensile fractures are present in a well, S_{Hmax} is determinable from the occurrence of tensile fractures alone. Thus, the width of breakouts allows us to estimate the strength of the rock *in situ*. For example, in the case illustrated in Figure 7.10, had the strength been appreciably less than 138 MPa, the breakout width would have been greater. Had the strength been greater than 138 MPa, the breakouts would have been narrower (or absent altogether). Because the width of breakouts can be measured accurately (Chapter 6), the occurrence of drilling-induced tensile fractures not only yields estimates of S_{Hmax}, in the presence of breakouts, they provide a means to obtain a direct *in situ* estimate of rock strength, as well.

Estimating S_{Hmax} from breakout rotations

In areas of active faulting, wells penetrate formations where there are localized stress perturbations due to slip on faults. These perturbations are manifest as rotations of breakout (and/or drilling-induced tensile fracture) azimuth along the wellbore as a function of depth. These have been seen in oil and gas wells in several parts of the world and this subject is briefly revisited in Chapter 11.

One example of a breakout rotation can be seen in the ultrasonic televiewer in the left panel of Figure 7.14 from the KTB scientific research well in Germany. As first

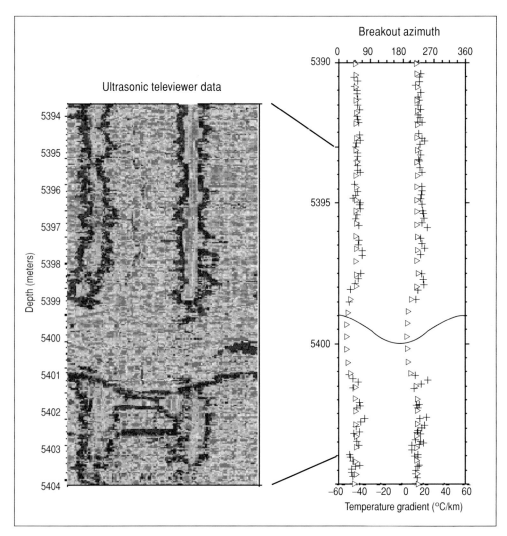

Figure 7.14. Rotation of wellbore breakouts near a fault at 5399 m in the KTB borehole that can be modeled as the result of a perturbation of the stress field induced by slip on the fault (Barton and Zoback 1994). This is illustrated on the right (see text).

illustrated in Figure 6.4, the dark bands on opposite sides of the well correspond to breakouts. Near the fault at 5399 m there is both an absence of breakouts immediately above the fault and an apparent rotation of breakout rotation immediately above and below the fault.

Breakout rotations were first noted in a scientific research well drilled near the San Andreas fault in southern California (Shamir and Zoback 1992). In that study, it was shown that slip on active faults was the most likely cause of the breakout rotations.

If the different mechanical properties of the faults were the cause of the rotation, the orientation of the breakouts would be perturbed over a much greater length of the wellbores than that observed. Brudy, Zoback *et al.* (1997) studied breakout and tensile fracture orientation with depth in the ultradeep KTB research borehole using wellbore image data and the interactive analysis technique referred to above. They documented the fact that while the average stress orientation to ~8 km depth was quite consistent, numerous relatively minor perturbations of stress orientation (at various wavelengths) are superimposed on the average orientation due to slip on faults at various scales.

Through interactive analysis of the shape of the wellbore at various depths, the orientation of the breakouts can be accurately determined as a function of depth using ultrasonic imaging data (Barton, Tessler *et al.* 1991). The orientation of the breakouts on the left side of Figure 7.14 are shown by the + symbols in the image on the right. Barton and Zoback (1994) used dislocation modeling to replicate the observed breakout rotations in the KTB wellbore at 5.4 km depth (Figure 7.14) and showed how modeling could be used to constrain the magnitude of S_{Hmax} based on knowing (i) the magnitudes of S_{hmin} and S_v, (ii) the unperturbed orientation of S_{Hmax} and (iii) the strike and dip of the causative fault. Note that the modeling results (triangles on right hand image) were used to replicate the breakout rotation observed in televiewer data (left image) which are shown by + symbols in the right image. The breakouts do not form right next to the fault that slipped due to the stress drop on the fault (Shamir and Zoback 1992). There is also a temperature gradient anomaly at the position of this fault due to fluid flow into the borehole along this fault (see Barton, Zoback *et al.* (1995). The magnitude of S_{Hmax} determined from modeling the breakout rotation was consistent with the range of values obtained from analysis of drilling-induced tensile fractures and breakouts by Brudy, Zoback *et al.* (1997) in this well (Barton and Zoback 1994) as shown by the + at 5.4 km depth in Figure 7.7.

Summary

In the sections above, we outlined two techniques for determination of the maximum horizontal principal stress, S_{Hmax}, which use observations of compressive and tensile wellbore failure in vertical wells. This requires independent knowledge of the least horizontal principal stress, S_{hmin}, which is determinable from carefully conducted leak off tests, or mini-fracs. These techniques are extended to deviated wells in Chapter 8.

In numerous field studies, the techniques described above have yielded consistent values at various depths in a given well and multiple wells in a given field (*e.g.* Zoback, Barton *et al.* (2003). In cases where the wells are drilled in areas of active faulting, the values obtained for S_{Hmax} are consistent with predictions of frictional faulting theory (Zoback and Healy 1984; Zoback and Healy 1992; Brudy, Zoback *et al.* 1997; Townend and Zoback 2000). In many case studies, the stress states determined have been used to successfully predict compressive and tensile failures in subsequently drilled wells (of varied orientation). Hence, from an engineering perspective, predictions of wellbore stability based on stress fields determined using the techniques described in this chapter have proven to yield useful and reliable results. This will be illustrated in the case studies discussed in Chapter 10.

8 Wellbore failure and stress determination in deviated wells

Many wells being drilled for oil and gas production are either horizontal, highly deviated from vertical or have complex trajectories. Because of this, it is necessary to understand the factors that control the occurrence of compressive and tensile failures in wells with arbitrary orientation. In this section, I generalize the material presented in Chapters 6 and 7 for wells of any orientation. I begin the chapter by considering the ways in which compressive failures (breakouts) and drilling-induced tensile fractures occur in arbitrarily oriented wells in normal, strike-slip and reverse faulting environments. The basic principles of failure of wells with arbitrary orientation will be utilized extensively when applied to wellbore stability in Chapter 10.

Most analyses of breakouts and tensile fracture compilations do not include observations of failures in inclined boreholes. There are three reasons for this. First, four-arm caliper logs (typically used to study breakouts) frequently track *key seats* as discussed in Chapter 6. This makes it difficult to detect breakouts with caliper data. Second, breakout directions in deviated holes vary significantly from what would occur in vertical holes (*e.g.* Mastin 1988; Peska and Zoback 1995). The same is true of drilling-induced tensile fractures. Thus, it is not straightforward to relate breakout and tensile fracture observations to stress directions. Third, drilling-induced tensile fractures in deviated wells occur in an *en echelon* pattern at an angle to the wellbore axis and can be difficult to distinguish from natural fractures in the formation. I demonstrate theoretically (and illustrate through several examples) how to distinguish the two.

Because of the complexities associated with breakout and tensile fracture occurrence in deviated wells, it has been typical to ignore data from deviated wells when assessing *in situ* stress orientations. In this chapter we take the opposite approach. I show that the details of failure of deviated wells are sensitive to the exact stress conditions *in situ*, so the study of such failures can provide important insight into stress orientations and magnitudes. In this chapter we demonstrate that the analysis of wellbore failures in deviated wells makes possible several new techniques for stress determination.

I will present the material in this chapter in the context of deviated wells and principal stresses acting in horizontal and vertical directions. As alluded to in Chapter 1, this is the usual case world-wide. However, complex stress fields, such as near salt bodies (as illustrated in Figure 1.10) require knowing the orientation of the stress tensor when the principal stresses are not in horizontal and vertical planes. The generalized theory presented in this chapter allows us to do so.

Finally, I address the issue of determination of stress orientation from cross-dipole sonic logs. I begin by reviewing the basic idea of deriving stress orientations from this data in vertical wells when bedding is sub-horizontal and go on to illustrate how the technique can work in deviated wells when the bedding (or aligned fractures) is likely to be at an oblique angle to the well trajectory.

State of stress surrounding an arbitrarily deviated well

Various authors have addressed different aspects of wellbore failure in deviated wells. Bradley (1979) was the first to model for compressive well failure of a deviated well for the purpose of recommending proper mud weights to prevent borehole failure. However, he did all of his analyses for the rare case where the two horizontal stresses are equal and less than the vertical stress. Daneshy 1973; Richardson (1981), Roegiers and Detournay (1988), Yew and Li (1988) and Baumgärtner, Carvalho *et al.* (1989) have done numerical and experimental analyses of hydraulic fracture formation in wells at various orientations to principal stresses, although only several specific borehole orientations and stress states were considered. In this chapter, we present a systematic analysis of wellbore stability (including both compressive and tensile failures) for arbitrarily inclined boreholes in a wide variety of stress states ranging from normal faulting, to strike-slip to reverse faulting environments. We also consider the likelihood of compressive and tensile borehole failure as a function of rock strength and borehole fluid pressure over a wide range of conditions.

In a deviated well, the principal stresses acting in the vicinity of the wellbore wall are generally not aligned with the wellbore axis (Figure 8.1a). To consider failure in a well of arbitrary orientation, we must define three coordinate systems (Figure 8.1b): (1) a geographic coordinate system, X, Y and Z oriented north, east and vertical (down); (2) a stress coordinate system, x_s, y_s and z_s (corresponding to the orientations S_1, S_2, and S_3) and (3) the wellbore coordinate system x_b, y_b and z_b where x_b is radial, pointing to the bottom of the well, z_b is down along the wellbore axis and y_b is orthogonal in a right-hand coordinate system. To most easily visualize wellbore failure we will always *look down* deviated wells and evaluate wellbore failure as a function of angle, θ, from the bottom of the well in a clockwise direction. Despite the complexities associated with such cases, to analyze whether (and how) failure might initiate at the wellbore wall, we simply need to consider whether the principal stresses acting in a plane tangential to the wellbore wall, σ_{tmax} and σ_{tmin} (and σ_{rr} acting normal to the wellbore wall) are such that that they exceed the strength of the rock. I define the angle between the axis of the wellbore and the plane normal to σ_{tmin} as ω (Figure 8.1a), and consider stress

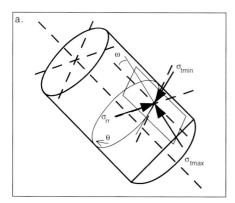
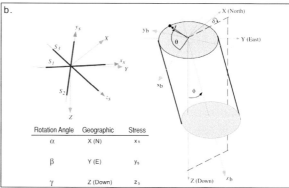
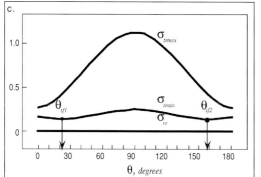
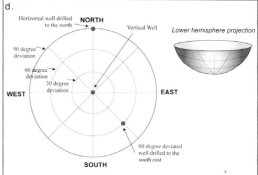

Figure 8.1. (a) Stresses acting in the wall of an arbitrarily oriented well. (b) Coordinate system used to transform knowledge of principal stresses, to that around the wellbore. (c) Variation of principal stresses around an arbitrarily oriented wellbore. (d) Lower hemisphere projection used to display relative stability of wells with different deviations and azimuth. Modified from Peska and Zoback (1995).

variations as a function of position around the well going clockwise from the bottom (Figure 8.1c).

In the case of an arbitrarily deviated well there is no simple relation between the orientation of far-field stresses and the position around the well at which either compressive or tensile failure might possibly occur. Thus, while breakouts in a vertical well *always* form at the azimuth of S_{hmin}, regardless of stress magnitude or rock strength (as long as the principal stresses are vertical and horizontal), this is not the case for a well that is arbitrarily oriented with respect to the *in situ* principal stresses. In this case, the position of the breakouts depends on the magnitude and orientation of principal stresses as well as the orientation of the well with respect to the stress field. This will be illustrated below.

Following Peska and Zoback (1995), we utilize tensor transformations to evaluate stress in the three coordinate systems of interest. In tensor notation, the principal stresses are given by

$$
\mathbf{S}_s = \begin{pmatrix} S_1 & 0 & 0 \\ 0 & S_2 & 0 \\ 0 & 0 & S_3 \end{pmatrix} \tag{8.1}
$$

To rotate these stresses into a wellbore coordinate system we first need to know how to transform the stress field first into a geographic coordinate system using the angles α, β, γ (Figure 8.1b). This is done using

$$
\begin{pmatrix} x_s \\ y_s \\ z_s \end{pmatrix} = \mathbf{R}_s \begin{pmatrix} X \\ Y \\ Z \end{pmatrix} \tag{8.2}
$$

where,

$$
\mathbf{R}_s = \begin{pmatrix} \cos\alpha\cos\beta & \sin\alpha\cos\beta & -\sin\beta \\ \cos\alpha\sin\beta\sin\gamma - \sin\alpha\cos\gamma & \sin\alpha\sin\beta\sin\gamma + \cos\alpha\cos\gamma & \cos\beta\sin\gamma \\ \cos\alpha\sin\beta\cos\gamma + \sin\alpha\sin\gamma & \sin\alpha\sin\beta\cos\gamma - \cos\alpha\sin\gamma & \cos\beta\cos\gamma \end{pmatrix} \tag{8.3}
$$

To transform the stress field from the geographic coordinate system to the borehole system, we use

$$
\begin{pmatrix} x_b \\ y_b \\ z_b \end{pmatrix} = \mathbf{R}_b \begin{pmatrix} X \\ Y \\ Z \end{pmatrix} \tag{8.4}
$$

where,

$$
\mathbf{R}_b = \begin{pmatrix} -\cos\delta\cos\phi & -\sin\delta\cos\phi & \sin\phi \\ \sin\delta & -\cos\delta & 0 \\ \cos\delta\sin\phi & \sin\delta\sin\phi & \cos\phi \end{pmatrix} \tag{8.5}
$$

With $\mathbf{R}_s$ and $\mathbf{R}_b$ defined, we can define the stress first in a geographic, $\mathbf{S}_g$, and then in a wellbore, $\mathbf{S}_g$, coordinate system using the following transformations

$$
\begin{aligned} \mathbf{S}_g &= \mathbf{R}_s^{\mathrm{T}} \mathbf{S}_s \mathbf{R}_s \\ \mathbf{S}_b &= \mathbf{R}_b \mathbf{R}_s^{\mathrm{T}} \mathbf{S}_s \mathbf{R}_s \mathbf{R}_b^{\mathrm{T}} \end{aligned} \tag{8.6}
$$

where we define effective stress using the generalized form of the effective stress law described above (equation 3.10). We go on to define individual effective stress components around the well (simplified here for the wellbore wall) as

$$
\begin{aligned} \sigma_{zz} &= \sigma_{33} - 2\nu(\sigma_{11} - \sigma_{22})\cos 2\theta - 4\nu\sigma_{12}\sin 2\theta \\ \sigma_{\theta\theta} &= \sigma_{11} + \sigma_{22} - 2(\sigma_{11} - \sigma_{22})\cos 2\theta - 4\sigma_{12}\sin 2\theta - \Delta P \\ \tau_{\theta z} &= 2(\sigma_{23}\cos\theta - \sigma_{13}\sin\theta) \\ \sigma_{rr} &= \Delta P \end{aligned} \tag{8.7}
$$

Note that there is a change of sign in equation (8.7) that corrects an error in Peska and Zoback (1995). The principal effective stresses around the wellbore are given by

$$
\sigma_{t\max} = \frac{1}{2}\left(\sigma_{zz} + \sigma_{\theta\theta} + \sqrt{(\sigma_{zz} - \sigma_{\theta\theta})^2 + 4\tau_{\theta z}^2}\right)
$$

$$
\sigma_{t\min} = \frac{1}{2}\left(\sigma_{zz} + \sigma_{\theta\theta} - \sqrt{(\sigma_{zz} - \sigma_{\theta\theta})^2 + 4\tau_{\theta z}^2}\right)
$$

(8.8)

Failure of arbitrarily deviated wells

To evaluate the stability of wells of any orientation we use a lower hemisphere diagram as illustrated in Figure 8.1d, where each point represents a well of a given azimuth and deviation. Vertical wells correspond to a point in the center, horizontal wells correspond to a point on the periphery at the appropriate azimuth and deviated wells are plotted at the appropriate azimuth and radial distance. Figure 8.2 shows the relative stability of wells of various orientations for normal, strike-slip and reverse faulting environments. The principal stresses are in vertical and horizontal planes although Figure 8.2 also could have been calculated for any arbitrary stress field (Peska and Zoback 1996). The stress magnitudes, pore pressure and mud weight assumed for each set of calculations are shown in each figure. The stresses and pore pressure correspond to a depth of 3.2 km and hydrostatic pore pressure. The mud weight is assumed to be equal to the pore pressure, for simplicity. The color shown in each figure represents *the rock strength required to prevent the initiation of breakouts*. A Mohr–Coulomb failure criterion was utilized in these calculations, but any of the failure criteria discussed in Chapter 4 could have been used in the calculations. Red colors represent relatively unstable well orientations as higher rock strength is required to prevent breakout initiation whereas dark blue represents relatively stable well orientations as failure is prevented by much lower rock strength.

Note that in normal faulting environments, breakout initiation is more likely to occur in wells that are highly deviated in the direction of maximum horizontal stress than for vertical wells. Conversely, wells that are highly deviated in the $S_{h\min}$ direction are more stable than vertical wells. This is easy to understand for the case of horizontal wells. Those drilled parallel to $S_{H\max}$ have a trajectory that results in the greatest principal stress, S_v, *pushing down* on the well and the minimum principal stress, $S_{h\min}$, acting in a horizontal direction normal to the well path. This yields the maximum possible stress concentration at the wellbore wall. For horizontal wells drilled parallel to $S_{h\min}$, S_v still *pushes down* on the well, but $S_{H\max}$ (which is only slightly below the value of S_v) acts in a horizontal plane normal to the well path, resulting in a lesser stress concentration (and much less stress anisotropy) on the wellbore wall. Note that the color patterns in

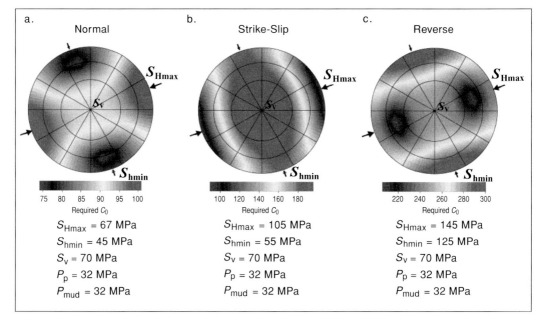

Figure 8.2. The tendency for the initiation of wellbore breakouts in wells of different orientation for normal, strike-slip and reverse faulting stress regimes. Similar to the figures in Peska and Zoback (1995). The magnitudes of the stresses, pore pressure and mud weight assumed for each case is shown. The color indicates the rock strength required to prevent failure, hence red indicates a relatively unstable well as it would take high rock strength to prevent failure whereas blue indicates the opposite. The strength scale is different for each figure as the stress magnitudes are progressively higher from normal to strike-slip to reverse faulting. Note that because these calculations represent the initiation of breakouts, they are not directly applicable to considerations of wellbore stability (see Chapter 10). (*For colour version see plate section.*)

Figures 8.2a, b and are quite different. Hence, no universal rule-of-thumb defines the relative stability of deviated wells with respect to the principal stress directions. In the case of strike-slip faulting, vertical wells are most likely to fail whereas horizontal wells drilled parallel to S_{Hmax} are most stable. In the case of reverse faulting environments, sub-horizontal wells drilled parallel to S_{hmin} are most unstable. Again, for vertical and horizontal wells, these general patterns are somewhat intuitive in terms of magnitudes of the principal stresses acting normal to the wellbore trajectory.

There are two important additional points to note about these figures. First, the strength scale is different for each figure. At a given depth, stress magnitudes are more compressive for strike-slip faulting regimes than for normal faulting regimes and more compressive still for reverse faulting regimes. Therefore, it takes considerably higher strengths to prevent breakout initiation in strike-slip regimes than normal faulting regimes and still higher strengths in reverse faulting regimes. Thus, for a given value of rock strength, wellbores are least stable in reverse faulting regimes and most

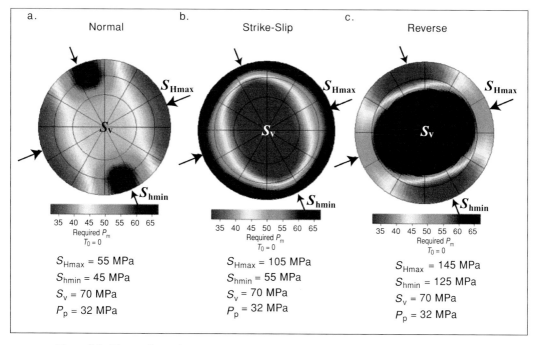

Figure 8.3. The tendency for the initiation of tensile fractures to form in wells of different orientation for normal, strike-slip and reverse faulting stress regimes. Similar to the figures in Peska and Zoback (1995). The magnitudes of the stresses, pore pressure and mud weight assumed for each case is shown. Note that the color indicates the mud pressure required to initiate tensile failure. Hence red indicates that tensile fractures are likely to form as little excess mud weight is required to initiate failure whereas blue indicates the opposite. (*For colour version see plate section.*)

stable in normal faulting regimes. Second, it should also be noted that these figures were constructed for the initiation of breakouts, not the severity of breakouts, which is addressed below. As discussed at length in Chapter 10, to drill stable wells it is not necessary to prevent breakout initiation; it is necessary to limit breakout severity.

Similar to Figure 8.2, Figure 8.3 represents the tendency for tensile fractures to occur. In this case, the assumed stress states and pore pressure are identical to those in Figure 8.2, but the colors now indicate the magnitude of mud weight required to induce tensile failure of the wellbore wall. Zero tensile strength was assumed although assuming a finite value of tensile strength would not have changed the results in any significant way. When drilling-induced tensile fractures are expected at mud weights close to the pore pressure, the figures are shaded red. When extremely high mud weights are required to initiate tensile fractures, the figures are shaded dark blue. In the cases for normal and strike-slip faulting (Figures 8.3a,b), the darkest blue corresponds to mud weights in excess of the least principal stress, which could likely not be achieved prior to losing circulation. Note that in strike-slip faulting areas (Figure 8.3b), all wells deviated less than 30° are expected to have drilling-induced tensile fractures, as previously discussed for the case of vertical wells in Chapters 6 and 7. Conversely, in normal and reverse faulting environments, drilling-induced tensile fractures are expected to occur at mud weights close to the pore pressure only in highly deviated wells. In normal faulting areas, such fractures are expected in sub-horizontal wells drilled parallel to S_{Hmax}, whereas in reverse faulting environments the same is true for sub-horizontal wells drilled parallel to S_{hmin}.

As mentioned above, the orientation of the wellbore breakouts in deviated wells depends on the orientation of the well with respect to the stress field and *in situ* stress magnitudes. Figure 8.4a shows the orientation of breakouts for deviated wells in a strike-slip faulting regime with S_{Hmax} acting in the NW–SE direction. The orientations of breakouts (if they were to occur) are shown in a *looking down the well* reference frame (see inset). Thus, wells deviated to the northeast or southwest would have breakouts on the top and bottom of the well whereas deviated wells drilled to the southeast or northwest would have breakouts on the sides. The orientations of tensile fractures (if they were to occur) are shown in Figure 8.4b. The two lines indicate the position of the tensile fractures around the well and the angle with respect to the wellbore axis (see Peska and Zoback 1995 and the inset). As noted by Brudy and Zoback (1993) and Peska and Zoback (1995), drilling-induced tensile fractures in deviated wells generally occur as *en echelon* pairs of fractures which are inclined to the wellbore wall at the angle ω, referred to above. In Chapter 10, we discuss how hydraulic fractures that form at the wellbore wall as *en echelon* tensile fractures propagate away from a well must coalesce (*link-up*) as they turn and become perpendicular to the least principal stress. A similar situation is discussed by Baumgärtner, Carvalho *et al.* (1989) for the case of hydraulically fracturing a vertical well when the least principal stress is vertical (reverse faulting regime). Axial tensile fractures form at the wellbore wall when the $\sigma_{\theta\theta}$ goes to zero (as a result of borehole pressurization), but the fractures *roll-over* into a horizontal plane as they propagate away from the well.

To make some of the previous calculations more relevant to wellbore stability, Figure 8.5 was calculated using the stress state used in the construction of Figure 6.3. This is a strike-slip stress state with S_{Hmax} acting in an E–W orientation. However, the colors in Figure 8.5a now indicate the width of breakouts for wellbores of any arbitrary orientation at the depth of interest in the prescribed stress state assuming a uniaxial compressive strength of 45 MPa, a coefficient of internal friction of 1.0 and a Mohr–Coulomb failure criterion. As can be seen by comparison with Figure 6.3, vertical wells are expected to have breakout widths of about 90°. Wellbore deviations up to about 30° (independent of azimuth) have a similar degree of instability, as do wells of any deviation drilled approximately east–west. Breakout orientations in east–west striking wells are expected on the sides of the hole whereas those trending north–south would be expected to have breakouts on the top and bottom (Figure 8.5b). Note that highly deviated wells drilled in the north–south direction are much more unstable as breakouts with much greater width would be expected to occur. In fact, such wells would undoubtedly be

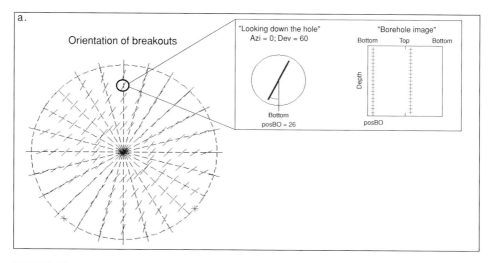

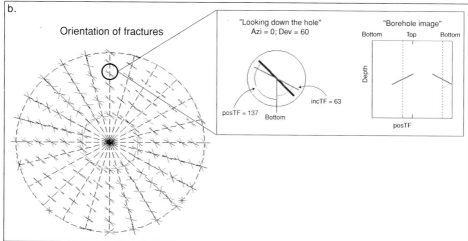

Figure 8.4. (a) The orientation of breakouts, if they were to form, in wells of different orientations. A looking-down-the-well convention is used as indicated in the inset. Similar to the figures in Peska and Zoback (1995). (b) The orientation of tensile fractures, if they were to form, in wells of different orientations is indicated by two angles that define the position of the tensile fracture around the wellbore's circumference as well as the orientation of the fracture trace with respect to the wellbore axis, as indicated in the inset. In both figures, a strike-slip faulting regime with S_{Hmax} acting in a NW–SE direction is assumed in the calculations. (*For colour version see plate section.*)

washed out as the breakouts subtend nearly the entire circumference of the well. The relationship between breakout width and wellbore stability is discussed at length in Chapter 10. In this stress state, drilling-induced tensile fractures are likely at mud weights close to the pore pressure in wells with a wide range of orientations (Figure 8.5c). Only in wells drilled approximately east–west (parallel to S_{Hmax}) would

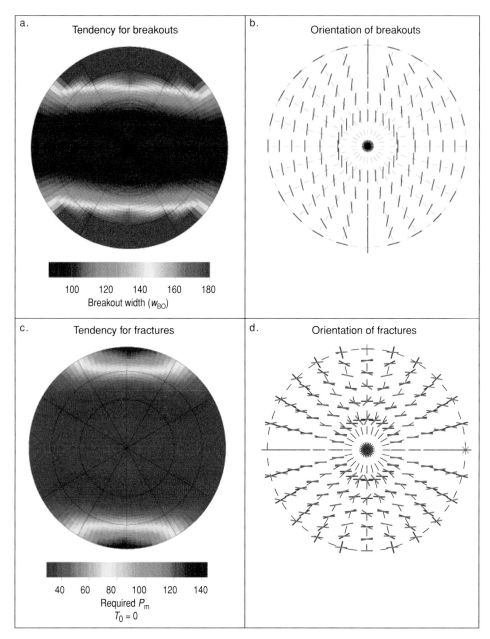

Figure 8.5. Lower hemisphere representations of the relative stability of wellbores of varied orientation with respect to the formation of wellbore breakouts (a, b) and drilling-induced tensile fractures (c, d). (a) The width of breakouts in red areas indicate the orientations of unstable wellbores as nearly the entire circumference of the well fails. (b) Orientation of wellbore breakouts (if they form) in a *looking-down-the-well* coordinate system. (c) The tendency of drilling-induced tensile fractures to form in terms of the magnitude of excess mud weight needed to initiate failure. (d) The orientation of induced tensile failures (colors are the same as in Figure 8.4b). (*For colour version see plate section.*)

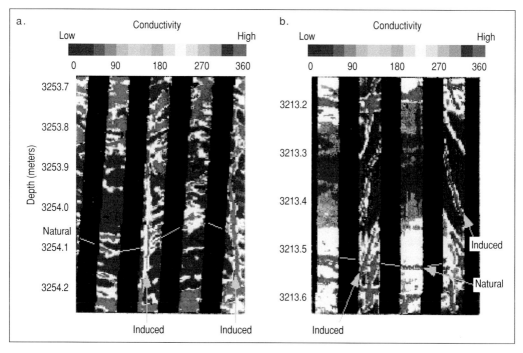

Figure 8.6. Electrical resistivity image of drilling-induced tensile fractures observed in the KTB pilot hole (after Peska and Zoback 1995).

the tensile fractures be axial and oriented at the S_{Hmax} direction as in a vertical well (Figure 8.5d). At all other well orientations, the fractures would be significantly inclined with respect to the wellbore axis.

An example of *en echelon* drilling-induced tensile fractures observed in the vertical KTB pilot hole in Germany is shown in Figure 8.6b (after Peska and Zoback 1995). The pilot hole was continuously cored and these fractures are not present in the core. Nearly all the tensile fractures were axial (Brudy, Zoback *et al.* 1997) as shown in Figure 8.6a. The sinusoidal features in Figure 8.6a represent foliation planes in granitic gneiss. In intervals where the stress field is locally perturbed by slip on active faults (see Chapter 9), the drilling-induced fractures that form occur at an angle ω to the wellbore axis because one principal stress is not vertical.

Figure 8.7 is intended to better illustrate how *en echelon* drilling-induced tensile fractures form. It is obvious that the tensile fracture will first form at the point around the wellbore where the minimum principal stress, σ_{tmin}, is tensile. Because the wellbore is deviated with respect to the principal stresses, ω is about 15° and 165° in the sections around the wellbore where the borehole wall is locally in tension (Figure 8.7a). The fractures propagate over a span of the wellbore circumference, θ_t, where tensile stress exists (Figure 8.7b). The fractures do not propagate further because as the fracture

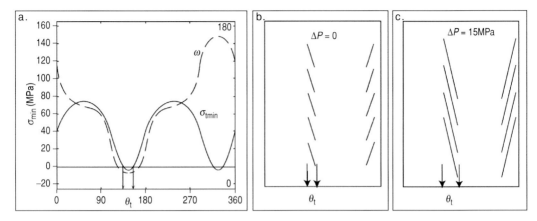

Figure 8.7. Theoretical illustration of the manner of formation of *en echelon* drilling-induced tensile fractures in a deviated well. (a) The fracture forms when σ_{tmin} is tensile. The angle the fracture makes with the axis of the wellbore is defined by ω, which, like σ_{tmin} varies around the wellbore. (b) The *en echelon* fractures form over the angular span θ_t, where the wellbore wall is in tension. (c) Raising the mud weight causes the fractures to propagate over a wider range of angles because σ_{tmin} is reduced around the wellbore's circumference.

grows, σ_{tmin} becomes compressive. Raising the wellbore pressure (Figure 8.7c) allows the fracture to propagate further around the wellbore because the σ_{tmin} is decreased by the amount of ΔP, thus increasing the angular span (θ_t), where tensile stress are observed around the wellbore circumference.

Confirming that S_{Hmax} and S_{hmin} are principal stresses

Drilling-induced tensile fractures were ubiquitous in the KTB pilot hole and main borehole. As mentioned above, along most of the well path, the tensile fractures are axial. As this is a near-vertical borehole, it indicates that there is a near-vertical principal stress. However, in a few sections of the wellbore, the state of stress is locally perturbed by slip on faults and is rotated away from a horizontal and vertical orientation (Brudy, Zoback *et al.* 1997). Over the entire depth interval studied in detail in the KTB boreholes (from $\sim$ 1 km to $\sim$8 km depth), axial drilling-induced tensile fractures indicate that one principal stress is nearly always vertical and the cases where this is not true is limited to zones of locally anomalous stress (Brudy, Zoback *et al.* 1997). Near vertical drilling-induced tensile fractures were observed to $\sim$7 km depth in the Siljan wells drilled in Sweden (Lund and Zoback 1999). A similar situation was encountered by Wiprut, Zoback *et al.* (2000) who documented both axial and *en echelon* fractures in a suite of five oil wells in the Visund field of the northern North Sea. While the stress field is well-characterized by a near-vertical and two horizontal principal stresses, there are zones of anomalous stress where principal stresses deviate from their average orientation by ±10°.

As a brief historical note, it is worth pointing out that at the time when the integrated stress maps presented in Chapter 1 were initially compiled by Zoback and Zoback (1980; 1989) and Zoback (1992) it was assumed that one principal stress was vertical. The rationale for this assumption was that at relatively shallow depth in the crust, the presence of a sub-horizontal free surface would require one principal stress to be approximately vertical. At greater depth, it was argued that for nearly all intraplate crustal earthquakes for which there are reliable focal mechanisms, either the P-, B- or T-axis was sub-horizontal. While this is a relatively weak constraint on stress orientation, the consistency of the apparent horizontal principal stresses in the compiled maps indicated that the assumption appeared to generally correct. To date, drilling-induced tensile fractures have been identified in scores of near-vertical wells around the world. As such fractures are nearly always axial, it provides strong confirmation that the assumption that principal stresses *in situ* are vertical and horizontal is generally valid.

Estimating S_{Hmax} from breakouts and tensile fractures in deviated wells

Several authors have addressed the subject of the relationship between the failure of inclined holes and the tectonic stress field. Mastin (1988) demonstrated that breakouts in inclined holes drilled at different azimuths were expected to form at various angles around a well bore. Qian and Pedersen (1991) proposed a non-linear inversion method to attempt to extract information about the *in situ* stress tensor from breakouts in an inclined deep borehole in the Siljan impact structure in Sweden. Qian, Crossing *et al.* (1994) later presented a correction of their results because of errors in the published equations of Mastin (1988) (although the figures in Mastin's paper are correct). Zajac and Stock (1992) suggested that it is possible to constrain stress magnitudes from breakout azimuths if there are observations from a number of inclined holes drilled at various azimuths in a uniform stress field. This technique is conceptually similar to a technique reported by Aadnoy (1990a,b) to estimate *in situ* stress from leak-off test data in a number of inclined boreholes that assumes that peak pressures from leak-off tests are hydrofrac fracture initiation pressures. For the reasons outlined in Chapter 7, this is likely a questionable assumption. Using formal geophysical inversion theory with observations of breakouts or tensile failures presumes that there will be data available from multiple wells of varied orientation that are sampling a uniform stress field. Whether such data are likely to be available is a questionable assumption.

There are several straightforward ways to use observations of tensile fractures and breakouts in deviated wells for determination of the magnitude and orientation of S_{Hmax} at depth assuming that the vertical principal stress, S_v, and minimum horizontal

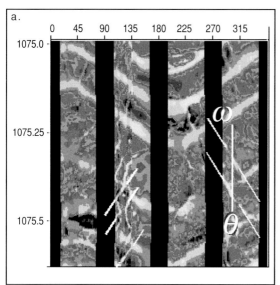
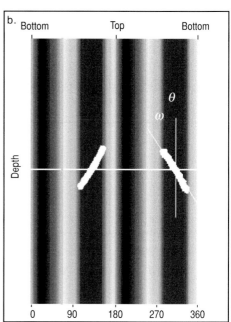

Figure 8.8. (a) Drilling-induced tensile fractures in a geothermal well in Japan make an angle ω with the axis of the wellbore and are located at position indicated by the angle θ from the bottom of the wellbore. (b) Theoretical model of the observed fractures in (a) replicate both ω and θ for the appropriate value of the magnitude and orientation of S_{Hmax}.

stress, S_{hmin}, are determined in the manner described previously. For example, iterative forward modeling of the *en echelon* fractures seen in Figure 8.8a in a moderately deviated geothermal well in Japan yielded knowledge of the magnitude and orientation of S_{Hmax}. Three observations constrain the modeling – the position of the fractures around the wellbore, θ, their deviation with respect to the wellbore axis, ω, and the very existence of the tensile fractures. The orientation of S_{Hmax} was found to be about N60°W and S_{Hmax} was found to be slightly in excess of the vertical stress (because of severe wellbore cooling that commonly occurs in geothermal wells, drilling-induced tensile fractures are induced even if there are relatively modest stress differences).

In the two cases considered below, the wells are significantly deviated but the occurrence of axial drilling-induced tensile fractures in vertical sections of the wells indicates that the principal stresses are vertical and horizontal. In the generalized case of a deviated well in a deviated stress field, LOT's will provide information on the magnitude of the least principal stress, but it would be necessary to do iterative forward modeling of observed wellbore failures to determine principal stress orientations and magnitudes.

Wiprut, Zoback *et al.* (2000) observed drilling-induced tensile fractures in the vertical section of a well in the Visund field of the northern North Sea. Below 2600 m depth, the well gradually increased in deviation with depth with a *build-and-hold* trajectory at

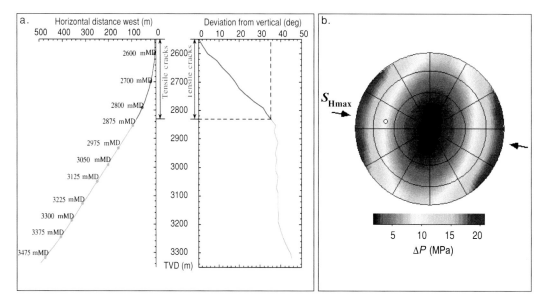

Figure 8.9. Drilling-induced tensile fractures were observed in the near-vertical portion of a well in the Visund field in the northern North Sea which abruptly ceased when the well deviated more than 35° (center). As shown in the figure on the right, this result is predicted by the stress state shown in Figure 8.10 (after Wiprut, Zoback *et al.* 2000). *Reprinted with permission of Elsevier. (For colour version see plate section.)*

an azimuth of N80°W (Figure 8.9a). Numerous drilling-induced tensile fractures were observed in the near-vertical section of this well and in other near-vertical wells in the field. The depth interval over which the fractures are observed is shown in red. Note that the occurrence of drilling-induced tensile fractures stops abruptly at a measured depth of ~2860 m, or equivalently, when the deviation of the well reached 35°. Electrical image data quality was excellent in all sections of the well. Hence, the disappearance of the fractures was not the result of poor data quality. In addition, there was no marked change in drilling procedure, mud weights, etc. such that disappearance of the fractures with depth does not appear to be due to a change in drilling procedures.

In a manner similar to that illustrated for vertical wells in Figure 7.10, it is possible to constrain the magnitude of S_{Hmax} after taking into account the ECD and thermal perturbation of the wellbore stress concentration as accurately as possible. The vertical stress, pore pressure and magnitude of least principal stress (from leak-off test data) for the Visund field were presented in Figure 7.4. Careful note was taken where bit trips and wash and ream operations may have perturbed the mud pressure in the well at given depths (Figure 7.12). As illustrated in Figure 8.10, the heavy black line indicates the magnitude of S_{Hmax} as a function of S_{hmin} to cause drilling-induced tensile fractures in a well with the appropriate deviation, ECD and amount of cooling. Because of this well's deviation, the line that defines the magnitude of S_{Hmax} required to explain the occurrence of drilling-induced tensile fractures is no longer nearly coincident with the strike-slip faulting condition as was the case for vertical wells as illustrated in Chapter 7.

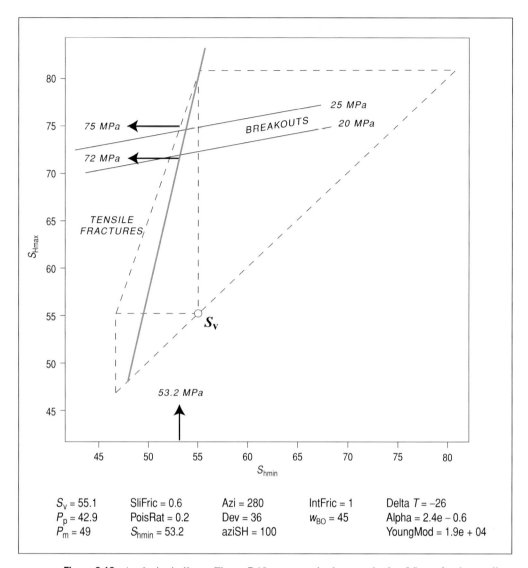

Figure 8.10. Analysis similar to Figure 7.10 to constrain the magnitude of S_{Hmax} for the tensile fractures observed in the deviated well shown in Figure 8.9. The heavy black line indicates the magnitude of S_{Hmax} as a function of S_{hmin} to cause drilling-induced tensile fractures in a well with the appropriate deviation, ECD and amount of cooling. Modified from Wiprut, Zoback *et al.* (2000). *Reprinted with permission of Elsevier.*

Note that for a measured value of S_{hmin} of 53.2 MPa that is appropriate for the depth of interest, an S_{Hmax} value of 72–75 MPa is indicated (other parameters used in the modeling are discussed by Wiprut, Zoback *et al.* 2000). This value was incorporated in the compilation of S_{Hmax} magnitudes compiled for the Visund field (mostly from vertical wells) presented in Figure 7.13.

Of course, the interesting observation in this well is that the occurrence of tensile fractures abruptly ceased when the well reached a deviation of 35° (Figure 8.9a). In fact, this is exactly what is expected for the stress field determined in Figure 8.10. As shown in Figure 8.9b, near-vertical wells are expected to fail in tension at mud weights just a few MPa above the pore pressure, in contrast to wells deviated more than 35° which require excess wellbore pressures over 9 MPa to initiate tensile failure. As the ECD was approximately 6 MPa above the pore pressure in this well, there was sufficient mud weight to induce tensile fractures in the near vertical section of the well, but insufficient mud weight to do so in the more highly deviated sections.

This type of forward modeling is quite useful in putting constraints on the magnitude and orientation of S_{Hmax} when observations of wellbore failure are available in deviated wells. As we often have knowledge of the vertical stress and least principal stress, we can use iterative forward modeling to constrain values of S_{Hmax} magnitude and orientation that match the inclination of *en echelon* tensile failures with respect to the wellbore axis, ω, and their position around the wellbore circumference. As was the case with vertical wells, the absence of drilling-induced tensile fractures in a deviated well allows us to put upper bounds on the magnitude of S_{Hmax}.

Because the position of both tensile fractures and breakouts around a deviated wellbore depends on the magnitude and orientation of all three principal stresses (as well as the orientation of the wellbore), independent knowledge of S_v and S_{hmin} enables us to constrain possible values of the orientation and magnitude S_{Hmax}. This technique was used by Zoback and Peska (1995) to model the position of breakouts around a deviated well in the Gulf of Mexico to determine the magnitude and orientation of the maximum horizontal stress. In this case the position of the breakouts was determined from multi-arm caliper data, the magnitude of least principal stress was known from mini-frac data, and the vertical stress was obtained from integration of density logs, leaving the magnitude and orientation of S_{Hmax} as the two unknowns. As illustrated in Figure 8.11a, an iterative grid search technique was used to find the range of values of S_{Hmax} magnitude and orientation compatible with the observations cited above. The breakouts will only occur at the position around the wellbore in which they were observed if the orientation of S_{Hmax} is at an azimuth of about $136 \pm 8°$. This corresponds to a direction of S_{hmin} that is orthogonal to the strike of a nearby normal fault (Figure 8.11b), exactly as expected from Coulomb faulting theory. The estimate of S_{Hmax} obtained from this analysis ranges between 39.5 and 43 MPa (Figure 8.11a). Predictions of wellbore stability based on such values are consistent with drilling experience.

As a practical point, it turns out to be quite difficult to utilize the technique illustrated in Figure 8.11 with highly deviated wells in places like the Gulf of Mexico where the sediments are extremely weak. The problem with using multi-armed caliper data is that because deviated wells are usually key-seated, the caliper arms usually get locked in the key seats and will not detect breakouts. In a study of approximately 40 wells in the South Eugene Island area of the Gulf of Mexico, Finkbeiner (1998) was only able to

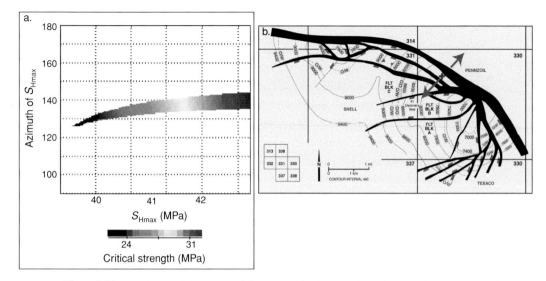

Figure 8.11. (a) The possible values of S_{Hmax} magnitude and orientation consistent with wellbore breakouts in a deviated well in the Gulf of Mexico (after Zoback and Peska 1995). © *2002 Society Petroleum Engineers.* (b) The stress orientation determined in this analysis indicates extension orthogonal to the strike of a major normal fault penetrated by the well. Original map after Holland, Leedy *et al.* (1990). *AAPG©1990 reprinted by permission of the AAPG whose permission is required for futher use.* (*For colour version see plate section.*)

document the occurrence of breakouts in five wells. In all other cases, the caliper logs were dominated by key seats. In another study, Yassir and Zerwer (1997) used four-arm calipers to map stress in the Gulf of Mexico. There was a great deal of scatter in their results, some of which was undoubtedly caused by extensive key seating. Hence, unless image logs are available to unequivocally identify wellbore failures, it may be difficult to employ the technique illustrated in Figure 8.11 if extensive key seating has occurred.

Distinguishing drilling-induced tensile fractures from natural fractures

In this section I briefly consider the problem of distinguishing *en echelon* drilling-induced tensile fractures from natural fractures in image logs. It is important *to get this right* for two reasons: First, misidentification of drilling-induced tensile fractures as natural fractures would lead to a misunderstanding the fractures and faults that actually exist in a reservoir at depth. Second, mistaking natural fractures for drilling-induced fractures would deny one the ability to utilize such fractures in a comprehensive stress analysis.

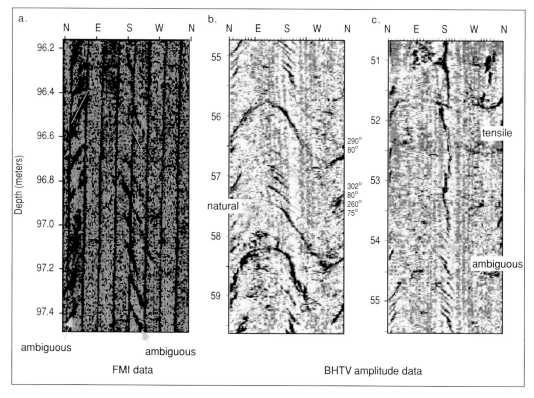

Figure 8.12. Three wellbore images from the Soultz geothermal well in eastern France. (a) Features indicated as *ambiguous* might be *en echelon* drilling induced tensile fractures or partial sinusoids associated with natural that cannot be seen all around the well. (b) Partial sinusoids have the same orientation as natural fractures. (c) Axial drilling-induced tensile fractures and possible partial sinusoids or *en echelon* drilling-induced fractures at approximately the same position around the wellbore (courtesy J. Baumgärtner).

At first glance, this would seem to be a trivial problem. As introduced in Chapter 5, natural fractures appear as sinusoids on an image log whereas *en echelon* drilling-induced tensile fractures have a distinctly different appearance. However, two points must be kept in mind. First, the combination of poor data quality and small aperture features sometimes makes it difficult to trace the sinusoid associated with a natural fracture all the way around a well. An example of this is Figure 8.12b, an ultrasonic televiewer log from the Soultz geothermal well in eastern France. Second, the *en echelon* drilling-induced tensile fractures discussed above are not linear and can be curved a significant amount. This is illustrated in the modeling shown in Figure 8.13a. For the combination of stress magnitude and orientation (and wellbore orientation, of course) used in the calculation, ω varies rapidly in the region where the wellbore wall is in tension. Hence, the drilling-induced tensile fractures will have a curved appearance in wellbore image data. A pronounced example of this is shown in Figure 8.13b, an

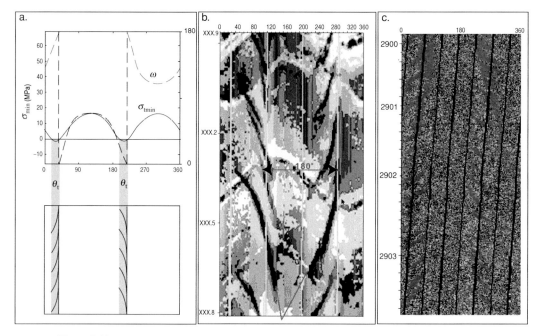

Figure 8.13. (a) Theoretical illustration of the formation of curved drilling-induced tensile fractures. Note that the fractures curve because ω varies over the angular span of the wellbore wall where the σ_{tmin} is negative. (b) Image log showing curved drilling-induced tensile fractures in a well in Argentina. Note that the fractures on the left side of the image are concave down whereas those on the right are concave up. These fractures are clearly not natural fractures which would have a sinusoidal trace on the wellbore wall. (c) Image log showing curved drilling-induced fractures in the Soultz geothermal well in eastern France.

image log from a fractured reservoir in Argentina. Such fractures are sometimes called *fish-hook* or *J-fractures*.

Returning to Figure 8.12a one can see that it is not immediately clear whether the fractures labeled *ambiguous* are *en echelon* drilling-induced tensile fractures or whether they are segments of natural fractures as seen in Figure 8.12b. This is especially problematic in this case because near axial drilling-induced fractures are present at some depths in this well (Figure 8.12c). An important aspect of tensile fracture initiation that aids in the correct interpretation of wellbore images is that curved drilling-induced tensile fractures have an opposite sense of curvature on either side of the well and thus do not define sections of a sinusoidal fracture trace (Barton and Zoback 2003). This is illustrated in the calculations shown in Figure 8.13a as well as the examples shown in the image logs in Figures 8.13b,c. The reversed curvature of the fractures is quite clear in Figure 8.13b, but is more subtle in Figure 8.13c, which is also from the Soultz well.

With this in mind, it can be seen in Figure 8.12a that the apparent *en echelon* fractures have the same, downward curvature on both sides of the well and could be *fit by a* sinusoid. Hence, these are segments of pre-existing fractures that cannot be seen around the entire wellbore, similar to those in Figure 8.12b. One interesting aspect of the *partial sinusoids* seen in Figures 8.12a,b is that they tend to occur at the azimuth of minimum compression around the wellbore. Note there are partial sinusoids in Figure 8.12c that are at the same azimuth as the axial drilling-induced tensile fractures. We refer to these as *drilling-enhanced* fractures (Barton and Zoback 2002). We interpret them to be small-aperture natural fractures that would normally be too fine to be seen on image logs if it were not for the preferential spalling of the fracture as the well is being drilled in the portions of the wellbore circumference where σ_{tmin} is minimum. While there has been no specific modeling to confirm this interpretation, we have noted a strong correlation between drilling-enhanced natural fractures and the orientation of S_{Hmax} in vertical wells.

Determination of S_{Hmax} orientation from shear velocity anisotropy in deviated wells

There are numerous observations of shear wave anisotropy (the polarization of shear waves in an anisotropic medium into fast and slow components) in the upper crust. The mechanisms proposed to explain them fall into two general categories. First is stress-induced anisotropy in response to the difference among the three principal stresses. In this case, vertically propagating seismic waves will be polarized with a fast direction parallel to the open microcracks (Crampin 1985), or perpendicular to closed macroscopic fractures (Boness and Zoback 2004). In both cases, the fast shear direction is polarized parallel to S_{Hmax}. Second is structural anisotropy due to the alignment of sub-parallel planar features such as aligned macroscopic fractures or sedimentary bedding planes. In this case the propagating shear waves exhibit a fast polarization direction parallel to the strike of the structural fabric or texture. In geophysical exploration, shear velocity anisotropy is commonly modeled with a transversely isotropic (Maxwell, Urbancic *et al.* 2002) symmetry where the shear waves are polarized parallel and perpendicular to the planes normal to the formation symmetry axis (Thomsen 1986).

Utilization of cross-dipole sonic logs in vertical wells has been used to determine stress orientation from the *fast* shear polarization direction when bedding planes are sub-horizontal or aligned fractures are not likely to influence the polarization of the shear waves. We limit our discussion to cross-dipole sonic shear wave logs (Kimball and Marzetta 1984; Chen 1988; Harrison, Randall *et al.* 1990) in order to compare and contrast the results from utilization of this technique with the other techniques discussed in this book. The sondes used to obtain these data have linear arrays of transmitters and receiver stations commonly spaced at 6 inch intervals and the transmitter on these dipole sonic tools is a low-frequency dipole source operating in the frequency range of

~1–5 kHz. A dispersive flexural wave propagates along the borehole wall with a velocity that is a function of the formation shear modulus. The dispersive nature of the flexural wave is used to filter out the high frequencies corresponding to short wavelengths that sample the rocks subjected to the stress concentration around the borehole (Sinha, Norris *et al.* 1994). In fact, at low frequencies, the flexural wave velocity approximates the shear velocity of the formation and has a depth of investigation of approximately 1.5 m into the formation such that it should be insensitive to the stress concentration around the wellbore. When the sondes are oriented (either geographically or with respect to the top or bottom of the well) the polarization direction of the fast and slow shear velocity directions can be obtained as measured in a plane perpendicular to the wellbore. The observations of interest are thus shear velocities that correlate with low frequencies that penetrate deeper into the formation beyond the altered zone around the wellbore. In addition, borehole ovality is known to bias the results of a shear wave splitting analysis with dipole sonic logs (Leslie and Randall 1990; Sinha and Kostek 1996), and care must be taken not to mistake shear polarization in the formation for the effects of ovality.

An example of cross-dipole data in vertical wells is shown in Figure 8.14 (after Yale 2003). Using cross-dipole data in vertical wells, they showed the direction of maximum horizontal stress in the Scott field of the North Sea was equally well determined from shear velocity anisotropy (solid arrows) and wellbore breakouts (dashed lines). Note that while the stress orientations obtained from the breakout data compare extremely well with that implied from the cross-dipole analysis, overall the stress orientations in the field seem to be quite heterogeneous with no overall trend apparent. In fact, the stress orientations seem to follow the trend of faults in the region. We return to this case study in Chapter 12 where we offer a model to explain these varied stress orientation observations.

A second example is shown in Figure 8.15 from an oil field in Southeast Asia. Fifteen vertical wells with dipole sonic logs were used to determine the direction of maximum horizontal compression (Figure 8.15a). This analysis delineates subtle differences in stress orientation in the northwest, southwest and southeastern parts of the field. Ten vertical wells with electrical image data that detected wellbore breakouts (Figure 8.15b) show the same stress directions in the northwest and southwest parts of the field. Although no breakout data are available in the southeast part of the field, the direction implied by the dipole sonic log can be used with confidence to determine the stress orientation. It seems clear that with appropriate quality control (see below), shear velocity anisotropy obtained from dipole sonic data in vertical wells can provide useful information about stress orientation if structural sources of velocity anisotropy can be ruled out.

Following closely Boness and Zoback (2006), I briefly address the topic of stress orientation determination from shear velocity anisotropy in this chapter to look at the more complex problem of determining stress orientation from cross-dipole sonic data when wells are highly deviated. In this case, the potential influence of bedding planes

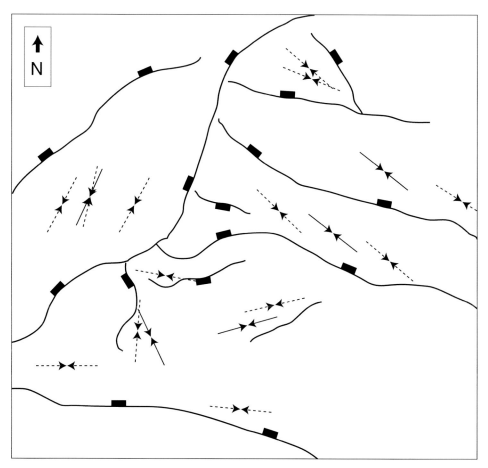

Figure 8.14. Map of S_{Hmax} orientations in vertical wells of the Scott Field of the North Sea utilizing both wellbore breakouts (solid arrows) and the fast shear direction in dipole sonic logs (dashed arrows) (after Yale 2003).

(or aligned fractures) on velocity anisotropy needs to be taken into account because aligned features might be encountered at a wide range of orientations to that of the wellbore and result in a fast direction that is difficult to discriminate from that induced by stress. As mentioned above, the shear waves generated and received by the dipole sonic tools are recorded in the planes normal to the axis of the borehole. Thus, the minimum and maximum shear velocities observed (and used to compute the amount of anisotropy) are not necessarily the absolute minimum and maximum velocities in the earth, which may exist in planes that are not perpendicular to the borehole axis. We define the true fast direction as the orientation in the earth with the absolute fastest shear velocity (a series of parallel planes described by a dip and dip direction) and the apparent fast direction as the fastest direction in a plane perpendicular to the borehole.

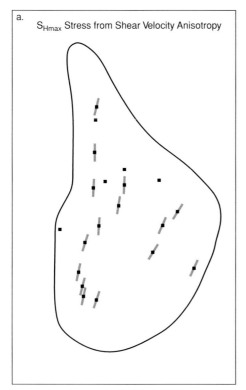
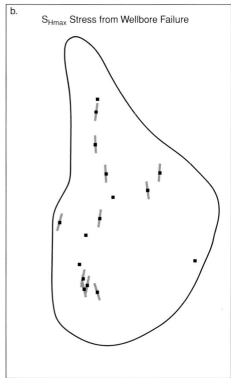

Figure 8.15. Stress maps of an oil field in Southeast Asia determined from (a) analysis of fast shear wave polarizations in dipole sonic logs in vertical wellbores and (b) breakouts detected in electrical image data.

Sinha, Norris *et al.* (1994) and Boness and Zoback (2006) modeled elastic wave propagation in a borehole with an axis at a range of angles to the formation symmetry axis. They demonstrated how the amount of anisotropy varies as the borehole becomes more oblique to the symmetry axis of the formation and that the maximum anisotropy is recorded at a 90° angle.

The geometry of the borehole relative to the formation will not only dictate the amount of anisotropy observed but also the apparent fast direction that is recorded by the tool. In the case of an arbitrarily deviated wellbore, it is probable that the borehole will be at some oblique angle to the symmetry axis (Figure 8.16a) and more generally, that neither the borehole nor the formation will be aligned with the cartesian coordinate axes (Figure 8.16b).

A case history that illustrates the controls on shear wave velocity anisotropy in a highly deviated well is that of the dipole sonic logs obtained in the SAFOD (San Andreas Fault Observatory at Depth) boreholes between measured depths of 600 m and 3000 m (Boness and Zoback 2006). Two boreholes were drilled at the SAFOD site,

a.

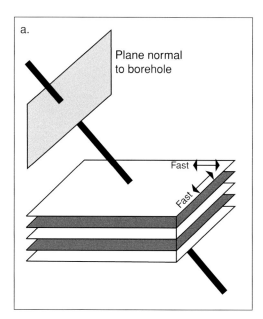

Plane normal
to borehole

Fast

Fast

b.

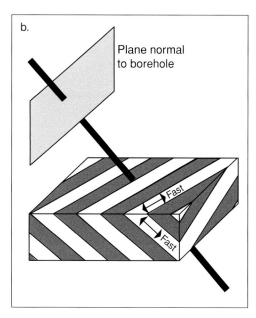

Plane normal
to borehole

Fast

Fast

Figure 8.16. (a) Geometry of a borehole at an oblique angle to a vertically transverse formation and (b) the general case when a borehole is oblique to a formation with a symmetry axis that is not aligned with one of the cartesian coordinate axes. After Boness and Zoback (2006).

a vertical pilot hole drilled to 2.2 km depth in granodiorite and a well deviated ~55° from vertical below 1.5 km depth that encountered an alternating sequence of sedimentary rocks below 1920 m depth (measured depth). Independent data are available on the orientations of bedding planes and fractures from electrical imaging data as well as breakouts and drilling-induced tensile fractures in the vertical pilot hole (Hickman and Zoback 2004). Boness and Zoback (2004) showed that in the vertical pilot hole, the fast shear velocity anisotropy direction was parallel to the direction of maximum horizontal compression obtained from wellbore failures apparently caused by the preferential closure of fractures in response to an anisotropic stress state.

Figure 8.17 (from Boness and Zoback 2006) illustrates the three-dimensional model for computing the apparent fast direction that will be recorded on the dipole sonic tools for any arbitrary orientation of the borehole and dipping bedding planes. In the case of stress-induced anisotropy, the true fast direction is parallel to the maximum compressive stress, oriented across the closed fractures. Thus, the apparent fast direction is not described by a plane but rather a line that lies within the plane normal to the borehole, in the direction normal to the fracture opening direction. The apparent fast direction is the vertical projection of the maximum compressive stress on the plane perpendicular to the borehole and will have the same azimuth as S_{Hmax} (Figure 8.17b), with a dip that depends on the orientation of the borehole. In the case of structural anisotropy, the true fast direction is oriented along the planes (be they fractures/bedding/aligned minerals)

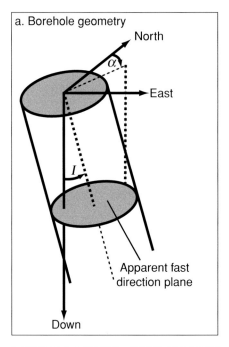

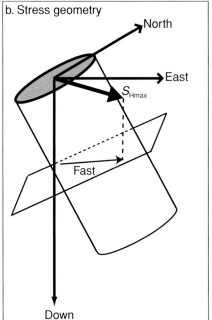

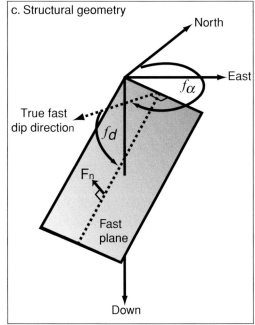

Figure 8.17. (a) Figure illustrating the geometry of the borehole with the plane in which the apparent fast direction is measured with the sonic logs. (b) In the case of stress-induced anisotropy, the apparent fast direction in the plane perpendicular to the borehole has an azimuth equivalent to that of S_{Hmax}, although the dip depends on the borehole trajectory. (c) Geometry used to compute the apparent fast direction that will be observed on the dipole sonic tool for structural anisotropy when the fast direction lies in an arbitrarily oriented plane. The angles f_α, f_d, F_n are defined in the text. After Boness and Zoback (2006).

and the orientation will be dependent on the propagation direction. However, the apparent fast direction has to be in the plane perpendicular to the borehole. Therefore, the apparent fast direction that is observed with the sonic tool will be a line that lies in both the true fast plane and the plane normal to the borehole, *i.e.* a line that marks the intersection of both planes (Figure 8.17c). The goal of this formalism is to allow one to either determine the true fast direction in the earth given an observed apparent fast direction or if the formation geometry is known one can predict the apparent fast direction that will be recorded by the dipole sonic tool for known transverse anisotropy.

For a borehole with azimuth from north, α, and inclination from the vertical, I, the vector, B_n that defines the axis of the borehole from an arbitrary origin is given by:

$$
B_n = \left[\sin\alpha \sqrt{1 + \left(\sin\left(\frac{\pi}{2} - I\right)\right)^2} \right.
$$
$$
\left. \times \cos\alpha \sqrt{1 + \left(\sin\left(\frac{\pi}{2} - I\right)\right)^2} - \sin\left(\frac{\pi}{2} - I\right) \right]
\tag{8.9}
$$

where all angles are in radians. Given the dip, f_d, and dip direction, f_α, of the true fast plane we compute three discrete points, F_1, F_2 and F_3, in the fast plane that has a corner at the origin used to define the borehole. The normal to the fast plane, F_n, may now be computed using $A = F_1 - F_2$ and $B = F_2 - F_3$, thus giving $F_n = A \times B$. The vector defining the apparent fast direction, f^a, is then found by computing the vector that is both in the true fast plane and perpendicular to the borehole such that $f^a = B_n \times F_n$.

For the arbitrary case of a well with an azimuth of 45° (i.e. northeast) and an inclination of 45°, Figure 8.18 shows the apparent fast direction and dip that will be measured in the borehole for true fast directions dipping to the north, east, south and west (i.e. 0°, 90°, 180° and 270°) over a range of true fast dip angles from horizontal to vertical (i.e. 0° to 90°). Typically the azimuth of the fast direction is reported (as a direction between −90° west and 90° east) but the dip of the fast direction is omitted as only a vertical T.I. symmetry is considered. However, the dip of the apparent fast direction can easily be computed given the orientation of the borehole as the observed azimuth lies in a plane normal to the borehole. For completeness we present both the azimuth (as an angle between −180° and 180° in the direction of dip) and the dip of the apparent fast direction. The dip of the fast azimuth provides valuable information about the true orientation of the fast direction within the formation.

Figure 8.18 illustrates that the apparent fast direction strongly depends on the relative geometry of the borehole and true fast direction (shown here as a bedding plane). In this example with a northeast trending borehole, one can see that if the beds dip to the north the apparent fast direction will be southwest. However, if the beds dip to the east, the apparent fast direction is southeast. For this borehole trajectory, the dip of the true fast direction (or bedding planes) has the biggest effect on the apparent fast direction when the beds are dipping to the south and west, *i.e.* away from the direction of penetration.

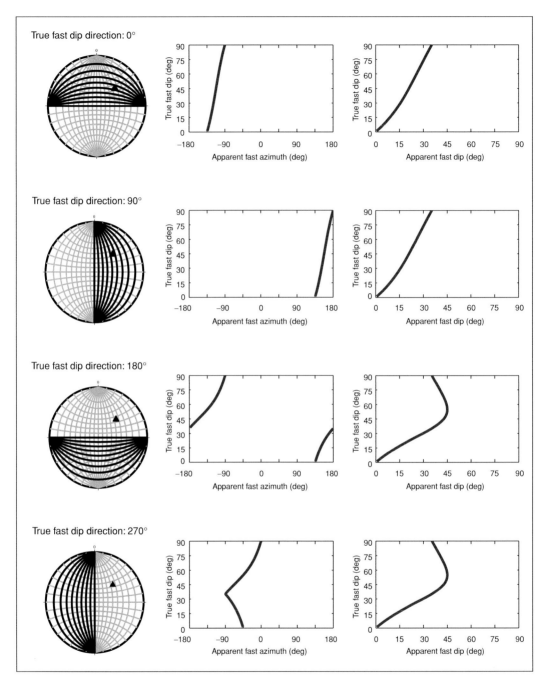

Figure 8.18. Model results for the arbitrary case of a borehole with an azimuth of 45° inclined at 45° (shown as a triangle on the stereonets), for four true fast dip directions of 0°, 90°, 180° and 270°, at a full range of dips from 0–90° (shown as great circles) (after Boness and Zoback 2006).

The true fast direction is most closely approximated by the apparent fast direction when the formation axis is close to being perpendicular to the borehole. This corresponds to the results of Sinha, Norris *et al.* (1994) showing the amount of anisotropy will also be at a maximum when the formation axis is normal to the borehole.

After applying quality control measures to the dipole sonic data collected in the SAFOD boreholes, Boness and Zoback (2006) computed the mean fast direction of the shear waves over 3 m intervals using Bingham statistics (Fisher, Lewis *et al.* 1987) because the fast directions are of unit amplitude (*i.e.* they are not vectors). The normalized eigenvalues give a measure of the relative concentration of orientations about the mean and we discard any mean fast direction over a 3 m interval with a normalized eigenvalue of less than 0.9. In the granite at depths shallower than 1920 m in both the pilot hole and main hole the faults and fractures observed on the image logs show no preferential orientation. However, there is an excellent correlation between fast directions in granite section and the direction of S_{Hmax} from a wellbore failure analysis in the pilot hole (Hickman and Zoback 2004) (Figure 8.19). The fact that the fast shear waves were found by Boness and Zoback (2006) to be polarized parallel to the stress in the shear zones encountered in the two boreholes indicates that this is not structural anisotropy but is instead directly related to perturbations in the stress state. Active fault zones are often frequently associated with a localized rotation of S_{Hmax} and a localized absence of breakouts (see Chapter 11).

Boness (2006) analyzed the electrical image log from 2000 m to 3000 m in discrete intervals of 10 m to compute the mean bed orientation using Fisher vector distribution statistics (Fisher, Lewis *et al.* 1987) to compute the mean bed orientations. We can then use the theoretical formulation presented above to compute the apparent fast direction for each discrete 10 m interval that would be observed in the SAFOD borehole if the shear waves were being polarized with a fast direction parallel to the bedding planes. Between 2000 m and 3000 m the borehole has an average azimuth and deviation from vertical of 35° and 54°. Within the massively bedded sandstones (2170 m to 2550 m), Figure 8.19 shows that the sonic log exhibits a northeast fast polarization direction consistent with observations in the granite at shallower depths, but that do not correlate with the theoretical fast directions if bedding planes were polarizing the shear waves (Figure 8.19). However, in the finely laminated, clay-rich shale and siltstone units below 2550 m the northwest fast direction of the sonic shear waves generally correlates well with the theoretical fast directions for structural anisotropy. We interpret the seismic anisotropy within these finely bedded stratigraphic layers to be controlled by the alignment of clay and mica platelets in the strike direction of the bedding planes. The electrical image log indicates that the bedding within most of the sandstone units is spaced at much larger intervals on the order of 0.5 m to 2 m. The spacing of these bedding planes is comparable to the 1.5 m wavelength of the sonic waves at the low frequencies of interest, which explains why we only observe structural anisotropy

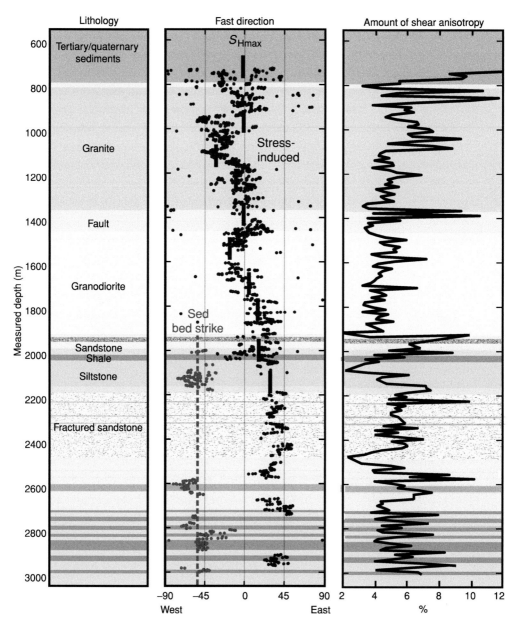

Figure 8.19. Observations of shear wave velocity anisotropy from the dipole sonic logs in the pilot hole and main hole. The direction of the sedimentary bedding planes is the mean strike determined in the electrical conductivity image and the black bars in the middle plot indicate the orientation of S_{Hmax} in the pilot hole (after Boness and Zoback 2006).

within the shale despite the sub-parallel bedding planes being present within all the sedimentary units.

In summary, in deviated wells (or vertical wells encountering bedding or aligned fractures at a high angle to the wellbore), the theory developed by Boness and Zoback (2006) allows one to separate stress-induced from structurally induced anisotropy (assuming the orientation of the structures are known) thus providing another technique for determining stress orientation.

9 Stress fields – from tectonic plates to reservoirs around the world

In this chapter, I discuss *in situ* stress fields at a wide variety of scales – from global patterns of tectonic stress (with a brief discussion of the sources of large-scale tectonic stress) to examples of normal, strike-slip and reverse faulting stress states in different sedimentary basins around the world. The purpose of this review is to (i) illustrate the robustness of the stress measurement techniques discussed in Chapters 6–8, (ii) emphasize the fact that sedimentary basins are, in fact, found in normal, strike-slip and reverse faulting environments (as discussed in Chapter 1) and (iii) demonstrate that critically stressed faults are found in many sedimentary basins such that stress magnitudes are often found to be consistent with those predicted on the basis of frictional faulting theory (as discussed near the end of Chapter 4).

In this chapter I also review empirical methods used for stress magnitude estimation at depth. Specifically, I provide an overview of some of the techniques being used for estimating the magnitude of the minimum principal stress in normal faulting environments (such as the Gulf of Mexico) for cases where direct measurements of the least principal stress from extended leak-off tests and mini-fracs are not available. I discuss in detail one particular model, the *bilateral constraint*, which has been widely used for stress estimation at depth using values of Poisson's ratio from geophysical logs. As discussed in this section, this model is not based on sound physical principles and leads to erroneous values of the horizontal principal stresses. Finally, because stress magnitude information is needed as a continuous function of depth to address problems such as wellbore stability during drilling (as discussed in Chapter 10), in the final section of this chapter, I discuss a method for interpolation and extrapolation of measurements of stress magnitude at selected depths based on the principal of constant effective stress ratios.

Global stress patterns

Knowledge of the magnitude and distribution of stress in the crust can be combined with mechanical, thermal and rheological constraints to examine a broad range of geologic processes. For example, such knowledge contributes to a better understanding of the processes that drive (or inhibit) lithospheric plate motions as well as the forces responsible for the occurrence of crustal earthquakes – both along plate boundaries and in intraplate regions. While such topics are clearly beyond the scope of this book, they are briefly addressed here to provide a broad-scale context for the discussions of stress at more regional, local, field and well scales that follow in Chapters 10–12.

Figure 9.1 is a global map of maximum horizontal compressive stress orientations based on the 2005 World Stress Map data base. As with Figure 1.5, only data qualities A and B are shown and the symbols are the same as those in Figure 1.5. While global coverage is quite variable (for the reasons noted above in North America), the relative uniformity of stress orientation and relative magnitudes in different parts of the world is striking and permits mapping of regionally coherent stress fields. In addition to the paucity of data in continental intraplate regions, there is also a near-complete absence of data in ocean basins.

Figure 9.2 presents a generalized version of the global stress map that is quite similar to the map presented by Zoback and others (1989) and showing mean stress directions and stress regime based on averages of the data shown in Figure 9.1. Tectonic stress regimes are indicated in Figure 9.2 by color and arrow type. Black inward pointing arrows indicate S_{Hmax} orientations in areas of compressional (reverse) stress regimes. Red outward pointing arrows give S_{hmin} orientations (extension direction) in areas of normal faulting stress regimes. Regions dominated by strike-slip tectonics are distinguished with thick inward-pointing and orthogonal, thin outward-pointing arrows.

A number of first-order patterns can be observed in Figures 9.1 and 9.2:

1. In many regions a uniform stress field exists throughout the upper brittle crust as indicated by consistent orientations from the different measurement techniques sampling very different rock volumes and depth ranges.
2. Intraplate regions are dominated by compression (reverse and strike-slip stress regimes) in which the maximum principal stress is horizontal. Such stress states are observed in continental regions throughout the world and likely exist in regions where data are absent. The intraplate compression seen in several ocean basins (the northeast Indian Ocean and just west of the East Pacific rise, for instance) are indicated by rare intraplate oceanic earthquakes.
3. Active extensional tectonism (normal faulting stress regimes) in which the maximum principal stress is the vertical stress generally occurs in topographically elevated areas of the continents. The areas of extensional stress near mid-ocean ridges in the Indian Ocean are likely the result of cooling stresses in the crust near, but not along, the spreading centers.
4. Regional consistency of both stress orientations and relative magnitudes permits the definition of broad-scale regional stress provinces, many of which coincide with physiographic provinces, particularly in tectonically active regions. These provinces may have lateral dimensions on the order of 10^3–10^4 km, many times the typical lithosphere thickness of 100–300 km. These broad regions of the earth's crust

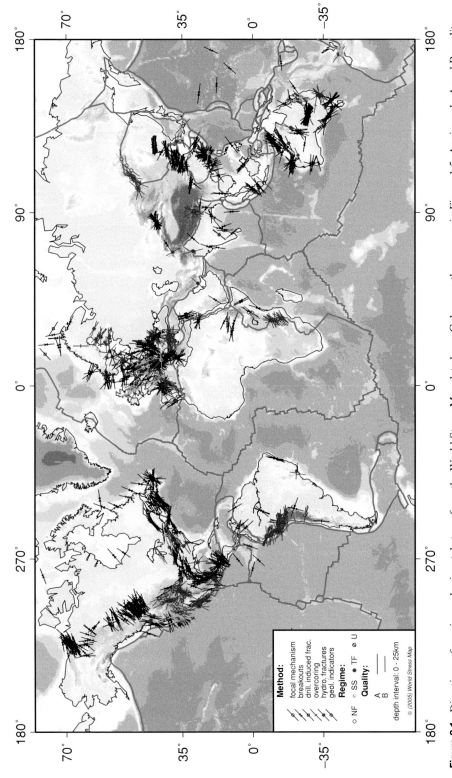

Figure 9.1. Directions of maximum horizontal stress from the World Stress Map data base. Colors are the same as in Figure 1.5. Again, only A and B quality data are shown. (http://www-wsm.physik.uni-karlsrohe.de/) (*For colour version see plate section.*)

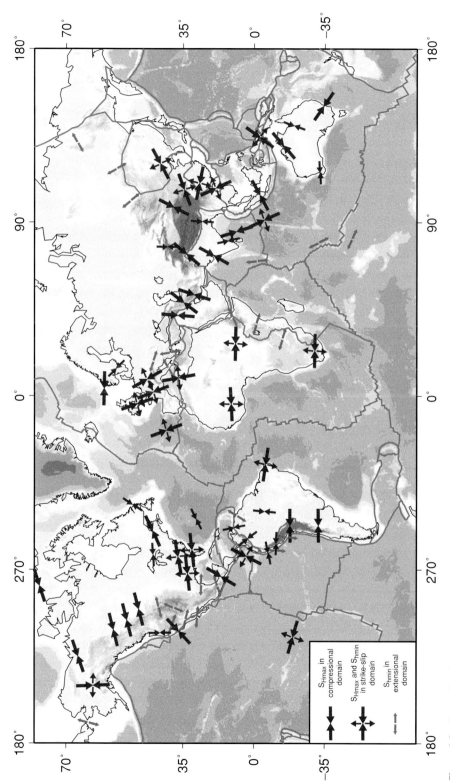

Figure 9.2. Generalized world stress map based on the data in Figure 9.1, similar to that of Zoback, Zoback *et al.* (1989). Inward pointing arrows indicate high compression as in reverse faulting regions. Paired inward and outward arrows indicate strike-slip faulting. Outward directed red arrows indicate areas of extension. Note that the plates are generally in compression and that areas of extension are limited to thermally uplifted areas. (*For colour version see plate section.*)

subjected to uniform stress orientation or a uniform pattern of stress orientations (such as the radial pattern of stress orientations in China) are referred to as *first-order* stress provinces (Zoback 1992).

Sources of crustal stress

As alluded to above, stresses in the earth's crust are of both tectonic and non-tectonic, or local, origin. The regional uniformity of the stress fields observed in Figures 1.5, 9.1 and 9.2 clearly demonstrate the tectonic origins of stress at depth for most intraplate regions around the world. For many years, numerous workers suggested that residual stresses from past tectonic events may play an important role in defining the tectonic stress field (*e.g.*, Engelder 1993). We have found no evidence for significant residual stresses at depth. If such stresses exist, they seem to be only important in the upper few meters or tens of meters of the crust where tectonic stresses are very small.

In the sections below the primary sources of tectonic stress are briefly discussed. Although it is possible to theoretically derive the significance of individual sources of stress in a given region, because the observed tectonic stress state at any point is the result of superposition of a variety of forces acting within the lithosphere, it is usually difficult to define the relative importance of any one stress source.

Plate driving stresses

The most fundamental sources of the broad-scale regions of uniform crustal stress are the forces that drive (and resist) plate motions (Forsyth and Uyeda 1975). Ultimately, these forces arise from lateral density contrasts in the lithosphere. Lithospheric plates are generally about 100 km thick, are composed of both the crust (typically about 40 km thick in continental areas) and the upper mantle, and are characterized by conductive heat flow. They are underlain by the much less viscous asthensophere.

The most important plate-driving processes resulting in intraplate stress is the ridge push compressional force associated with the excess elevation (and hot, buoyant lithosphere) of mid-ocean ridges. Slab pull (a force resulting from the negative buoyancy of down-going slabs) does not seem to be transmitted into plates as these forces appear to be balanced at relatively shallow depths in subduction zones. Both of these sources contribute to plate motion and tend to act in the direction of plate motion. If there is flow in the upper asthenosphere, a positive drag force could be exerted on the lithosphere that would tend to drive plate motion, whereas if a cold thick lithospheric roots (such as beneath cratons) this may be subject to resistive drag forces that would act to inhibit plate motion. In either case the drag force would result in stresses being transferred upward into the lithosphere from its base. There are also collisional resistive forces resulting either from the frictional resistance of a plate to subduction or from the collision of two continental plates. As oceanic plates subduct into the viscous lower mantle additional resistive forces add to the collision resistance forces acting at shallow depth. Another force resisting plate motion is that due to transform faults, although, as discussed below, the amount of transform resistance may be negligible.

While it is possible to specify the various stresses associated with plate movement, their relative and absolute importance in plate movement are not understood. Many researchers believe that either the ridge push or slab pull force is most important in causing plate motion, but it is not clear that these forces are easily separable or that plate motion can be ascribed to a single dominating force. This has been addressed by a detailed series of finite element models of the stresses in the North American plate (Richardson 1992).

Topography and buoyancy forces

Numerous workers have demonstrated that topography and its compensation at depth can generate sizable stresses capable of influencing the tectonic stress state and style (Artyushkov 1973). Density anomalies within or just beneath the lithosphere constitute major sources of stress. The integral of anomalous density times depth (the density moment of Fleitout and Froidevaux 1983) characterizes the ability of density anomalies to influence the stress field and to induce deformation. In general, crustal thickening and lithospheric thinning (negative density anomalies) produce extensional stresses, while crustal thinning and lithospheric thickening (positive density anomalies) produce compressional stresses. In more complex cases, the resultant state of stress in a region depends on the density moment integrated over the entire lithosphere. In a collisional orogeny, for example, where both the crust and mantle lid are thickened, the presence of the cold lithospheric root can overcome the extensional forces related to crustal thickening and maintain compression (Fleitout and Froidevaux 1983). Zoback and Mooney (2003) showed that regional intraplate relative stress magnitudes are generally predictable from buoyancy forces derived from lateral variations in the density and structure of the lithosphere.

Lithospheric flexure

Loads on or within an elastic lithosphere cause deflection and induce flexural stresses which can be quite large (several hundred MPa) and can perturb the regional stress field with wavelengths as much as 1000 km (depending on the lateral extent of the load (e.g. McNutt and Menard 1982). Some potential sources of flexural stress influencing the regional stress field include sediment loading along continental margins and the upwarping of oceanic lithosphere oceanward of the trench, the "outer arc bulge" (Chapple and Forsythe 1979). Sediment loads as thick as 10 km represent a potentially significant stress on continental lithosphere (e.g. Cloetingh and Wortel 1986; Turcotte

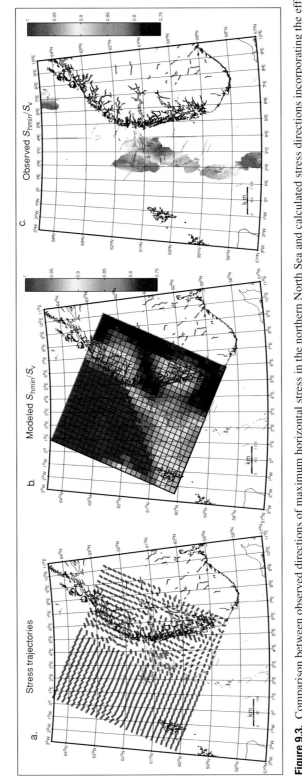

Figure 9.3. Comparison between observed directions of maximum horizontal stress in the northern North Sea and calculated stress directions incorporating the effects of lithospheric flexure due to removal of the Fennoscandian ice sheet (Grollimund and Zoback 2003). (a) The model accurately reproduces stress orientations in the northern North Sea. (b) The calculated stress magnitudes are shown as the ratio of the minimum horizontal stress to the overburden. (c) The magnitude of the minimum horizontal stress as determined from leak-off tests in wells is quite similar to that predicted by the model. *AAPG© 2003 reprinted by permission of the AAPG whose permission is required for futher use. (For colour version see plate section.)*

and Schubert 2002). Zoback (1992) suggested that a roughly 40° counter-clockwise rotation of horizontal stresses on the continental shelf offshore of eastern Canada was due to superposition of a margin-normal extensional stress derived from sediment load induced flexure.

Another illustration of the influence of lithospheric flexure on crustal stress is that associated with post-glacial rebound in offshore areas at relatively high latitudes. Grollimund and Zoback (2003) modeled the stress state in this portion of the North Sea region to assess the affect of deglaciation on regional stresses. Figure 9.3 shows the result of three-dimensional modeling of the stress field in this area. Figure 9.3a shows that stresses induced by lithospheric flexure do a very good job of explaining both the average E–W stress orientation observed in the northern North Sea as well as the subtle *swing* of maximum horizontal stress orientations from WNW–ESE on the west side of the Viking graben to ENE–WSW on the east side of the graben. We have also considered data on the magnitude of the least horizontal principal stress obtained from a study of approximately 400 wells offshore Norway. Note that the modeled ratio of the magnitude of minimum horizontal compression to the vertical stress (Figure 9.3b) compares favorably with measured values of the least horizontal stress magnitudes observed in the northern North Sea (Figure 9.3c). Thus, regional variations of the magnitude of the least principal stress also appear to support the hypothesis that the stress field offshore Norway has been strongly affected by deglaciation.

In the sections below, I review stress states in some sedimentary basins around the world to characterize stress magnitudes at depth in normal, strike-slip and reverse faulting regions. As discussed in Chapter 1, Anderson's faulting scheme defines the relative magnitudes of principal stress. In Chapter 4, we saw that one can predict stress magnitudes at depth through utilization of simplified two-dimensional Mohr–Coulomb failure theory and the concept of effective stress. As reviewed by McGarr and Gay (1978), Brace and Kohlstedt (1980), Zoback and Healy (1984), Brudy, Zoback *et al.* (1997) and Townend and Zoback (2000) numerous *in situ* stress measurements in areas of active faulting have proven to be consistent with Coulomb faulting theory assuming coefficients of friction in the range consistent with laboratory-determined values of 0.6–1.0 (Byerlee 1978). As discussed at the end of Chapter 4, this implies that the state of stress in the areas in which the measurements were made is controlled by the frictional strength of pre-existing faults. In other words, the state of stress in these areas is in frictional failure equilibrium – there are pre-existing faults just at the point of frictional sliding that control *in situ* stress states.

Normal faulting stress fields in sedimentary basins

Normal faulting stress states are observed in many parts of the world including the Gulf of Mexico region of the United States (both onshore and offshore) and the central

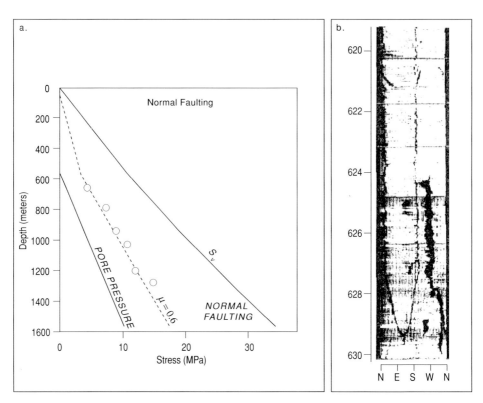

Figure 9.4. (a) Least principal stress measurements in the Yucca Mountain area of the Nevada Test Site (hole USW-G1) indicate a normal faulting stress regime with least principal stress directions consistent with frictional faulting theory for a coefficient of friction of ~0.6 (after Zoback and Healy 1984). Note the extremely low water table. (b) Drilling-induced hydraulic fractures imaged with a borehole televiewer explain the total loss of circulation during drilling (after Stock, Healy *et al.* 1985).

graben of the North Sea. One of the first places where frictional faulting theory was demonstrated to be clearly applicable to faulted crust *in situ* was the Yucca Mountain area of the Nevada Test Site. This site is located in the Basin and Range province of the western U.S., a region of high heat flow and active extensional tectonics. As illustrated in Figure 9.4a, the magnitudes of the least principal stress, S_{hmin}, obtained at various depths from mini-frac tests are consistent with the magnitudes predicted using Coulomb faulting theory for a coefficient of friction of ~0.6 (Zoback and Healy 1984). In other words, at the depths at which the measurements were made (~600–1300 m), the measured magnitude of S_{hmin} was exactly that predicted by equation (4.45) for a coefficient of friction of 0.6 and appropriate values of S_v and P_p (dashed line in Figure 9.4a). In addition, the direction of least principal stress in the region (WNW–ESE) is essentially perpendicular to the strike of the normal faults in the region (Stock, Healy *et al*. 1985), as expected for a normal faulting regime (Figure 5.1b).

When pore pressure is sub-hydrostatic, normal faulting occurs at a value of least principal stress that is lower than would be found at higher pore pressures. Because pore pressure is significantly below hydrostatic in this part of the Nevada Test Site, the process of drilling the well (and filling the borehole with drilling fluid) caused drilling-induced hydraulic fracturing and lost circulation to occur. Following drilling, the drilling-induced hydraulic fractures were observed on borehole televiewer logs (Stock, Healy *et al*. 1985) as shown in Figure 9.4b. Because this well was continuously cored, it is known that the fractures shown were not present in the formation prior to drilling. Apparently, all of the drilling fluid (and cuttings) went out into hydraulic fractures (such as the one shown in Figure 9.4b) as the hole was being cored. The explanation of this phenomenon is quite straightforward. Thus, raising the fluid level in the hole during drilling causes the pressure at ∼600 m depth to exceed the least principal stress and induces hydraulic fracturing. Hydraulic fracture propagation occurs when the fluid height in the borehole is ∼200 m below ground level. Stock, Healy *et al*. (1985) showed that the same phenomenon can be seen in three different holes drilled in this area. Interestingly, these wells could be successfully cored with 100% lost circulation.

In east Texas, another area characterized by active normal faulting, measurements of the least principal stress made in various lithologies of the Travis Peak formation are also consistent with frictional faulting theory. The abscissa of Figure 9.5 is the measured least principal stress value in different lithologies in the study area whereas the ordinate is that predicted by equation (4.45) for the appropriate depth and pore pressure (and coefficients of friction of ∼0.6). Note that regardless of whether the formation is sandstone, shale, siltstone or limestone, frictional faulting theory incorporating a coefficient of friction of 0.6 accurately predicts the measured stress values over a significant range of stresses.

Normal faulting is also seen in the central graben area of the North Sea. Figure 9.6 shows least principal stress values at the crest of the Valhall anticlinal structure (modified from Zoback and Zinke 2002). In this figure, the measured value of the least principal stress is shown at various pore pressures as depletion occurred over time (the approximate date of the measurements is also shown). Note that the measured values of the least principal stress in this weak chalk reservoir is predicted well by frictional faulting theory with a coefficient of friction of 0.6 (the solid line passing through the data) obtained from equation (4.45). The importance of normal faulting for formation permeability is discussed in Chapter 11. The evolution of the state of stress with depletion in the Valhall field (and faulting on the flanks of the reservoir induced by depletion) will be revisited in Chapter 12. However, it is important to note that for all the rock types in the case studies considered so far, the fact that a coefficient of friction of ∼0.6 is applicable is still further support for the applicability of this friction coefficient to

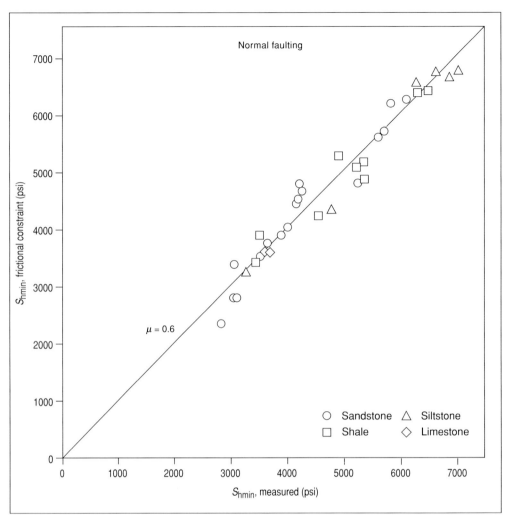

Figure 9.5. Comparison of calculated least principal stress magnitudes (ordinate) using Coulomb faulting theory (normal faulting) and a coefficient of friction of 0.6 with measured values (abscissa) in all lithologies of the Travis Peak formation in east Texas (see text).

faults *in situ*. Had lower coefficients of friction been applicable, the magnitude of the least principal stress could not have been as low as the values measured.

It is well known that active normal faulting is pervasive throughout the offshore Gulf of Mexico area. Because it is frequently important to be able to predict the magnitude of the least principal stress in advance of drilling, a number of researchers have proposed empirical methods to predict stress magnitudes from logs. These techniques are discussed later in this chapter. Figure 9.7 allows us to examine how well Coulomb faulting theory works in the South Eugene Island, Block 330 area. In this field, oil and gas are produced from a number of extremely young (Plio-Pleistocene)

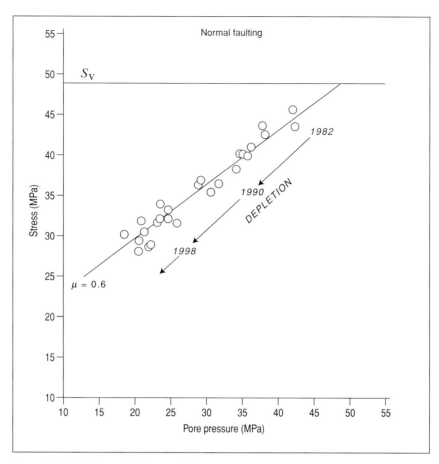

Figure 9.6. Evolution of the least principal stress with depletion at the crest of the anticlinal structure in the Valhall field of the North Sea (after Zoback and Zinke 2002). Note that the normal faulting theory explains the measured least principal stress magnitudes (for a coefficient of friction of 0.6) for the weak chalks in this field before, during and after depletion.

sand reservoirs separated by thick accumulations of shale (Alexander and Flemings 1995). Finkbeiner (1998) determined the magnitude of the least principal stress from frac pack completions – hydraulic fractures made in reservoir sands for the purpose of minimizing sand production. What makes these data important is that most published least principal stress data for the Gulf of Mexico were compiled from leak-off tests made at various depths in the region. These tests are usually made in shales (where the casing is set). In many published studies, these data have been interpreted with pore pressure measurements that were made in sands encountered at various depths in the region. Hence, many published empirical studies juxtapose stress and pore pressure measurements made in different wells and different lithologies. The advantage of using frac pack data is that at any given depth, the values of both the least principal stress

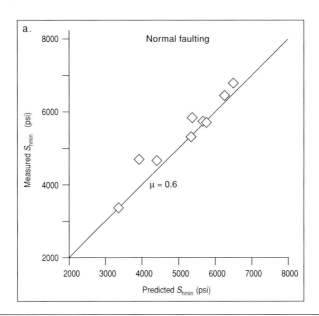

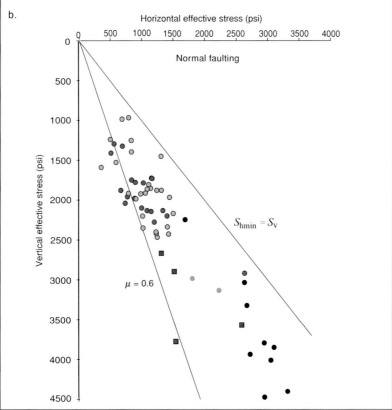

Figure 9.7. (a) Measured values of the least principal stress in the South Eugene Island Field of the Gulf of Mexico region compared with the predicted values based on Coulomb faulting theory using a coefficient of friction of 0.6 (straight line). The predicted stress magnitudes agree with some of the measured values but generally provide a lower bound. (b) Measured least principal effective stress magnitudes as a function of the vertical effective stress for several deep-water fields in the Gulf of Mexico prior to depletion. Again, predicted stress magnitudes using Coulomb faulting theory with a coefficient of friction of 0.6 generally provide a lower bound for the measured values.

and pore pressure are obtained in the same well, the same formation and at the same depth.

Figure 9.7a is similar to Figure 9.5 as it compares predicted (using equation 4.45) and measured (from frac pack completions) values of the least principal stress in producing formations at a variety of depths. All of the data shown are from formations not affected by depletion or stress changes associated with poroelastic effects (as discussed in Chapter 12). The figure illustrates frictional faulting theory (diagonal line) and provides a reasonable fit to some of the data, and a good lower bound for all of the data. Figure 9.7b shows measured least principal effective stress data as a function of the vertical effective stress for five deep-water fields in the Gulf of Mexico. As for the South Eugene Island field, the data come from undepleted sands from frac pack completions. The magnitude of the least principal effective stress using frictional faulting theory (equation 4.45) with a coefficient of friction of 0.6 yields the line shown. As in SEI, frictional faulting theory provides a lower bound for the measured values but there is considerable variability in the magnitude of the effective least principal stress as a function of the vertical effective stress. If one were using equation (4.45) to estimate the maximum mud weight, frictional faulting theory would yield a conservative estimate of the least principal stress. In other words, one could drill with a mud weight corresponding to this pressure without fear of hydraulic fracturing and loss of circulation. However, in many cases this may yield a mud weight that is too low to achieve the desired degree of wellbore stability. This is discussed at greater length in Chapters 8 and 10.

The general observation, that the least principal stress measurements in the young, relatively uncemented sands of offshore Gulf of Mexico, are frequently higher than predicted by Coulomb faulting theory using $\mu = 0.6$, was noted by Zoback and Healy (1984). There are several possible explanations of this. One possibility is that a coefficient of friction of 0.6 is too high for the faults in the region as the lithology is dominated by shale (e.g. Figure 2.6a). As mentioned in Chapter 6, Ewy, Stankowich et al. (2003) present laboratory data indicating that the effective coefficient of friction of shaley rocks is 0.2–0.3. While the measurements shown in Figure 9.7 were made in relatively clean sands, if the state of stress at depth was dominated by the frictional strength of the adjacent shales (because they compose most of the lithologic section) a lower coefficient of friction might be applicable.

Perhaps a more likely explanation of the relatively high values of the least principal stress observed in Figure 9.7 is related to the fact that the majority of these sands are essentially uncemented. Thus, if these sands exhibit the type of creep discussed in Chapter 4 for uncemented sands (Figure 3.11a, Table 3.2), it would be logical for creep to reduce the difference between the vertical stress and least principal stress over time. One reason this explanation is appealing with respect to the data shown in Figure 9.7 is that one would expect stress magnitudes to be consistent with Coulomb faulting theory in slightly cemented sands (where creep does not occur), but higher (even approaching the vertical stress) due to stress relaxation in uncemented sands.

Methods for approximating S_{hmin} in normal faulting areas

In Chapter 7 we discussed how mini-frac and leak-off tests can be used to accurately determine the magnitude of the least principal stress, S_{hmin}. Knowledge of the least principal stress is not only a critical step in determining the full stress tensor (as discussed at length in Chapters 7 and 8), it also provides important information for drilling stable wells. During drilling, mud weights must be kept below S_{hmin} to prevent accidental hydraulic fracturing and lost circulation, but above both the pore pressure (to avoid *taking a kick*) and the minimum mud weight required to prevent excessive wellbore failure (*i.e.* the *collapse* pressure) as discussed in Chapter 10. Because of this, a number of empirical techniques have been proposed for estimating the least principal stress in the absence of direct measurements. This issue is particularly important in normal faulting areas (such as the Gulf of Mexico) where overpressure is present at depth. As illustrated in Figure 1.4d, in overpressured normal faulting regions, there can be extremely small differences between P_p and S_{hmin}, which define the *mud window*, or the safe range of pressures to use while drilling.

In the sections below, the techniques summarized in Table 9.1 for estimating the least principal stress (or least principal effective stress) in the Gulf of Mexico are briefly discussed.

In their classic paper on hydraulic fracturing, Hubbert and Willis (1957) proposed an empirical expression for the magnitude of the least principal stress as a function of depth in the Gulf of Mexico region,

$$S_{hmin} = 0.3(S_v - P_p) + P_p \tag{9.1}$$

where the constant 0.3 was empirically determined from the analysis of hydraulic fracturing data. The scientific basis for this constant can be understood in terms of frictional faulting theory (Zoback and Healy 1984), as equation (4.45) produces essentially the same equation for a coefficient of friction of 0.6. However, as additional data became available for the offshore Gulf of Mexico area, Hubbert and Willis later adopted an empirical coefficient of 0.5 indicating that observed values for the least principal stress in the Gulf of Mexico generally exceed the values predicted using equation (4.45) with a coefficient of friction of 0.6.

Matthews and Kelly (1967) proposed a similar relation for the *fracture pressure*, or the magnitude of the pore pressure at which circulation is lost. As this requires propagation of a hydraulic fracture away from the wellbore, this value is essentially equivalent to the least principal stress. Thus, they proposed

$$S_{hmin} = K_i(S_v - P_p) + P_p \tag{9.2}$$

where K_i is a function of depth, z. Using this relation, functions for the Louisiana gulf coast and south Texas gulf coast region were proposed that varied in a non-linear fashion

Table 9.1. *Empirical methods for estimation of minimum stress in the gulf of mexico*

Method	Proposed equation	Effective stress ratio	Comments
Hubbert and Willis (1957)	$S_{hmin} = 0.3\,(S_v - P_p) + P_p$	$\dfrac{\sigma_{hmin}}{\sigma_v} = 0.3$	After first proposing this relation (which often underpredicted measured values) they modified the empirical constant to 0.5
Mathews and Kelly (1967)	$S_{hmin} = K_i(z)\,(S_v - P_p) + P_p$	$\dfrac{\sigma_{hmin}}{\sigma_v} = K_i(z)$	Requires an estimate of pore pressure at depth as well as empirically determined functions for $K_i(z)$.
Eaton (1969)	$S_{hmin} = \left(\dfrac{\nu}{1-\nu}\right)(S_v - P_p) + P_p$	$\dfrac{\sigma_{hmin}}{\sigma_v} = \left(\dfrac{\nu}{1-\nu}\right)$	While this equation is the same as the bilateral constraint discussed in the text, Eaton replaced Poisson's ratio, ν, with an empirical value that is a function of depth. The values used increase from ~ 0.25 at shallow depth (~ 1000 ft) to unreasonably high values of ~ 0.45 at depths of 10,000 ft and more.
Breckels and van Eekelen (1981)	$S_{hmin} = 0.197 z^{1.145} + 0.46(P_p - P_h)$ For $z < 11{,}500$ ft $S_{hmin} = 1.167z - 4596 + 0.46(P_p - P_h)$ For $z > 11{,}500$ ft		P_h is hydrostatic pore pressure at the depth, z, of interest. This expression is for pressure in psi.
Zoback and Healy (1984)	$\dfrac{S_{hmin} - P_p}{S_v - P_p}$ $= \left[(1+\mu^2)^{1/2} + \mu\right]^{-2}$	$\dfrac{\sigma_{hmin}}{\sigma_v}$ $= \left[(1+\mu^2)^{1/2} + \mu\right]^{-2}$	Based on frictional equilibrium. For $\mu = 0.6$, the effective stress ratio is 0.32.
Holbrook (1990)	$S_{hmin} = (1-\phi)(S_v - P_p) + P_p$	$\dfrac{\sigma_{hmin}}{\sigma_v} = 1 - \phi$	Replaces empirical constant, with function of porosity ϕ. Note that for reasonable porosities of 35%, it would yield a constant of 0.65.

from 0.4 and 0.48 at 2000 ft to values exceeding 0.7 at depths greater than 10,000 ft (see also Mouchet and Mitchell 1989).

Eaton (1969) suggested a physically based technique for determination of the least principal stress based on Poisson's ratio, ν.

$$S_{\text{hmin}} = \left(\frac{\nu}{1 - \nu}\right)\left(S_{\text{v}} - P_{\text{p}}\right) + P_{\text{p}} \tag{9.3}$$

This relation is derived from a problem in linear elasticity known as the *bilateral constraint*, which is discussed in more detail below. Despite the widespread use of this relation, even the author recognized that it was necessary to use an empirically determined effective Poisson's ratio which had to be obtained from calibration against least principal stress measurements obtained from leak-off tests. To fit available LOT data in the Gulf Coast, the effective Poisson's ratio must increase from 0.25 at $\sim$1000 ft to unreasonably high values approaching 0.5 at 20,000 ft. In other words, it was necessary to replace the Poisson's ratio term in equation (9.3) with a depth-varying empirical constant similar to equation (9.2). It is noteworthy that in west Texas, where pore pressures are essentially hydrostatic, Eaton argued that a constant Poisson's ratio of 0.25 works well. This is equivalent to the term $\left(\dfrac{\nu}{1 - \nu}\right)$ being equal to 0.33, a value quite similar to that derived from equation (4.45) for a coefficient of friction of 0.6.

Equation (9.3) is based on solving a problem in elasticity known as the *bilateral constraint* which has been referred to previously as a common method used to estimate the magnitude of the least principal stress from logs. Fundamentally, the method is derived assuming that the only source of horizontal stress is the overburden. If one applies an instantaneous overburden stress to a poroelastic half-space, rock will experience an equal increase in horizontal stress in all directions, S_{h}, as defined by equation (9.3), noting, of course, that ν is rigorously defined as Poisson's ratio, and not an empirical coefficient. The reason horizontal stress increases as the vertical stress is applied is that as a unit volume wants to expand laterally (the Poisson effect), the adjacent material also wants to expand, such that there is no lateral strain. Hence, the increase in horizontal stress results from the increase in vertical stress with no lateral strain.

An example of the bilateral constraint being used to estimate stress magnitude in the Travis Peak formation of east Texas is illustrated in Figure 9.8 (after Whitehead, Hunt *et al.* 1986). In this case, the predicted values of the least principal stress can be compared directly with a series of values determined from mini-fracs. One can see that that the three mini-fracs at depths of $\sim$9200–9350 ft show relatively low values of the least principal stress whereas the three at $\sim$9550–9620 ft show higher values. Both sets of measurements seem to fit well from the *stress log* (the right-hand side of Figure 9.8a), calculated using the bilateral constraint. If hydraulic fracturing were to be planned for the sand at $\sim$9400 ft that is surrounded by shales (see the gamma log in Figure 9.16a), it is obviously quite helpful to know that the magnitude of the least principal stress is

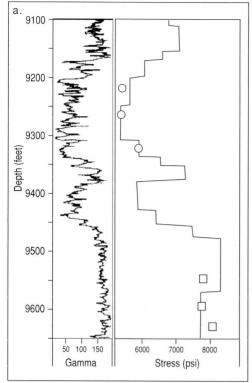
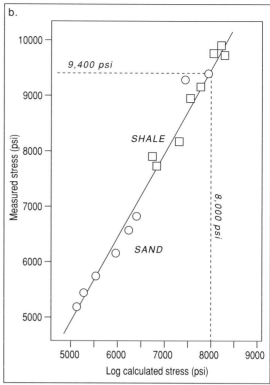

Figure 9.8. Comparison of log-derived least principal stress measurements using equation (9.3) with measured values (data after Whitehead, Hunt *et al*. 1986). (a) The gamma log helps distinguish the sands from shales. (b) Comparison of predicted and measured values of least principal stress. While both panels suggest that predictions of the least principal stress based on equation (9.3) match the observed data quite well, it was necessary to use an empirically determined Poisson's ratio and a *tectonic stress* that varies with depth in order for the predicted values to match the measured ones.

about 6000 psi in the sand at ~9400 ft (where the frac would be made) and the least principal stress in the adjacent shales (≥7000 psi) to know the pressure at which the hydrofrac might propagate vertically through the bounding shales.

The apparently good correlation between measured and predicted values of the least principal stress is even more dramatic in Figure 9.8b as one sees a linear relationship between the two sets of values in both the sands and shales. While this would make it appear that the bilateral constraint does a very good job of predicting the magnitude of the least principal stress in this case, it turns out that an empirically determined *effective* Poisson's ratio was used to match the log-determined values to the measurements. Moreover, while the measured and predicted values are linearly related, they are not equal at all stress values. Using equation (9.3) with an empirically correlated value of Poisson's ratio matches the measured values at low stress (~5500 psi) but underpredicts the measured stress values by 1400 psi at higher stress levels. Hence, the line fitting the data in Figure 9.8b also incorporates a depth-varying tectonic stress.

While using empirically determined Poisson's ratios of tectonic stress to characterize stress magnitudes in a given region may have some local usefulness, such methods clearly have little or no predictive value (see also Warpinski, Peterson *et al.* 1998). In fact, the data shown in Figure 9.8 are part of the same data set presented in Figure 9.5 for the Travis Peak formation where it was seen that frictional faulting theory (equation 4.45) does an excellent job of predicting the magnitude of the least principal stress – without empirically determining an effective Poisson's ratio and tectionic stress.

With respect to the Gulf of Mexico, to fit the deep water stress magnitude data presented in Figure 9.7b with a $\left(\dfrac{\nu}{1 - \nu} \right)$ model, ν would need to range from 0.2 (to fit the lowest values of the least principal stress) to 0.47 (to fit the highest). As was the case just described for the Travis Peak, however, these values do not match values measured with sonic logs, but represent *effective* values of ν, known only after the measurements were made.

Breckels and Van Eekelen (1981) proposed a number of empirical relations between the magnitude of the least principal stress and depth (in units of psi and feet) for various regions around the world. For the Gulf of Mexico region, they argue that if pore pressure is hydrostatic

$$S_{hmin} = 0.197z^{1.145} \tag{9.4}$$

fits the available data and for depths $z < 11{,}500$ ft and for $z > 11{,}500$ ft, they argue for the following

$$S_{hmin} = 0.167z - 4596 \tag{9.5}$$

Implicitly, this assumes a specific increase in the rate of pore pressure change with depth as well as an average overburden density. To include overpressure, they add a term equal to $0.46\,(P_p - P_h)$ to each equation, where P_h is hydrostatic pore pressure.

Zoback and Healy (1984) analyzed *in situ* stress and fluid pressure data from the Gulf Coast in an attempt to show that, as illustrated above for the Nevada test site, Travis Peak formation in east Texas and the Ekofisk and Valhall crests, the state of stress in the Gulf of Mexico is also controlled by the frictional strength of the ubiquitous active normal faults in the region. We rewrite equation (4.45) to compare it with equations (9.1) and (9.2):

$$S_{hmin} = 0.32(S_v - P_p) + P_p \tag{9.6}$$

Compilations of data on the magnitude of the least principal stress in the Gulf of Mexico show that at depths less than about 1.5 km (where pore pressure is hydro-static), S_{hmin} is about 60% of the vertical stress (as previously discussed in context of equation 4.45). In other words, equation (9.6) (or 4.45) is essentially identical to the original data compiled by Hubbert and Willis and equation (9.1) because the original data available to them were from relatively shallow depths where pore pressures are hydrostatic. The same thing is true for the data cited in west Texas by Eaton (1969). At greater depth, however, where overpressure is observed in the Gulf of Mexico, frictional faulting theory tends to underpredict measured values of the least principal stress, as mentioned previously. This can be seen by comparing equation (9.6) with the K_i values used in equation (9.2) to fit data from great depth. As K_i values get as high as 0.8–0.9 at great depth, it is clear that the theoretical value of the least principal stress predicted by equation (9.6) is less than the data indicate.

As previously mentioned, one possible explanation for this is that if the coefficient of friction of faults in smectite-rich shales is lower than 0.6, higher stress values would be predicted using equation (9.6) and hence better fit the observed data at depth. For example, a coefficient of friction as low as 0.2 results in an empirical coefficient in equation (9.6) of 0.67, closer to the values for K_i that should be used at depth as argued by Matthews and Kelly (1967).

Finally, Holbrook, Maggiori *et al.* (1993) proposed a porosity based technique for estimation of the least principal stress based on a force-balance concept:

$$S_{hmin} = (1 - \phi)(S_v - P_p) + P_p \qquad (9.7)$$

As porosity of overpressured shales is typically ~35%, it yields similar values to that predicted with $K_i \sim 0.65$ in the Matthews and Kelly (1967) relation for overpressured shales at depth, but would seriously overestimate the least principal stress in the cases presented in Figures 9.4–9.7.

Figure 9.9 presents calculated values (in units of equivalent mud weight in ppg) for the magnitude of the least principal stress for an offshore well in the Gulf of Mexico using the the formulae presented in Table 9.1 and discussed above. Input data include the vertical stress (calculated from integration of the density log), pore pressure and Poisson's ratio (determined from P- and S-wave sonic velocity measurements). Curve **a** illustrates the technique of Zoback and Healy (1984), curve **b** that of Breckels and Van Eekelen (1981), curve **c** that of Hubbert and Willis (1957) using the modified empirical coefficient of 0.5, curve **d** is that of Holbrook, Maggiori *et al.* (1993) and curve **e** is that of Eaton (1969). At a depth of 4000 feet, where pore pressure is hydrostatic, there is a marked variation between the predictions of the various techniques with the method of Zoback and Healy (1984) (or that of Hubbert and Willis 1957 with an empirical constant of 0.3) which yields the lowest values of S_{hmin} (slightly in excess of 11 ppg). Recall from the discussion above that where pore pressures are hydrostatic, the lower estimates of the least principal stress seem to be more representative of measured values. Also, as illustrated above in Figures 9.4–9.6, this technique seems to predict the least principal stress well in cases of cemented rocks in normal faulting environments. Note also that at this depth, the technique of Eaton (1969) predicts much higher values (about 14.5 ppg). However, where pore pressure is elevated, these techniques consistently underestimate

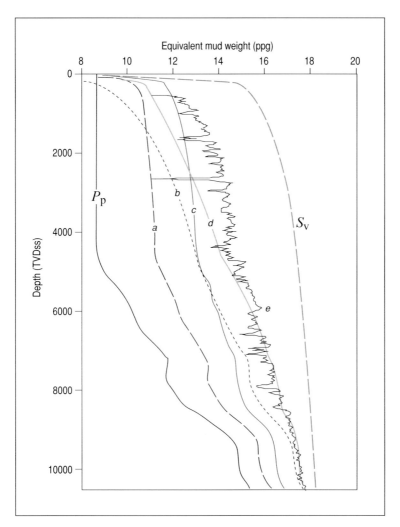

Figure 9.9. Different types of empirical techniques for determination of S_{hmin} from pore pressure and vertical stress data sometimes yield significantly different predictions, as shown for this well in the Gulf of Mexico. Curve **a** utilized equation (9.6) based on the technique proposed by Zoback and Healy (1984), curve **b** is based on equations (9.4) and (9.5) of Breckels and Van Eekelen (1981), curve **c** is from equation (9.1) after Hubbert and Willis (1957) utilizing the modified empirical coefficient of 0.5, curve **d** is based on equation (9.7) after Holbrook, Maggiori *et al.* (1993) and curve **e** is after equation (9.3) of Eaton (1969).

the magnitude of the least principal stress. As mentioned above, in overpressured shale-rich rocks in the Gulf of Mexico, the Matthews and Kelly (1967) technique with a constant value K_i equal to 0.6 seems to work reasonably well.

To summarize, it is perhaps appropriate to consider the empirical techniques presented in Table 9.1 and illustrated in Figure 9.9 in a somewhat analogous manner as the empirical techniques presented in Chapter 4 for estimating rock strength from logs. The practical importance of these techniques is clear as well as their local usefulness, when appropriatedly calibrated. However, it needs to be remembered that the use of such empirical techniques in areas where they have not yet been calibrated has appreciable uncertainty. The data presented in Figure 9.7b show this clearly as there is considerable variance in the magnitude of least principal stress in the deep water Gulf of Mexico for a given value of the vertical effective stress.

Compressional stress states in sedimentary basins

The stress magnitude data and stress estimation techniques discussed in the previous two sections of this chapter focused on normal faulting enviroments where both S_{hmin} and S_{Hmax} are less than the vertical stress, S_v. In this section, I present stress magnitudes at depth from more compressional environments, in order of increasing stress magnitudes at depth.

Normal/strike-slip

These areas are those where S_{hmin} is less than the vertical stress but $S_{Hmax} \approx S_v$. If the state of stress is in frictional equilibrium (as predicted by Coulomb faulting theory using $\mu = 0.6$), equation (4.45) will accurately describe the state of stress. In other words, S_{hmin} will be significantly below S_v, at the value predicted by equation (4.45) and $S_{Hmax} \approx S_v$. One such area is shown in Figure 9.10a, for a field in southeast Asia. The way to think about this type of *transitional* stress state is that $S_{max} \approx S_v \equiv S_1$ and $S_{hmin} \equiv S_3$. In this case, both equations (4.45) and (4.46) could be satisfied and both normal and strike-slip faults could be critically stressed, assuming, of course, that they have the appropriate orientation to the principal stress axes (as discussed in Chapter 4). This type of stress state is found in many parts of the world, including much of western Europe (Zoback 1992).

Strike-slip faulting

As discussed previously in Chapters 1 and 4, these stress states are those in which the vertical stress is the intermediate stress. If the state of stress is in frictional equilibrium, the difference between the horizontal stresses are described by equation (4.46), and S_{Hmax} will be appreciably greater than S_{hmin}. An example is shown in Figure 9.10b, from an area of the Timor Sea (Castillo, Bishop *et al.* 2000). Note that in this area, the values of both S_{hmin} and S_{Hmax} are elevated with respect to the area represented by the data in Figure 9.10a. At depths between 1.6 and 4.2 km, the magnitude of the maximum horizontal stress is exactly what would be predicted from equation (4.46)

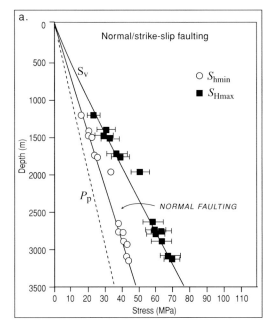
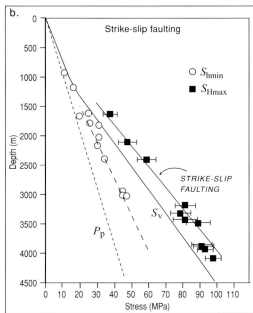

Figure 9.10. (a) Stress measurements in a well in southeast Asia are characterized by a normal/strike-slip stress state. Coulomb faulting theory with a coefficient of friction of 0.6 fits the measured values of S_{hmin} quite well for normal faulting as in the cases shown above. However, because $S_{Hmax} \sim S_v$, the difference between S_{hmin} and S_{Hmax} would also predict strike-slip motion on well-oriented planes. Note that the pore pressure is essentially hydrostatic. (b) Stress measurements in a portion of the Timor Sea indicating a strike-slip faulting stress state (after Castillo, Bishop *et al.* 2000).

for the measured magnitude of the least principal stress for a coefficient of friction of 0.6.

Figure 9.11 is the same as Figure 7.13 (for the Visund field of the northern North Sea) but is shown again to illustrate that the difference between the magnitudes of S_{hmin} and S_{Hmax} is consistent with equation (4.46). In other words, the line fitting the values of S_{Hmax} was derived from equation (4.46) based on the lines fitting the S_{hmin} values and the pore pressure. Note that this is a more compressional strike-slip stress state than that shown in Figure 9.10b. In fact, because the magnitude of S_{hmin} is only slightly below the magnitude of S_v, this is almost a strike-slip/reverse faulting stress state.

Strike-slip/reverse faulting

These are areas where S_{Hmax} is considerably greater than the vertical stress but $S_{hmin} \approx S_v$. If the state of stress is in frictional equilibrium as predicted by Coulomb

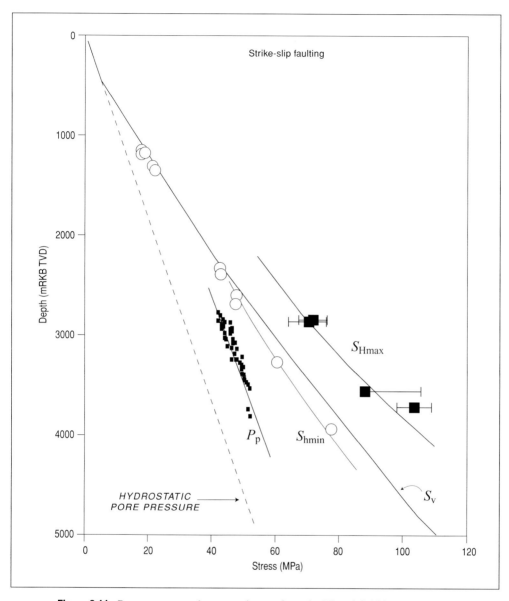

Figure 9.11. Pore pressure and stress estimates from the Visund field in the northern North Sea (after Wiprut, Zoback *et al.* 2000). This is a strike-slip stress state but one that is close to strike-slip/reverse as S_{hmin} is only slightly below S_v. *Reprinted with permission of Elsevier.*

faulting theory using $\mu = 0.6$, S_{Hmax} will be significantly greater than S_v (at the values predicted by equations 4.46 and 4.47) and $S_{hmin} \approx S_v$. Data from one such area are shown in Figure 9.12a, for a well in central Australia. Note that the mini-frac data indicate that the least principal stress is the vertical stress. This implies either a reverse or strike–slip reverse stress state. The reason we know the latter is the case is because of the presence

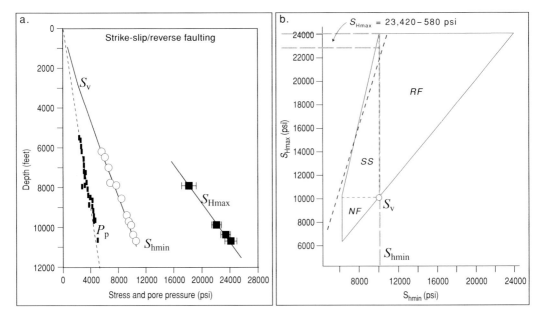

Figure 9.12. (a) Highly compressional (reverse/strike-slip) stress state observed in central Australia. Even though least principal stress values correspond to the overburden, the occurrence of drilling-induced tensile fractures requires S_{hmin} to be approximately equal to S_v as indicated in (b), but S_{Hmax} to be much higher (courtesy D. Castillo).

of drilling-induced tensile fractures. Figure 9.12b (similar to Figure 7.10) indicates that the only way tensile fractures can form (i.e. the stress state is above the dashed diagonal line) and $S_{hmin} \geq S_v$ is in the upper-left corner of the diagram, the region corresponding to a strike-slip/reverse faulting regime.

Reverse faulting

Areas where both S_{hmin} and S_{Hmax} exceed S_v are relatively rare and would be difficult to document because mini-frac tests would not yield reliable values of S_{hmin}. Figure 9.13 illustrates the stress state at a site of shallow reverse faulting earthquakes in central Connecticut (Baumgärtner and Zoback 1989). In this case, very careful open-hole hydraulic fracturing tests were used to determine the magnitude of the minimum horizontal stress as well as S_{Hmax}. The reason that S_{hmin} could be determined from the hydraulic fracturing data is that vertical hydraulic fractures initiate at the wall of a vertical well pressurized between two inflatable packers, even if the least principal stress is vertical. Immediate shut-in of the well prior to fracture propagation makes it possible to determine S_{hmin} before the hydraulic fracture *rolls over* into a horizontal plane as it propagates away from the wellbore. Note that the magnitude of S_{Hmax} is appreciably above S_v,

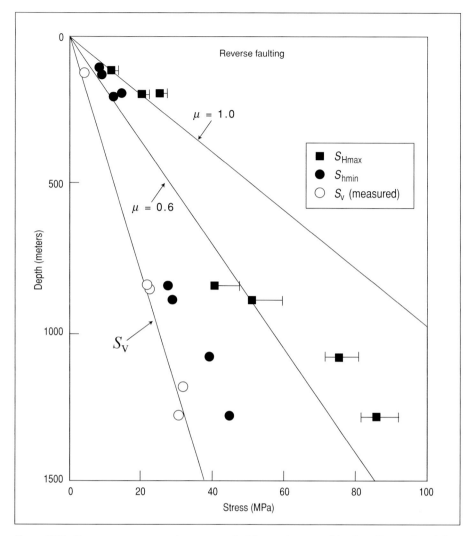

Figure 9.13. Stress measurements in a reverse faulting regime near Moodus, Connecticut (after Baumgärtner and Zoback 1989). Although the least principal stress was vertical, S_{hmin} was determined from *early* shut-in pressures and S_{Hmax} was estimated from *classical* hydraulic fracturing theory as discussed in Chapter 7. The error bars indicate whether pore pressure and effective stress are assumed in the equation for these extremely low-porosity crystalline rocks. *Reprinted with permission of Elsevier.*

consistent with equation (4.47) for coefficients of friction about 0.7. Corroborating evidence of this stress state comes from a sequence of shallow (1–2 km deep) micro-earthquakes that occurred very near the borehole. These earthquakes had reverse faulting focal mechanisms with east–west P axes. Breakouts in the vertical borehole were found on the N and S sides of the hole, thus also indicating E–W compression.

A few more comments about the bilateral constraint

The section above that discussed empirical methods for estimating stress magnitudes at depth focused on normal faulting areas and the determination of the minimum principal stress. This is principally because of the need of drillers to estimate permissible mud weights during drilling – an issue that is most important in normal faulting areas where S_{hmin} has the smallest values. In point of fact, techniques such as the bilateral constraint (equation 9.3) are used more broadly for estimating stress magnitude at depth used with tectonic stress added empirically to match measured values when available. So what is wrong with using the bilateral constraint for predicting the least principal stress at depth? First, diagenesis occurs over geologic time and stresses in the earth, originating from a variety of tectonic processes (as summarized earlier in this chapter), will act on rock to the degree that the rock can support such stresses. Thus, it is geologically somewhat naïve to view diagenesis as occurring in the absence of either gravitational or tectonic stress such that there is, at some point, an elastic half-space and gravitational forces can be instantaneously applied. Second, there is appreciable horizontal strain in the earth, especially in extending sedimentary basins. Third, the two horizontal stresses are rarely equal as a result of the wide variety of tectonic sources of stress acting on rock at depth (Chapter 1). The existence of consistent directions of principal stresses over broad regions is an obvious manifestation of anisotropic magnitudes of the horizontal stress. Moreover, these tectonic sources of stress often result in one (or both) of the horizontal stresses exceeding the vertical stress, as required in areas of strike-slip or reverse faulting and demonstrated previously in this chapter. Attempts to correct for this by adding arbitrary tectonic stresses only make the matter worse by adding more empirically determined parameters.

Figure 9.14 (after Lucier, Zoback *et al.* 2006) illustrates a lithologic column for a well drilled in the central U.S. V_p, V_s and density logs and log-derived elastic moduli (using equations 3.5 and 3.6 and the relations presented in Table 3.1) are also shown. Hydraulic fracturing of the Rose Run sandstone at $\sim$2380 m was being considered to stimulate injectivity. As a result, a series of mini-frac measurements were made within the Rose Run and in the formations immediately above and below (Figure 9.15). Moreover, as shown in the figure, estimates of S_{hmin} and S_{Hmax} magnitudes at other depths were made from analysis of tensile and compressive wellbore failures in the manner described in Chapter 7. In general, a strike-slip faulting stress state is seen. However, note the unusually low magnitudes of S_{hmin} and S_{Hmax} at the depth of the Rose Run. This indicates that this would be a particularly good interval for hydraulic fracturing as a relatively low pressure would be needed to exceed the least principal stress ($\sim$35 MPa) and as long as the frac pressure did not exceed $\sim$42 MPa, the fracture would not grow vertically out of the injection zone.

The data presented in Figures 9.14 and 9.15 make it possible to test the applicability of equation (9.3), although as S_{Hmax} is significantly greater than S_{hmin} (and is mostly

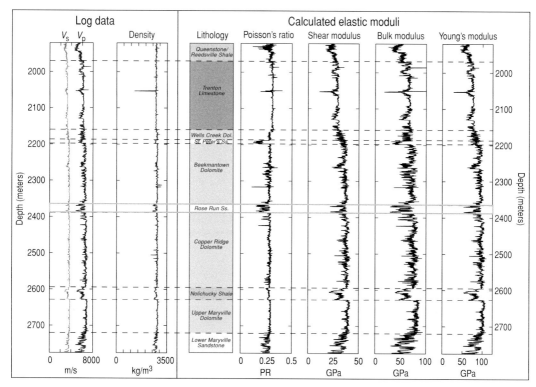

Figure 9.14. Several geophysical logs, lithology and log-derived elastic moduli from the AEP Mountaineer drillhole in W. Virginia (after Lucier, Zoback *et al.* 2006). *AAPG© 2006 reprinted by permission of the AAPG whose permission is required for futher use.*

greater than S_v), it is obvious that it will be necessary to add a tectonic stress, at least to S_{Hmax}. The result of this is illustrated in Figure 9.16 for the depth interval between 2050 and 2450 where the mini-fracs were made. The log-derived Poisson's ratio values are shown on the left and the values of S_{hmin} derived from Poisson's ratio (a stress log) using equation (9.3) is shown on the right. Note that measured stress values at 2060 m and 2425 m are appreciably above the predicted value (by over 5 MPa) whereas the three "low" stress measurements in the vicinity of the Rose Run (2340–2420 m) are 3–5 MPa less than the predicted values. Equation (9.3) obviously does a poor job of matching the measured values. As it is sometimes argued that an empirically calibrated *effective* Poisson's ratio should be used in equation (9.3), if we multiply the log-derived values of v by 1.25 prior to utilizing equation (9.3), the measured least principal stress at 2060 and 2525 m can be fit, but the low stress values between 2340 and 2420 m are fit even more poorly with a $\sim$10 MPa misfit between the measured and predicted values.

    In summary, no evidence has been found that indicates that horizontal principal stresses result simply from the weight of the overlying rock. Forcing this method upon data is, in general, both unwarranted and unwise.

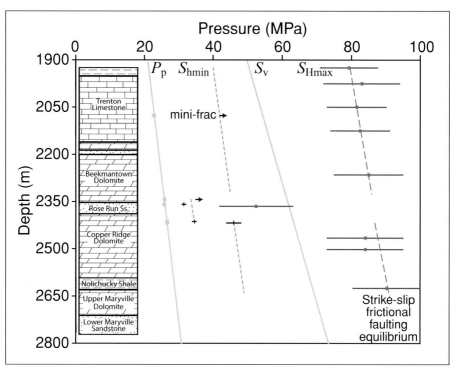

Figure 9.15. Measured values of pore pressure and S_{hmin} from mini-frac stress measurements in the AEP Mountaineer drill hole along with stimated values of S_{Hmax} values derived from wellbore failure analysis (after Lucier, Zoback *et al.* 2006). Note the low values of S_{hmin} in the Rose Run sandstone, making it ideal for hydraulic fracturing. *AAPG© 2006 reprinted by permission of the AAPG whose permission is required for futher use.*

Interpolation and extrapolation of stress magnitude data

The final topic considered in this chapter is related to creating stress profiles from measurements at specific depths. Obviously, leak-off tests or mini-fracs are made at specific depths to yield the magnitude of the least principal stress, and the techniques described in Chapters 7 and 8 for determination of S_{Hmax} are applied at specific depths where wellbore breakouts and/or drilling-induced tensile fractures are observed. As illustrated in the chapters that follow, many applications of stress data require relatively continuous knowledge of stress magnitude with depth. This is especially true in cases of wellbore stability.

In the cases where point measurements of stress are consistent with the predictions of equations (4.45), (4.46 or 4.47) (such as the cases cited earlier in this chapter), it is straightforward (and physically reasonable) to interpolate, or extrapolate, some of the point measurements to create a stress profile. For example, in a normal faulting case,

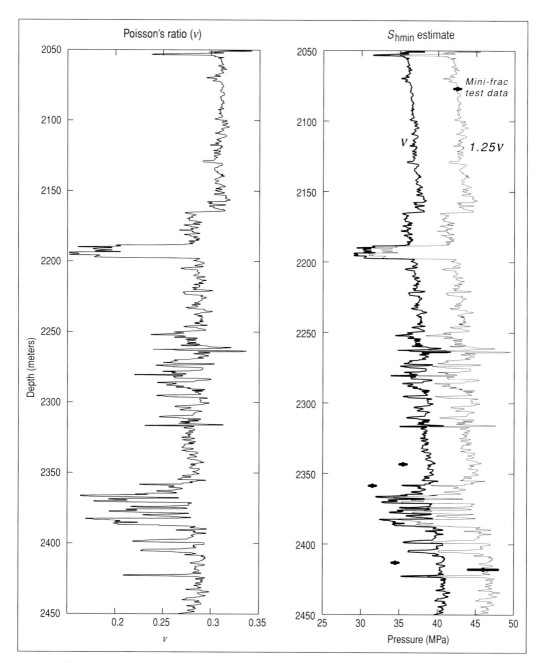

Figure 9.16. The section of the Mountaineer well between 2050 and 2450 m depth where the mini-frac measurements were made. The left column shows the log-derived values of Poisson's ratio. The right column shows a comparison of the measured values of S_{hmin} with those derived from equation (9.3). Note that using the log-derived values of Poisson's ratio significantly under-predicts the measured values at 2075 and 2425 m and over-predicts the three stress measurements in the Rose Run between 2340 and 2420 m. Using a multiplication factor of 1.25 for Poisson's ratio results in a better fit of the data at the highest and lowest depths, but still does not fit the Rose Run data.

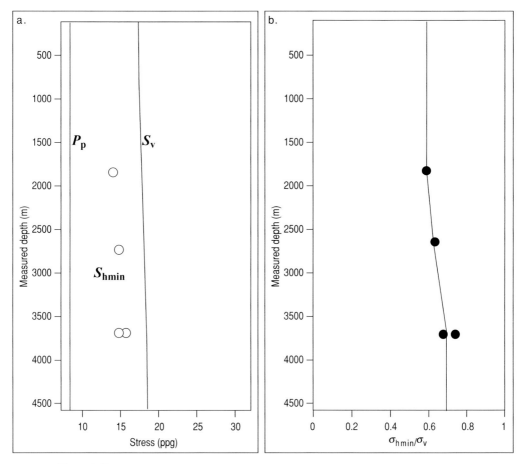

Figure 9.17. (a) Measurements of stress at specific depths need to be extended over depth in order to be used for many applications such as wellbore stability. (b) One method for interpolation and extrapolation of the measured stress values is to utilize a smoothly varying effective stress ratio with depth. In this case with four measurements of S_{hmin} and continuous profiles of P_p and S_v, one could estimate S_{hmin} at other depths by utilizing a function of σ_{hmin}/σ_v similar to what is shown.

it would be reasonable to interpolate or extrapolate the S_{hmin} measurements, but not necessarily S_{Hmax}.

One reasonable approach to the problem of extrapolation of measured stress values is based on relatively constant effective principal stress ratios. An example of this is illustrated in Figure 9.17, where pore pressure is hydrostatic. The measurements of S_{hmin} in Figure 9.17a indicate values that are not in frictional equilibrium for normal faulting. Hence, one cannot use equation (4.45) to estimate values at depths less than or deeper than the measured values. By first calculating the ratio of the minimum effective horizontal stress to the vertical effective stress, we can utilize these data to extrapolate beyond the range of measured values and establish a continuous profile of stress with depth (Figure 9.17b). In the case shown with four measurements of S_{hmin} and continuous profiles of P_p and S_v, one could estimate S_{hmin} at other depths by utilizing a function of σ_{hmin}/σ_v similar to what is shown. As an aside, this case study comes from an area suspected to be a strike-slip faulting region. This implies a lower limit of S_{Hmax} equal to S_v. Because there were no image logs available (or other data related to wellbore failure) that could be used to constrain the magnitude of S_{Hmax}, once a profile of the least principal effective stress was obtained, an upper bound for S_{Hmax} was calculated using equation (4.46). As the pore pressure is elevated, this corresponds to a value of $\sim$31.5 ppg at a measured depth of 3700 m.

Part III Applications

10 Wellbore stability

In this chapter I address a number of problems related to wellbore stability that I illustrate through case studies drawn from a variety of sedimentary basins around the world. In this chapter we focus on wellbore stability problems associated with mechanical failure of the formations surrounding a wellbore. Failure exacerbated by chemical reactions between the drilling mud and the formation is addressed only briefly. I make no attempt to discuss a number of critically important issues related to successful drilling such as hole cleaning, wellbore hydraulics, mechanical vibrations of the drilling equipment, etc. and refer readers to excellent texts such as that of Bourgoyne Jr., Millheim *et al.* (2003).

In each case study considered in this chapter, a comprehensive geomechanical model was developed utilizing the techniques described in previous chapters. The problems addressed fall into two general categories: Preventing significant wellbore instability during drilling and limiting failure of the formation surrounding the wellbore during production. The latter problem is sometimes referred to as sand (or solids) production as significant formation failure during production results in fragments of the formation being produced from the well along with hydrocarbons. Another aspect of wellbore failure with production, the collapse of well casings due to depletion-induced compaction and/or the shearing of wells by faults through injection- (or depletion-) induced faulting, will be discussed in Chapter 12.

When considering the types of wellbore stability problems that could occur while drilling, we must first define what we mean as a *stable* well and an *optimal* mud weight. The practical manifestation of this is related to the concept of the safe *mud window* for drilling, a term referring to the difference between the minimum and maximum mud weight one should use when drilling at given depth. If wellbore stability is not a concern in a given area, the minimum mud weight is usually taken to be the pore pressure so that a well does not flow while drilling. When wellbore stability is a consideration, the lower bound of the mud window is the minimum mud weight required to achieve the desired degree of wellbore stability. In both cases, the upper bound of the mud window is the mud weight at which lost circulation occurs due to hydraulic fracturing of the formation.

The next topics I consider are how well trajectory affects wellbore stability (first alluded to in Chapter 8) and the feasibility of underbalanced drilling. Underbalanced drilling refers to intentionally drilling with mud weights that are less than the pore pressure. This is done to prevent high mud weights from damaging formation permeability, to increase the rate-of-penetration (ROP) and to prevent mud losses in permeable intervals. We introduce utilization of Quantitative Risk Assessment (QRA) for problems of wellbore stability (Ottesen, Zheng *et al.* 1999; Moos, Peska *et al.* 2003) in the context of underbalanced drilling, although it is applicable to a variety of problems in geomechanics, wellbore stability, in particular. QRA allows one to predict the probability of drilling a successful well in the context of uncertainties in the various parameters utilized in wellbore stability analysis (stress magnitudes, pore pressure, rock strength, mud weight, etc.) as well as to evaluate which of the parameters are most important in the assessment.

Three specialized topics related to wellbore stability during drilling are also considered in this chapter. First is the influence of weak bedding planes on wellbore stability as they introduce anisotropic rock strength which can affect the stability of wells drilled at particular angles to the bedding planes. We illustrate the importance of this in two very different cases: one where near-vertical wells were being drilled in steeply dipping strata and the other where highly deviated wells were being drilled through relatively flat-laying shales. Next, I briefly consider the complex topic of chemical effects on rock strength and wellbore stability when drilling through *reactive* shales. By reactive shales we mean those that chemically react with drilling mud in such a way as to weaken a formation leading to wellbore instability. Drilling with chemically inert fluids (such as oil-based mud) is one way to deal with this problem, but there are often environmental and cost considerations that make this option undesirable. While the complex topic of chemical interactions between drilling fluids and formations is generally beyond the scope of this book, we do discuss a case in this chapter in which utilizing a higher mud weight can offset the chemical weakening of mud/shale interactions. Third, I consider the often serious drilling difficulties associated with very high pore pressure environments. As illustrated in Chapters 2 and 4, in areas of highly elevated pore pressure (especially in normal faulting areas), there is a very small difference between the pore pressure and least principal stress. This can result in an extremely small mud window such that even when using pressure-while-drilling measurements to achieve maximum control of mud weights, it can be an extremely challenging problem. Finally, I briefly illustrate the problem of time-dependent wellbore failure due to fluid penetration into fractured rock surrounding a wellbore that was alluded to in Chapter 6.

It should be noted that the calculations of wellbore stability related to mechanical failure usually assume a perfect *mud cake*. In other words, the full difference between P_m and P_p acts to raise σ_{rr} and decrease $\sigma_{\theta\theta}$, as discussed in Chapter 6. This may not always be the case, especially in fractured formations, unless one takes special measures to assure that the mud pressure does not penetrate the formation (Labenski, Reid *et al.* 2003). Detailed discussion of the types of mud types and additives that might be used to assure this is beyond the scope of the book. Nonetheless, one needs to be cognizant of this problem and take steps to assure that the mud pressure does not penetrate into the formation.

In the second part of this chapter we address some of the types of problems associated with minimizing sand production. Minimizing sand production is a complex problem which is the subject of a number of specialized publications (Ott and Woods 2003), and generally requiring comprehensive numerical modeling. We illustrate here approaches to three specific problems. The first is how reservoir depletion and drawdown (related to the rate of production) affect open-hole completions in wells of different orientations. Now that horizontal drilling is widely used, open-hole completions are much more common. A variant of this problem occurs with multi-lateral wells (secondary wells drilled through the casing in a new direction). We consider a case study where the question was whether or not uncased multi-laterals would stay open with depletion and drawdown, resulting in considerable cost savings. The final topic related to sand production considered here is how oriented perforations provide an effective means of sand control in some cases and why sand production sometimes increases with the amount of water being produced.

Preventing wellbore instability during drilling

Perhaps the first issue to address when considering the topic of wellbore stability during drilling is to define what is meant by drilling a stable well. An unstable well is one in which excess breakout formation produces so much failed material from around the wellbore that the total volume of cuttings and failed material in the hole cannot be circulated out by mud circulation. In fact, as a wellbore enlarges due to the excessive wall failure, the velocity of drilling mud in the annulus between the outside of the bottom hole assembly and the wall of the hole decreases. This, in turn, reduces the ability of the mud to clean the cuttings and debris out of the well. Together, the excessive failed rock and reduced cleaning capacity associated with mud circulation can cause the cuttings and failed rock to stick to the bottom hole assembly. This is sometimes called *wellbore collapse* because it seems as if the wellbore has collapsed in on the bottomhole assembly. The mud weight needed to stabilize the wellbore wall and prevent this from occurring is sometimes referred to as the collapse pressure.

So what is a stable well? As illustrated in Figure 10.1, it need not be a well in which no wellbore failure is occurring. As discussed in Chapters 6 and 7, as wellbore breakouts grow they deepen, but do not widen. Hence, a breakout with limited breakout width, $\leq 60°$ in the example shown in Figure 10.1a, produces a failed zone of limited size. Literally thousands of near-vertical wells have been studied in which breakouts of such size are present but there were no significant wellbore stability problems.

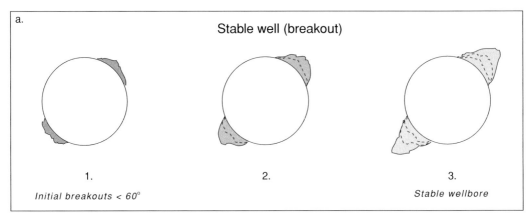

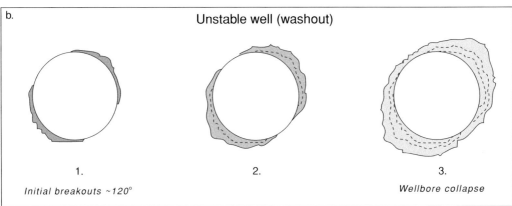

Figure 10.1. (a) Schematic representation of breakout growth when the initial breakout size is relatively small (<60°) and (b) when it is relatively large (~120°). When breakouts are narrow, they deepen as they grow, but do not widen (see Figure 6.15) such that a relatively small amount of failed material falls into the well and the diameter does not change markedly. Hence, stable wells can be drilled that allow a degree of wellbore failure to occur. Wide breakouts, however, can lead to washouts due to the lack of insufficient intact material around the wellbore wall to support the applied stresses.

In designing a stable wellbore it is only necessary to raise mud weight (and/or alter the well trajectory) sufficiently to limit the initial breakout width to an acceptable amount. Empirically, many case studies have shown that designing for maximum breakout widths of ~90° in vertical wells is often a reasonable, if somewhat conservative criterion for mud weight prediction. Intuitively one can see that breakout widths that exceed 90° correspond to failure of more than half of a well's circumference. In this case, the well could lack adequate *arch support, i.e.* sufficient unfailed formation to support the applied forces (Bratli and Risnes 1981). As illustrated schematically in Figure 10.1b, inadequate arch support could lead to failure all the way around a well. This leads to the formation of a washout (the well is enlarged in all directions) which can be quite

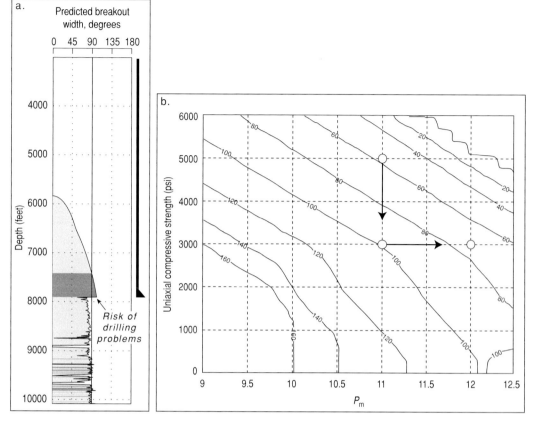

Figure 10.2. An example of how the simple empirical stability criterion of maintaining initial breakout widths <90° can be effective for achieving a desired degree of wellbore stability. (a) After development of a comprehensive geomechanical model, it was possible to demonstrate that the depths at which significant wellbore stability problems occurred (7500–7900 feet) was interval in which breakouts were expected to be occurring with widths greater than 90°. (b) By plotting breakout width (contours) as a function of mud weight (abscissa, in ppg) and rock strength (ordinate) one can see that for a UCS of 3000 psi, simply raising the mud weight from 11 to 12 ppg would have reduced the breakout width to an acceptable amount. Moreover, had the mud weight been raised, there would also have been narrower breakouts at greater depth after setting casing as the breakouts below 8000 feet were only slightly less than the 90°.

problematic because the well's diameter could simply keep increasing in size as failure progressed. The stress concentration around a well does not change as its diameter increases.

A practical example of this type of empirical stability criterion is illustrated in Figure 10.2a. Utilizing a comprehensive geomechanical model developed for a case study of wellbore stability, it was possible to demonstrate that the depths at which significant wellbore stability problems occurred (7500–7900 feet) was interval in which breakouts were expected to be occurring with widths greater than 90°. Casing was set at this depth to remedy this situation. However, as shown in Figure 10.2b, simply raising the mud weight from 11 to 12 ppg would have reduced the breakout width to an acceptable amount for the rock strength (and stress state) at that depth. Moreover, had the mud weight been raised, there would also have been narrower breakouts at greater depth, as well. As seen in Figure 10.2a, the predicted breakouts were only slightly less than the 90° after setting casing.

It is worth noting that while 90° has empirically proven to be an effective criterion for the upper limit of breakout width in vertical wells, it is common to be more conservative if it can be accomplished with reasonable mud weights, casing plans, etc. to allow for local variations of rock strength. It is also common to be more conservative in highly deviated, and especially horizontal, wells. The reason for this is that horizontal wells are more difficult to clean than vertical wells because gravity causes the cuttings to settle on the bottom of the hole. Hence, it is prudent to design horizontal wells (and highly deviated wells) with less wellbore failure than vertical wells.

There are, of course, a number of reasons not to raise mud weight any higher than necessary. The biggest potential problem with raising mud weight too high is associated with inadvertent hydraulic fracturing of the well and associated lost circulation. This is an extremely serious problem which, if it occurs while trying to deal with excessive wellbore failure, could lead to losing a well. As discussed at greater length below, this situation is especially problematic in cases of elevated pore pressure. As mentioned above, other problems associated with high mud weights are a decrease in the drilling rate, formation damage (decreases in permeability due to mud infiltration into the formation), mud losses into permeable zones and, if there isn't an adequate mud cake developed on the borehole wall, differential sticking can occur. Differential sticking occurs when the difference between the mud weight and pore pressure hydraulically clamps the bottom hole assembly to the wellbore wall.

Drilling with mud weights greater than either the pore pressure or the collapse pressure but less than the hydraulic fracture pressure (or frac gradient) is the principal consideration in well design and determination of casing set points. This is illustrated in Figure 10.3 (from Moos, Peska *et al.* 2003). In Figure 10.3a, the original well design is shown, using pore pressure as the lower bound and the frac gradient as the upper bound of the mud window. Based on a wellbore stability problem in a previously drilled well, the lower bound of the mud weight pressure was increased at depth, necessitating a number of casing strings and a very small window for the third string of casing (Figure 10.3b). When the well was drilled, it took two sidetracks to get through this interval. Development of a comprehensive geomechanical model allowed the upper two casing strings to be deepened, resulting in the mud window for the third casing string to be enlarged and for one less casing string to be used overall (Figure 10.3c). This simple case history demonstrates the importance of basing drilling decisions on a quantitative model of wellbore stability founded upon having a comprehensive geomechanical model. The case shown in Figure 10.3b is one in which

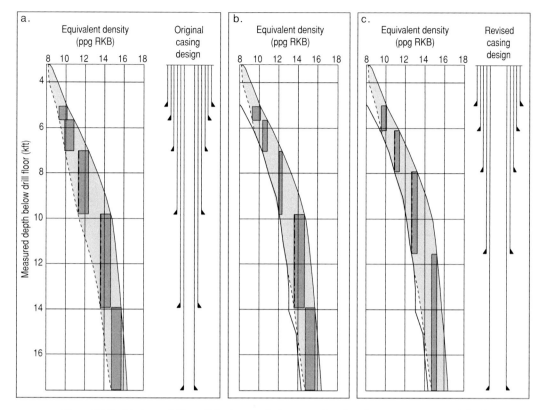

Figure 10.3. (a) A pre-drill well design, made by assuming that the pore pressure and the fracture gradient limit the mud window. (b) An illustration of the impact of considering the collapse pressure on the pre-drill design. There is an extremely narrow mud window for the third casing interval; and in fact, two sidetracks were required while drilling this section of the well. (c) A design made utilizing a comprehensive geomechanical model, which adjusts the positions of the first two casings to reduce the length of the third cased interval. Not only does this design avoid the extremely narrow mud window for the fourth casing that resulted in drilling problems, it also reduces the required number of casing strings (after Moos, Peska *et al.* 2003). *Reprinted with permission of Elsevier.*

millions of dollars were spent unnecessarily on an extra casing string and multiple sidetracks.

Figure 10.4 shows how a well's trajectory affects stability for normal, strike-slip and reverse faulting environments. The parameters used in this figure are the same as those used for the calculations shown in Figure 8.2, but there are two important differences. Rather than illustrate the rock strength needed to inhibit initial breakout formation in wells of any given orientation, Figure 10.4 shows the mud weight required to drill a stable well as a function of well orientation at a single depth. A relatively high uniaxial rock strength of 50 MPa was used for the calculations and a modified Lade failure criterion (Chapter 4) was adopted. For simplicity, a maximum breakout width of 30°

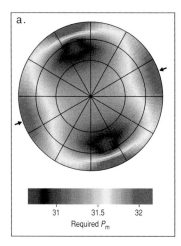
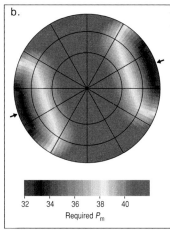
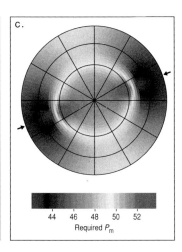

a.   b.   c.

Required P_m Required P_m Required P_m

Figure 10.4. The effects of wellbore trajectory and stress state on wellbore stability. The parameters used in this figure are the same as those used for the calculations shown in Figure 8.2. The figure shows the mud pressure (in ppg) required to drill a stable well (maximum breakout width 30° for a relatively strong rock (UCS ~50 MPa) as a function of well orientation at a depth of 3 km for hydrostatic pore pressure: (a) normal faulting, (b) strike-slip faulting and (c) reverse faulting. (*For colour version see plate section.*)

was used for all calculations, recognizing that in practice, this would be a very conservative approach. As for Figure 8.2, hydrostatic pore pressure (32 MPa) was used. For the case of normal faulting ($S_v = 70$ MPa, $S_{Hmax} = 67$ MPa, $S_{hmin} = 45$ MPa) shown in Figure 10.4a, a mud weight as low as ~30 MPa (slightly underbalanced) is sufficient to drill as stable well that is near vertical, but higher mud weights are needed to achieve well stability for deviated wells depending on orientation. However, the most unstable orientations (horizontal wells drilled parallel to either S_{Hmax} or S_{hmin}) require a mud weight that balances the pore pressure (32 MPa). In this case, the relatively high rock strength (50 MPa) and relatively low stresses associated with normal faulting environments combine to make drilling stable wells at almost any orientation easily achievable. For the case of strike-slip faulting ($S_{Hmax} = 105$ MPa, $S_v = 70$ MPa, $S_{hmin} = 45$ MPa) shown in Figure 10.4b, it is clear that for all orientations except highly deviated wells parallel to S_{Hmax}, mud weights of 40–42 MPa (corresponding to ~1.28 sg or ~10.7 ppg) are required to achieve the desired degree of stability. As the stresses are both larger in magnitude and more anisotropic than for the normal faulting case, the well trajectory has a more important effect on wellbore stability. For the case of reverse faulting ($S_{Hmax} = 145$ MPa, $S_{hmin} = 125$ MPa, $S_v = 70$ MPa) shown in Figure 10.4c, still higher mud weights are needed at all wellbore orientations because of the very high stress magnitudes. The most unstable wells (*i.e.* those requiring the highest mud weights) are near vertical (deviations <30°) and require ~52 MPa (~1.62 sg or ~13.7 ppg) to achieve the desired degree of wellbore stability. Lower mud weights

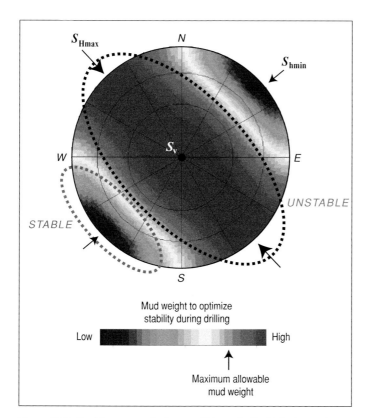

Figure 10.5. In this case study for an offshore Gulf of Mexico well, near vertical wells and those deviated to the northwest or southeast require unrealistically high mud weights (i.e. in excess of the frac gradient) to achieve an acceptable degree of wellbore stability. In contrast, wells that are highly deviated to the southwest or northeast are relatively stable. (*For colour version see plate section.*)

can be used to achieve the desired degree of stability whenever the well trajectories are more highly deviated. The most stable wells are those that are highly deviated and drilled in the direction of S_{Hmax}.

A practical example of how the principles illustrated in Figure 10.4 arise in practice can be seen for the case of a deviated well in the Gulf of Mexico that was being drilled with a build-and-hold trajectory to the southeast. At a certain depth, the well could not be drilled any further (and did not reach its intended target reservoir) because mud weights sufficient to stabilize the well exceeded the least principal stress. In other words, as the operator increased the mud weight in an attempt to stabilize failure of the wellbore wall, circulation was lost because the mud weight exceeded the least principal stress. Development of a geomechanical model enabled us to develop the figure shown in Figure 10.5 for the depth at which the wellbore stability problem was occurring. It is clear from this figure that it is not feasible to drill a highly deviated well with a southeast (or northwest) trajectory at this depth. In contrast, drilling a well highly deviated to the

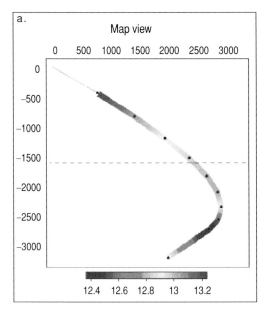
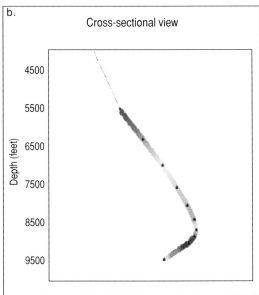

Figure 10.6. Two views of a well trajectory that takes advantage of the principles demonstrated in Figure 10.5. By turning the well to the southwest in the problematic area, wellbore stability could be achieved with a mud weight less than the fracture gradient. The color indicates the mud weight (in ppg) required to stabilize the well at a given depth. (*For colour version see plate section.*)

southwest or northeast would be possible. The reason drilling direction is so important in this case study is that at the depth of interest there is a significant difference between the maximum and minimum horizontal principal stresses. While S_{hmin} is well below S_v, as expected in this area of active normal faulting, S_{Hmax} was found to be approximately equal to S_v.

As there were no serious wellbore stability problems at shallower depth, a trajectory was designed for this well (shown in Figure 10.6a,b in both map view and cross-section) that would avoid drilling in a problematic direction at the depth where the wellbore became unstable. The key to deriving the appropriate trajectory was to keep the overall direction of the well to the southeast (the direction from the platform to the target reservoir), but the local trajectory at the depth where the wellbore stability problems were occurring (*i.e.* the depth at which Figure 10.5 was calculated) was to the southwest. A well was successfully completed with a trajectory similar to the one shown in Figure 10.6. A point to note is that the successful well trajectory was not significantly longer, and was not more strongly deviated, than the original well. It was simply necessary to think of the well trajectory in three dimensions. In a traditional build-and-hold trajectory, the well path is in a vertical plane. By rotating this plane, the path of the well could be essentially the same, but the azimuth of the well at the depth of concern different than that at shallow depth.

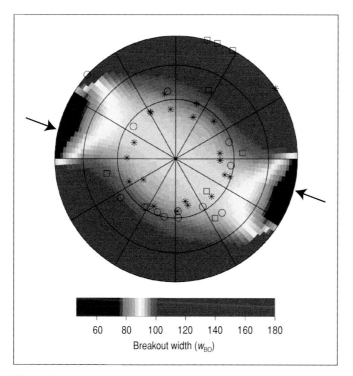

Figure 10.7. Illustration of the importance of drilling direction on success (after Zoback, Barton *et al.* 2003). The figure shows modeled breakout width in the shales in an oil field in South America at a depth of 2195 m TVD as a function of drilling direction assuming a mud weight of 10 ppg and $C_0 = 17.2$ MPa. The total circumference of the wellbore that fails is twice the breakout width. The symbols illustrate the number of days it took to drill the respective well, which is a measure of drilling problems associated with wellbore stability. The asterisks indicate wells that were not problematic (<20 days), the circles indicate wells that were somewhat problematic (20–30 days) and the squares indicate wells that were quite problematic (>30 days). The color scale ranges from an acceptable breakout width (70°, blue) in which less than half the wellbore circumference fails to an excessive amount failure corresponding to over half the circumference failing (breakout widths over 100°, dark red). *Reprinted with permission of Elsevier. (For colour version see plate section.)*

Another illustration of the importance of well trajectory on wellbore stability is shown in Figure 10.7 where we compare wellbore stability predictions based on the derived stress state to actual drilling experience in a field in a sub-Andean foreland basin in northwestern South America (Zoback, Barton *et al.* 2003). We divided the wells in the field into three categories depending on the time needed to drill each well: wells that were drilled in less than 20 days are considered not to be problematic; wells that took more than 20 days are considered to be problematic and those that required more than 30 days are considered to be quite problematic. In Figure 10.7 we compare the predicted failure width and the drilling experience (drilling time) as a function of drilling direction for all wells in the field.

The predicted failure widths displayed in Figure 10.7 correlate well with the drilling time. According to our predictions, near-vertical wells at deviations of less than $\sim 30°$ have breakout widths of less than $90°$ and should therefore pose only minor stability problems. Wells with high deviations drilled towards the NNE–SSW have the largest breakout widths, which explains why wells drilled in this direction were problematic (more than 30 drilling days). The most stable drilling direction is parallel to S_{Hmax} ($\sim 100°/280°$) with a high deviation (near-horizontal). The horizontal well drilled at an azimuth of N55°E would seem to be inconsistent in that it was drilled without problems yet is in an unstable direction. However, the mud weight used in this well was higher than the 10 ppg mud weight used in the calculations that was appropriate for the other wells. This decision was made after the determination of the orientation and magnitude of principal stresses in the reservoir using the techniques described herein. A great deal of money would have been saved had this been done earlier in the field's history.

While drilling with mud weights above formation pore pressure is common practice (even in cases where wellbore stability is not a serious problem), it would be appealing in many cases if wells could be drilled *underbalanced*, that is, with mud weights less than the pore pressure. This is most attractive for cases where there is the potential for damage to formation permeability due to infiltration of the mud filtrate. However, just as overbalanced drilling enhances wellbore stability, underbalanced drilling could seriously compromise wellbore stability if rock strength is low or the ambient stresses are high. Figure 10.8 demonstrates this for a well in South America where underbalanced drilling was being considered and pore pressure is hydrostatic (8.3 ppg). Figure 10.8a shows the mud weight required to achieve the desired degree of wellbore stability for a formation UCS of 7000 psi. Figures 10.8b,c show the same for formation strengths of 8000 and 9000 psi, respectively. Note in Figure 10.8c that if the strength where 9000 psi, mud weights appreciably lower than hydrostatic could be used for wells of any orientation without encountering wellbore stability problems. In fact, the most unstable drilling direction in this case is near-vertical and a mud weight of 7.3 ppg (1 ppg below hydrostatic) could be used and the well should be stable. For a formation strength of 8000 psi (Figure 10.8b), wells deviated less than 30° can be drilled successfully with mud weights slightly below hydrostatic. If the formation strength is 7000 psi, however, Figure 10.8a shows that most well trajectories require mud weights that are slightly in excess of hydrostatic pressure.

Quantitative risk assessment

While the type of trial and error calculation illustrated in Figure 10.8 is one way to assess the relationships among mud weight, well trajectory and wellbore failure, a more effective technique is to use Quantitative Risk Assessment (QRA). As described by Ottesen, Zheng *et al.* (1999) and Moos, Peska *et al.* (2003), QRA allows one to

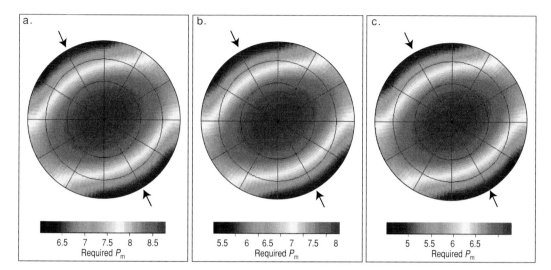

Figure 10.8. Wellbore stability as a function of mud weight. In each of these figures, all parameters are the same except rock strength. For a UCS of 7000 psi (a), a required mud weight of ~8.6 ppg (slightly overbalanced) is needed to achieve the desired degree of stability in near-vertical wells (the most unstable orientation). For a strength of 8000 psi (b), the desired degree of stability can be achieved with a mud weight of ~8 ppg (slightly underbalanced). If the strength is 9000 psi (c) a stable well could be drilled with a mud weight of ~7.3 ppg (appreciably underbalanced). (*For colour version see plate section.*)

consider how uncertainty of one parameter affects wellbore stability in terms of the mud weight required to achieve a desired degree of wellbore stability. In the case of underbalanced drilling, we are obviously interested in how much one could lower mud weight without adversely affecting well stability. More generally, QRA allows us to formally consider the uncertainty associated with any of the parameters affecting wellbore stability.

Figure 10.9 shows an example of the application of QRA, where the input parameter uncertainties are given by probability density functions (from Moos, Peska *et al.* 2003) that are specified by means of the minimum, the maximum, and the most likely values of each parameter. The probability density functions shown here are either normal or log-normal curves depending on whether the minimum and maximum values are symmetrical (*e.g.* S_v, S_{Hmax}, S_{hmin}, and P_p) or asymmetrical (as shown for C_0) with respect to the most likely value. In both cases, the functional form of the distribution is defined by the assumption that 99% of the possible values lie between the maximum and minimum input values.

Once the input uncertainties have been specified, response surfaces for the wellbore collapse (Figure 10.10) and the lost circulation pressures (not shown) can be defined. These response surfaces are assumed to be quadratic polynomial functions of the individual input parameters. Their unknown coefficients in the linear, quadratic and interaction terms are determined by a linear regression technique that is used to fit the

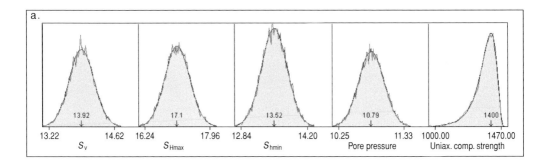

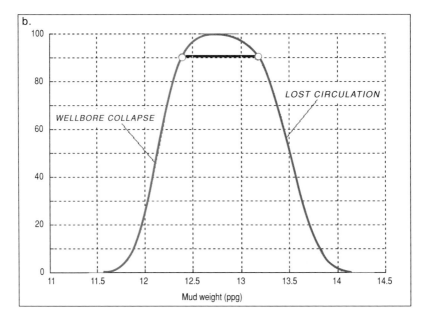

Figure 10.9. (a) Probability density functions (smooth, shaded curves) and the sampled values used in the QRA analysis (jagged lines) as defined by the minimum, most likely, and maximum values of the stresses, the pore pressure, and the rock strength. These quantify the uncertainties in the input parameters needed to compute the mud weight limits necessary to avoid wellbore instabilities. (b) Resulting minimum (quantified in terms of the likelihood of preventing breakouts wider than a defined collapse threshold) and maximum (to avoid lost circulation) bounds on mud weights at this depth. The horizontal bar spans the range of mud weights that ensure a greater than 90% likelihood of avoiding either outcome – resulting in a minimum mud weight of 12.4 ppg and a mud window of 0.75 ppg. After Moos, Peska *et al.* (2003). *Reprinted with permission of Elsevier. (For colour version see plate section.)*

surfaces to theoretical values of the wellbore collapse and lost circulation pressures. The theoretical values are calculated for multiple combinations of input values that are selected according to the representative design matrix based on the minimum, maximum and most likely values. The calculations assume that the rock behaves elastically

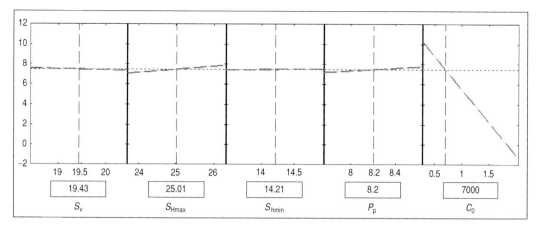

Figure 10.10. Response surfaces that illustrate the sensitivity of the mud weight predictions – expressed in ppg – associated with each parameter's uncertainty, as shown in Figure 10.9a (after Moos, Peska *et al.* 2003). *Reprinted with permission of Elsevier.*

up to the point of failure. The case considered here is the same as that considered in Figure 10.8. Note that for the range of uncertainties associated with S_v, S_{Hmax}, S_{hmin} and P_p, all have relatively little effect on the mud weight needed to achieve the desired degree of wellbore stability. The exact opposite is true for rock strength. As the rock strength increases, the value of the mud weight needed to drill a stable well decreases markedly. Note that if the rock strength is less than 7000 psi, underbalanced drilling cannot be considered for the well in question.

It should be noted that response surfaces such as those shown in Figure 10.10 are very case specific. While rock strength is always an important factor in cases of underbalanced drilling, in other cases the magnitude of the principal stresses or pore pressure might be quite important. Some of these dependencies are intuitive. The magnitude of S_v is very important when considering the stability of a horizontal well, but much less important for the case of a vertical well. If the horizontal well was being drilled in the direction of S_{Hmax}, its magnitude would have relatively little effect on the mud weight needed to stabilize the well. In the case of a vertical well, the magnitude of S_{Hmax} is always very important for wellbore stability, but not its orientation. In the case of deviated wells, both the magnitude and orientation of S_{Hmax} are important. Overall, sensitivity analyses such as are shown in Figure 10.10 are useful for guiding data collection (in future wells or in real-time) for reducing uncertainty in geomechanical models, assessing the importance of laboratory rock strength determinations, etc.

After the response surfaces have been determined, Monte Carlo simulations are performed to establish uncertainties in the wellbore collapse and the lost circulation pressures. Figure 10.9b shows the cumulative likelihood of avoiding wellbore collapse (the lower bound curve on the left) and the cumulative likelihood of avoiding lost circulation (the upper bound curve shown on the right) as a function of the mud weight

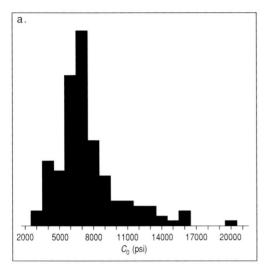
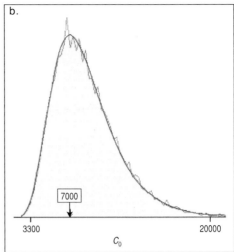

Figure 10.11. (a) Distribution of rock strengths (determined from log analysis) for the case study illustrated in Figure 10.8. Note that there is a wide distribution of strengths. The most likely strength is ~7000 psi, with the majority of rock strengths being higher. (b) A log-normal statistical representation of the rock strength data shown in (a). After Moos, Peska *et al.* (2003). *Reprinted with permission of Elsevier.*

at the depth of interest. The horizontal line illustrates the range of mud weights that will simultaneously provide at least a 90% certainty of avoiding both collapse and lost circulation (the operational mud window). This is because there is a greater than 90% certainty of avoiding collapse provided the mud weight is above 12.4 ppg (for example, a mud weight of 12.5 ppg provides a better than 95% certainty of avoiding collapse). At the same time, there is a 90% certainty of avoiding lost circulation provided the mud weight is less than 13.15 ppg (for example, for a mud weight of 13 ppg there is at least a 97% certainty of avoiding lost circulation). The analysis result suggests that optimum stability can be achieved utilizing a static mud weight close to the lower bound value of 12.4 ppg, and indicates that there is little likelihood of lost circulation so long as ECDs are below 13.1 ppg.

For the case illustrated in Figure 10.8, Figure 10.11 shows the distribution of the values of rock strength and the log-normal distribution function that fit those values. In this case, strength was determined from utilization of a log-based technique such as described in Chapter 4. The most likely value of strength is 7000 psi but there is quite a wide distribution of strengths implied by the algorithm used to estimate rock strength. Also, it can be somewhat dangerous to use the type of strength distribution shown in Figure 10.11 in a QRA analysis. The reason for this is that while ~75% of the interval of the well to be drilled underbalanced may have strengths greater than 7000 psi, a significant fraction of the well appears to have strengths that are less than 7000 psi. Hence, there could be significant problems associated with underbalanced

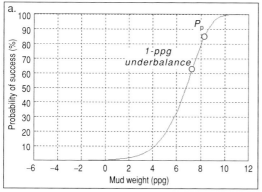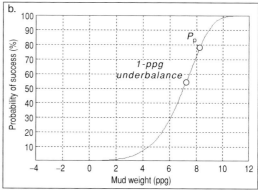

Figure 10.12. Cumulative probability distribution functions for wellbore stability for the reservoir section of the well proposed for underbalanced drilling, using a desired breakout width of (a) 60° and (b) 30°. The hydrostatic pore pressure in the reservoir is shown for comparison (after Moos, Peska *et al.* 2003). *Reprinted with permission of Elsevier.*

drilling in this case. In general, it is best to consider the statistical distribution of the strengths of the rocks in the weakest intervals in this type of analysis.

Utilizing the strength distribution in Figure 10.11 and the response surfaces shown in Figure 10.10, the probabilities of successfully drilling 1 ppg underbalanced are shown in Figure 10.12 for two different stability objectives – restricting breakout width to 60° in the first case and a more conservative 30° in the second. There is a 62% chance of success in the first case (*i.e.* restricting breakout width to 60°) when drilling 1 ppg underbalanced and a slightly lower (~55%) chance of success in the second (i.e. restricting breakout width to 30°) if drilled 1 ppg underbalanced. Taken together, the first case appears to be relatively more safe than the second, but this is simply because a less stringent stability criterion was used. In light of the comments above concerning the distribution of rock strengths inferred from the strength analysis (Figure 10.11) drilling with a slightly overbalanced mud system would be appreciably more stable than drilling underbalanced.

Role of rock strength anisotropy

In Chapter 4 we introduced the concept of anisotropic rock strength resulting from the presence of weak bedding planes in shales (Figures 4.12 and 4.13) and in Chapter 6 we illustrated how the presence of such planes cause broad *double-lobed* breakouts to form on each side of a well (Figure 6.16). In this section we consider two case studies where consideration of strength anisotropy has proven to be quite important in controlling wellbore stability.

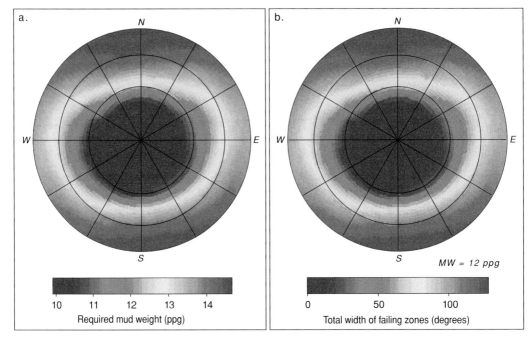

Figure 10.13. Drilling through sub-horizontal, weak bedding planes is only problematic in this case study when the wellbore deviation exceeds ~30°. Because there is little stress anisotropy, there are relatively minor differences in stability with azimuth. This can be seen in terms of the mud weight required to achieve an acceptable degree of failure (a) or the width of the failure zone at a mud weight of 12 ppg (b). (*For colour version see plate section.*)

In the first case considered, extended reach drilling was being done in an area with sub-horizontal bedding and relatively little stress anisotropy. Severe wellbore failures were encountered when drilling a highly deviated section of the well with 12 ppg mud through a thick shale unit immediately above the reservoir. It was initially recommended to switch from water-based mud to an oil-based mud because it was assumed that the problem was related to chemical interaction between the mud and shale (analogous to the case illustrated in Figure 6.18) because the activity of the drilling mud was too high with respect to that of the formation.

Developing a comprehensive geomechanical model for the formations that incorporates the effect of weak bedding planes on wellbore stability allows us to evaluate whether the observed wellbore failures in the highly deviated (~60°) wells could be understood in terms of mechanical wellbore failures and whether there was a combination of mud weights and wellbore trajectories that would allow the wells to be drilled through the unstable section. Figure 10.13 illustrates the role of trajectory and mud weight on wellbore stability. In Figure 10.13a, we show the mud weight required to limit breakout width to 60° in the problematic shale unit, a value that should allow a stable well to be successfully drilled. Because of the low stress anisotropy and modest dip of the formation, the figure is almost symmetrical. Thus, at almost any azimuth, as long the deviation is less than 30°, mud weights of only ∼10 ppg are sufficient to maintain the desired degree of wellbore stability. At 60°, however, mud weights of 13.5–14 ppg are needed. This is why severe wellbore stability problems were occurring in the highly deviated well when using 12 ppg mud. Because the bedding is sub-horizontal, the weak bedding planes only have an effect on wellbore stability when drilling at a high angle to the vertical.

The solution for drilling a stable well is thus fairly obvious. They could either maintain the trajectory they were using and raise the mud weight, or they could build angle at a shallower depth and *drop through* the problematic shale at lower angles. As it turned out, the operators did not want to raise mud weight any higher than 12 ppg. Figure 10.13b shows the impact of this with respect to the degree of wellbore instability to expect as a function of well trajectory. The take-away message from this figure is essentially the same as that in Figure 10.13a. If wells were drilled with relatively low deviation (less than 30°–40°) only a moderate degree of failure is expected (breakout widths of 50°–60°) whereas at higher deviations, wellbore failure is more severe.

Figure 10.14 illustrates how weak bedding planes can affect wellbore stability in a much different geologic environment. In this case, we consider drilling through steeply dipping shales in the Andean foothills of Colombia, an area of significant stress anisotropy (Willson, Last *et al.* 1999). A near-vertical well experienced severe mechanical failures accompanied by large increases in well diameter in a particularly problematic shale section. Oil-based mud was used so chemical effects on rock strength could be ruled out. Interestingly, it was noticed that the well diameter came *back-into-gauge* when a fault was crossed, even though it was the same formation. Above the fault the well was quite unstable, below the fault it was much more stable. After development of a comprehensive geomechanical model (through analysis of data from multiple wells in the field), a *post mortem* of the problematic well revealed that slip on weak bedding planes were responsible for the severe wellbore instabilities above the fault. The well became more stable after crossing the fault because the dip of the bedding changed dramatically, and no longer affected the degree of wellbore failure.

This is illustrated in Figure 10.14 which shows the mud weight required to limit breakout width to 60°. The strong asymmetry in these figures results from the steep dip of the bedding and the anisotropy of the stress field. The upper figure is for the formation above the fault where the bedding is dipping 60° to the SW. It shows that relatively low mud weights (less than 10 ppg) could be used only when drilling wells deviated more than 30° to the NW (or horizontal wells to the SE). This is because when drilling nearly orthogonal to bedding (the pole to the bedding planes is shown by the red dot), the weak bedding planes do not affect failure. As the well in question was being drilled nearly vertically (green dot), mud weights of at least 11.5 ppg would have been required to stabilize the well above the fault. As a lower mud weight was being used, the breakout width was much too large and the well became unstable. Below the fault (the lower

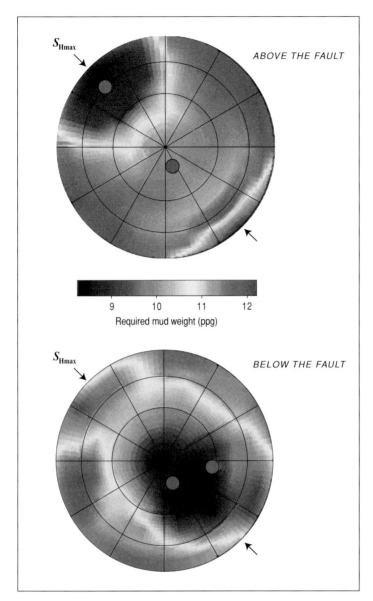

Figure 10.14. When bedding planes dip steeply, both the deviation and azimuth of wells have a strong effect on wellbore stability (similar to Willson, Last *et al*. 1999). (a) Wellbore stability diagram that shows the case above a fault at about 15,000 ft depth, where the bedding plane orientation (the red dot is the pole to the bedding planes) was such that drilling a near vertical well was quite problematic. Drilling orthogonal to the bedding planes (to offset the effect of bedding on strength) would require a steeply dipping well to the northwest. (b) Below the fault, the bedding orientation changes such that a near-vertical well is stable. (*For colour version see plate section.*)

part of Figure 10.14), the bedding is dipping about 30° to the west. For this situation, a near vertical well is in a relatively stable direction because the more flat-laying bedding planes did not affect wellbore stability.

Mud/rock interaction

As previously mentioned in Chapter 6, shales tend to be more unstable than sands or carbonates when drilling with water-based mud because chemical interactions can cause a significant reduction in the effective strength of some shales. Two effects contribute to this problem. The first is osmotic diffusion (the transfer of water from regions of low salinity to regions of high salinity), which causes water in low-salinity mud to diffuse across the membrane formed at the mud–rock interface and into the formation. The second is chemical diffusion (the transfer of specific ions from regions of high concentration to regions of low concentration).

When the salinity of the drilling mud water phase is lower than the salinity of the pore fluid in the formation, osmotic diffusion causes shales to swell and weaken due to elevated internal pore pressure. Consequently, one solution to shale instabilities is to increase the salinity of the water phase of the mud system, which works in some cases. Improving membrane efficiency helps limit the effects of chemical diffusion, which weakens shales through ion exchange with the mud system.

In calculating the magnitude of the pressure generated by osmotic diffusion, the parameter that is used to select the appropriate salinity is the so-called activity of the fluid. Activity (which is explicitly the ratio of the vapor pressure above pure water to the vapor pressure above the solution being tested) varies from zero to one. Typical water-based muds have activities between 0.8 and 0.9. Typical shales in places like the Gulf of Mexico have pore fluid activities between 0.75 and 0.85, based on extrapolations from laboratory measurements. Hence, the use of *typical* muds in *typical* shales is expected to cause an increase in the pore pressure within the shale, leading to shale swelling, weakening, and the development of washouts.

Mody and Hale (1993) published the following equation to describe the pore pressure increase due to a given fluid activity contrast:

$$\Delta P = E_m \times (^{RT}/_V) \times \ln(A_p/A_m) \tag{10.1}$$

If ΔP is negative, it indicates that water will be drawn into the shale. Here, R is the gas constant, T is temperature in kelvin, and V is the molar volume of the water (liters/mole). Decreasing the mud activity often alleviates shale swelling because ΔP is positive if A_p (the pore fluid activity) is larger than A_m (the mud activity), and water will be drawn out of the shale for this condition. The parameter E_m is the membrane efficiency, which is a measure of how close to ideal the membrane is. Explicitly, it is the ratio of the pressure change across an ideal membrane due to a fluid activity difference across the membrane, to the actual pressure difference across the membrane in question, and can be measured in the laboratory for various combinations of rock and mud. The membrane efficiency is affected by the mud chemistry, solute concentrations, and other properties; oil-based mud has nearly perfect efficiency, whereas water-based mud generally has very low efficiency. Some recently developed water-based synthetics have been designed to have high efficiencies, allowing instability-free drilling without the environmental issues associated with oil-based mud. Figure 6.18a shows the relationship between membrane efficiency, mud fluid activity, and degree of failure (quantified in terms of the widths of the failed regions) for shale with a nominal pore fluid activity of 0.8. Higher mud activities than the shale pore fluid cause an increase in breakout width, whereas predicted breakout width is less for muds with lower activities.

The model described by equation (10.1) is time-independent. However, some shales are known to develop instabilities over time, leading to their characterization as 5-day or 10-day shales, etc. Time-dependent models have been developed that predict variations in pore pressure as a function of time and position around the hole. These are explicitly both chemo-elastic and poro-elastic (that is, they account for interactions between the pore pressure and the stress as well as the chemical effects on the pore pressure). The results allow selection of mud weights for specific mud activities, or mud activities for specific mud weights. Figure 10.15a shows a plot of failure vs. time for a mud activity of 0.9. As can be seen, failure gets worse over time, and even a mud weight as high as the fracture gradient of 16 ppg maintains hole stability for less than one day. On the other hand, for a mud activity of 0.7 (Figure 10.15b), the time before failure begins to worsen is extended. It is possible to select a mud weight below the fracture gradient and yet still provide several days of working time. It is important to note that this example is markedly different than that shown in Figure 6.18 in that for the case shown in Figure 10.15, it is not straightforward to simply use increased mud weight to overcome the weakening effects of mud/rock interaction.

Maximizing the frac gradient

In Chapter 7 we discussed hydraulic fracturing and made the point that because of the extremely high stress intensity in the tip area of a propagating hydraulic fracture, once a fracture reaches a length of only $\sim$1 m, the strength of the formation had negligible influence on the pressure needed to propagate a fracture. Hence, except in the case when high viscosity fluids were being pumped at high flow rates, the hydraulic fracture extension should occur at a pressure very close to the magnitude of the least principal stress. When drilling in areas of severe overpressure, especially in normal faulting environments such as the Gulf of Mexico (shown schematically in Figure 1.4d, and for the Monte Cristo field in Figure 2.2), it can be quite difficult to keep the mud pressure at the bottom of the hole during drilling in the narrow window between pore pressure

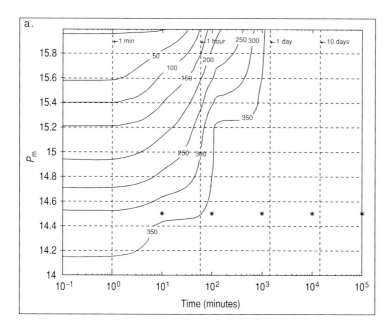

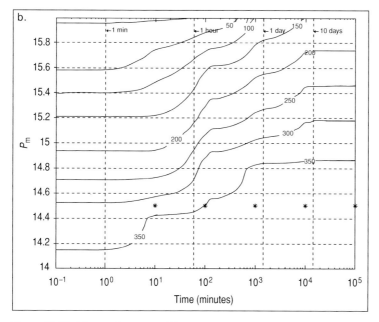

Figure 10.15. The total amount of failure (angular span of well's circumference) as a function of time and mud weight for a shale. (a) Case for a pore fluid activity of 0.8, subjected to a mud with a water phase activity of 0.9. (b) Same as (a) except the mud water phase activity is 0.7. When the mud activity is lower than the pore fluid in the shale, even very high mud weights (the fracture gradient is 16 ppg) stabilize the well for less than 1 day. By lowering the mud activity, the mud weight can be decreased while keeping failure under control and extending working time.

and S_{hmin}. A similar problem occurs when drilling through a highly depleted sand in order to reach a deeper horizon because of the decrease in the magnitude of the least principal stress accompanying depletion (see Chapter 12).

In this section, we follow Ito, Zoback *et al.* (2001) and address the theoretical possibility of drilling with mud weights in excess of the least principal stress for cases of particularly high pore pressure (or high minimum mud weights needed to maintain wellbore instability). In fact, there is empirical evidence that this can occur. The following was reported by a major oil company drilling in the Gulf of Mexico:

While drilling a highly deviated well at elevated pore pressure, lost circulation occurred at an ECD of 14.8 ppg caused by a pressure surge while attempting to free a stuck logging tool. The measured value of the least principal stress at this depth was 13.0 ppg. Lost circulation material (LCM) was used to establish circulation at 14.9 ppg (1.9 ppg over the least principal stress). Once circulation was re-established, the well drilled to TD with an ECD of ~14 ppg (1 ppg above the measured least principal stress).

We will return to this anecdotal account and offer one explanations of why it was possible to re-establish circulation and continue drilling with mud weights greater than the least principal stress.

We consider three critical wellbore pressures, p_{frac}, p_{link} and p_{grow}. For the general case of a well that is deviated with respect to the *in situ* stress field (discussed in Chapter 8), tensile fractures initiate at the wellbore wall at a pressure we will call p_{frac}. As these fractures grow away from the wellbore wall they will attempt to turn to be perpendicular to the least principal stress and link up at p_{link}. Once the fractures have linked up and turned to be perpendicular to the least principal stress, they propagate away from the wellbore at p_{grow}. It is obvious that lost circulation cannot occur if the wellbore pressure during drilling is below p_{frac}. However, even if p_{frac} is exceeded and tensile fractures initiate at the wellbore wall, fracture propagation (and hence lost circulation) will be limited as long as the wellbore pressure is below p_{link}, the pressure required for multiple tensile fractures to link up around the wellbore. Finally, if the wellbore pressure is greater than p_{link}, the fractures will not grow away from the wellbore (and significant lost circulation will not occur) if the wellbore pressure is below p_{grow}, which must exceed (if only slightly) the least principal stress. In general, our modeling shows that p_{frac} and p_{link} can be maximized by drilling the wellbore in an optimal orientation, and p_{grow} can be maximized by using "non-invading" drilling muds that prevent fluid pressure from reaching the fracture tip.

First, let us consider p_{grow}, the fluid pressure in the fracture necessary to cause fracture propagation once it has already propagated away from the wellbore. For simplicity, the fracture is modeled as a penny shaped fracture oriented normal to S_3. The pressure distribution in the fracture is assumed be uniform, as shown in Figure 10.16, which means that if there is a significant pressure gradient in the fracture, we will be calculating a lower bound value of p_{grow}. However, we must take into account the fact that drilling

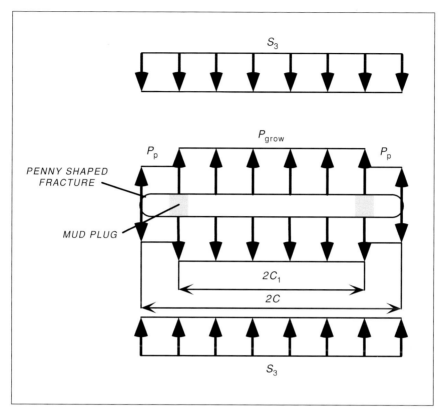

Figure 10.16. Definition of the invaded zone ($2c_1$) where the internal pressure is p_{grow} and the non-invaded zone ($c - c_1$) where the internal pressure is p_p during fracture growth. After Ito, Zoback *et al.* (2001). © *2001 Society Petroleum Engineers*

fluids containing solids may prevent pressure from reaching the fracture tip if the solids effectively plug the fracture due to its narrow width (Morita, Black *et al.* 1996). Thus, the pressure is assumed to be uniform (at p_{grow}) except for a zone at the fracture tip where the drilling fluid does not reach and pressure remains equal to the pore pressure P_p. We refer to these zones as the invaded zone and the non-invaded zone, respectively. For this case, Abé, Mura *et al.* (1976) derived a theoretical relationship between p_{grow}, P_p and S_3. The relationship is given by

$$\frac{p_{grow} - S_3}{S_3 - P_p} = \frac{1}{1 - \sqrt{1 - \left(\frac{c_1}{c}\right)^2}}\left[\sqrt{1 - \left(\frac{c_1}{c}\right)^2} + \sqrt{\frac{\pi}{4c}\frac{K_{IC}}{S_3 - P_p}}\right] \quad (10.2)$$

where K_{IC} is the fracture toughness of the rock, and c and c_1 are the radius of the fracture and that of the invaded zone, respectively. K_{IC} can be neglected for large size fractures such as the fracture which we consider here (see Figure 4.21). This leads to the

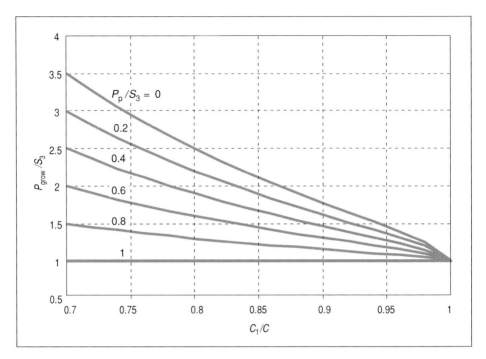

Figure 10.17. Variation of fracture propagation pressure with size of the non-invaded zone for different ratios of pore pressure to the least principal stress (after Ito, Zoback *et al.* 2001). Note that this is likely a lower-bound estimate of p_{grow}/S_3 as a uniform pressure distribution is assumed in the invaded zone (see text). © *2001 Society Petroleum Engineers*

following:

$$\frac{p_{grow}}{S_3} = \frac{1 - \frac{P_p}{S_3}\sqrt{1 - \left(\frac{c_1}{c}\right)^2}}{1 - \sqrt{1 - \left(\frac{c_1}{c}\right)^2}}$$ (10.3)

The relationship between p_{grow}/S_3 and c_1/c obtained from equation (10.3) is plotted in Figure 10.17. This figure shows that existence of the non-invaded zone contributes to maximize p_{grow}, the fracture propagation pressure. Note that for relatively low values of c_1/c (corresponding to significant non-invaded zone at the tip of the fracture) and low values of P_p/S_3, the fracture propagation pressure can appreciably exceed S_3. Recall that this is a lower limit of p_{grow}; if a pressure gradient exists in the fracture due to fluid flow, p_{grow} will be even larger. However, if the non-invaded zone represents only about 5% of the fracture length, and if P_p approaches the value of S_3 (in the case of overpressured formations), the increase in p_{grow} above S_3 is much more modest.

Fuh, Morita *et al.* (1992) argued that special loss prevention material can be utilized to enhance the non-invaded zone and report dramatically increased fracture propagation pressures – by as much as 8 ppg in one well and 3–6 ppg in others. They are presented in an equation which gives a theoretical relationship between p_{grow} and the fracture width at the inlet of the non-invaded zone, which is very similar to equation (10.2). The results reported by Fuh, Morita *et al.* (1992) are explainable in terms of the results presented in Figure 10.17. The use of lost circulation material during drilling should be most effective in cases of hydrostatic pore pressure when S_3 is relatively high. Figure 10.17 shows that the potential increase in p_{grow} above S_3 is much less in cases of highly elevated pore pressure, which are the cases we are principally interested in because it is in such cases that there are small differences between P_p and S_3.

In the anecdotal case mentioned above, for the appropriate values of P_p and S_3, the expected value of p_{grow} will be 15 ppg for $c_1/c = 0.95$. Recall that circulation was recovered at 14.9 ppg. Laboratory experiments by Morita, Black *et al.* (1996) indicate that p_{grow} is higher with water-base muds because the non-invaded zone was larger than with oil-base muds. This difference may arise from the fact that the size of the non-invaded zone is likely larger with water-based mud because filtrate loss from fracture surfaces allows mud cake within the fracture to form the non-invaded zone.

We next consider the pressure, p_{link}, at which inclined tensile fractures at the wellbore wall would be likely to link up to form large axial fractures. In principle, the link-up phenomena will be dominated by the stress state in the plane tangent to the wellbore (Figure 8.1a). We disregard the fracture toughness of rock, because the compressive stresses acting in the plane are so large that its effect on the fracture propagation is expected to be very large compared with the effect of the fracture toughness. In general, the fluid pressure in the fracture must be equal to or larger than s_{norm} in order for a fracture to grow. The stresses acting parallel and normal to the fracture in the plane Σ are denoted as s_{para} and s_{norm}, respectively.

The following two cases (i) and (ii) can be considered in terms of whether the fracture will grow:

- Case (i) $P_w = s_{norm} > s_{para}$
- Case (ii) $P_w \geq s_{norm}, s_{para} \geq s_{norm}$

where s_{para} and s_{norm} are taken to be approximated by

$$S_{para} = S_t|_{\omega=\omega_f} \tag{10.4}$$
$$S_{norm} = S_t|_{\omega=\omega_f+90°} \tag{10.5}$$

Note that s_{para} and s_{norm} will change with p_w in contrast to ω_f, the initial fracture angle at the wellbore wall, which is given by p_{frac}.

When p_w reaches s_{norm}, the fractures will start to grow. The fractures will grow by reorienting themselves to normal to s_{para} in this case, because fractures will tend to grow normal to the minimum compressive stress. As a result, the fractures will grow towards adjacent fractures and will link up finally to form the axial fractures. Therefore, the critical wellbore pressure at the link-up, p_{link}, can be estimated by solving the equation

$p_w = s_{norm}$. Under a stress state that leads to $s_{para} \geq s_{norm}$, the fracture will always grow in a direction parallel to the initial fracture. However, as there is the interference between adjacent fractures, the fractures will reorient themselves to deviate from the direction of the initial fracture line under a certain combination of s_{para}, s_{norm} and p_w. Weng (1993) carried out a 2D analysis of the link-up problem of *en echelon* fractures taking account of the interference between them, and obtained a criterion which defines the link-up phenomena. Although his analysis was conducted originally for the case of fractures that link up in a region away from a wellbore, we adopt here this criterion to approximate fracture link-up near the wellbore wall. The criterion is expressed as

$$\omega_f \leq \omega_{crit} \tag{10.6}$$

where

$$\omega_{crit} = \sin^{-1}\left[0.57 \left(\frac{\Delta s}{\Delta p} \right)^{-0.72} \right] \tag{10.7}$$

and

$$\Delta s = s_{para} - s_{norm} \tag{10.8}$$

$$\Delta p = p_w - s_{norm} \tag{10.9}$$

The numeric constants in equation (10.7) (*i.e.* 0.57 and 0.72) were obtained by numerical simulations of fracture link-up after Weng (1993). Figure 10.18 shows the critical angle ω_{crit} for fracture link-up calculated from equation (10.7). For the inclined fractures with the angle ω_f above the critical angle curve, the fractures will not link up. For the inclined fractures with ω_f below the critical angle curve, the fractures will link up. However, ω_{crit} is a function of p_w. For initial fractures with a given ω_f, the critical wellbore pressure p_{link} at which the initial fractures just link up to form the axial fractures can be estimated by substituting ω_{crit} with ω_f in equation (10.7) and solving the equation for p_w.

Thus, the procedure to estimate p_{link} is summarized as follows:
(a) For a given set of remote stresses and wellbore orientation, estimate ω_f using the methodology presented in Chapter 8.
(b) Estimate the wellbore pressure p_w which satisfies $p_w = s_{norm}$. Note that s_{norm} is a function of p_w. The estimated p_w is denoted here as p^*.
(c) If $s_{norm} > s_{para}$ at $p_w = p^*$, then $p_{link} = p^*$.
(d) If $s_{norm} \leq s_{para}$ at $p_w = p^*$, then p_{link} is estimated by substituting ω_{crit} with ω_f in equation (10.7) and solving the equation for p_w.

Again, returning to the anecdotal case above, the fact that drilling could be resumed with a mud weight of 14 ppg, about 1 ppg in excess of the least principal stress, is explainable in terms of the fact that for the angle at which the fracture forms at the wellbore wall (corresponding to the deviation of the well and stress state), the fracture link-up pressure was about 1 ppg above S_3.

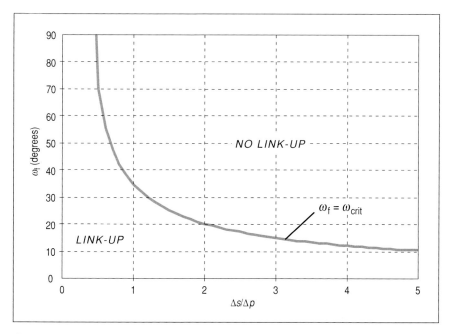

Figure 10.18. The critical fracture angle, ω_{crit}, at the wellbore wall that controls the link-up of inclined fractures (required to form large fractures propagating away from the wellbore) is a function of the initial fracture angle, ω_f, and Δs and Δp, as defined in the text (after Ito, Zoback *et al.* 2001). © *2001 Society Petroleum Engineers*

Another practical example of this theory is shown in Figure 10.19 for a deviated well being planned for the Caspian Sea following Ito, Zoback *et al.* (2001). In this case, the well plan utilizes the wellbore collapse pressure as the lower bound for the safe drilling mud window. In the case when the least principal stress was used as the upper bound of the mud window, four strings of casing would be required (Figure 10.19, left). However, taking into account the fact that there cannot be lost circulation if there is no fracture link-up and using the link-up pressure as the upper bound for the mud window allows the well to be drilled with one fewer casing strings (Figure 10.19, right). While the p_{link} is appreciably greater than the least principal stress over a considerable range of depths, the fact that p_{link} was greater by about 0.2 sg at ∼4350 m meant that it would be possible to run the 11.75 casing to TD.

Wellbore ballooning

Figure 7.6 introduced the subject of wellbore ballooning but did not offer a physical explanation of the phenomenon. The discussion above offers one possible explanation. Should wellbore pressure be high enough to induce *en echelon* drilling-induced fractures in the wellbore wall but not high enough for the fractures to link up and propagate away

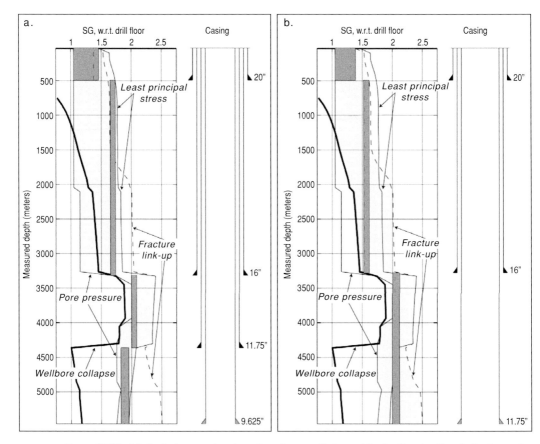

Figure 10.19. Mud windows and casing plans for a well planned in the Caspian Sea. In both panels the lower bound of the safe drilling window is the wellbore collapse pressure. The two cases shown consider the least principal stress as the upper bound of the safe drilling window (a) and p_{link} (b). By utilizing p_{link} we achieve a sufficient increase in the mud window to save one casing string. After Ito, Zoback *et al.* (2001). © *2001 Society Petroleum Engineers*

from the wellbore, there would be the potential to store an appreciable amount of drilling mud in the small fractures in the wellbore wall – explaining the balloon type behavior referred to in Figure 7.6. As mentioned previously, any indication of ballooning in PWD data should be taken as a warning that the downhole pressure is close to the pressure at which lost circulation could occur.

Mud penetration and time-dependent wellbore failure

As alluded to in Chapter 6, Paul and Zoback (2006) discussed the process of time-dependent wellbore failure of the SAFOD scientific research borehole in the terms of mud penetration into the rock surrounding the wellbore over time. A wellbore deviated 55° from vertical was being drilled through highly faulted rocks composed of arkosic sandstones and conglomerates with interbedded shales to the southwest of the San Andreas fault and siltstone and mudstones associated with the Great Valley formation on the northeast side of the fault. After successfully drilling through the fault during the summer of 2005, comprehensive geophysical logging showed the near-vertical San Andreas fault zone to be characterized by a broad ($\sim$250 m wide) zone of anomalously low P- and S-wave velocity and resistivity, with multiple discrete, narrower ($\sim$3–10 m wide) zones of even lower velocities and resistivities.

Paul and Zoback (2006) did a wellbore stability analysis of the section of the borehole penetrating the fault zone based on observations made at shallower depth. There was considerable uncertainties in both the stress state and rock strength. Quantitative risk assessment was used in a manner similar to that described earlier in this chapter. As shown in Figure 10.20a, logging-while-drilling (LWD) acoustic caliper data indicated that the mud weights used to drill the well (compared with the range of values of minimum mud weight based on the wellbore stability analysis) was adequate to keep the borehole diameter essentially in gauge during drilling. The acoustic caliper data indicate the hole diameter in a vertical and horizontal direction. The broad zone associated with anomalous geophysical properties alluded to above extends from 3190 to 3410 m and an actively creeping strand of the San Andreas has been observed at 3300 m. In the lower part of the interval logged (from 3630–3700 m) there was a moderate degree of failure as the wellbore was drilled. Everywhere else, the hole was essentially in gauge as it was drilled. In marked contrast, after drilling was completed (approximately three weeks later), six-arm caliper logs indicated substantial enlargement of the wellbore with time (Figure 10.20b). As this was observed in the arkosic, silty and shaley formations, it was not interpreted to be the result of water/rock chemical interactions lowering formation strength over time. In addition, as there has been a substantial amount of drilling in the Great Valley formation without significant problems associated with the use of water-based drilling muds, Paul and Zoback (2006) interpret the time-dependent failure of this hole in terms of time-dependent fluid penetration into the formation surrounding the wellbore because of the numerous fractures in these rocks. They argued that the increased pore pressure around the well with time reduced the effectiveness of mud weight to stabilize the hole and could cause cavitation of the rock surrounding the wellbore when the borehole pressure dropped due to shutting off the pumps or tripping out of the hole. Because the hole is highly deviated and the wellbore seems to be principally enlarged on the top, key seating could also be a source of erosion of the wellbore wall, especially on the top of the hole. In reality, it is likely a combination of these processes that was responsible for the time-dependent hole failure.

Preventing sand production

The final topic considered in this chapter is related to failure of the formation surrounding a wellbore during production. This is frequently referred to as sand production.

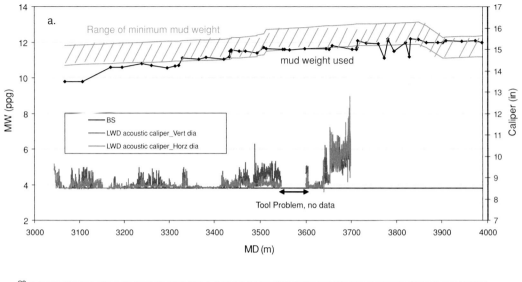

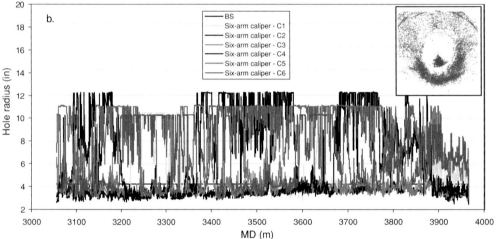

Figure 10.20. (a) Logging-while-drilling acoustic caliper data show relatively little borehole failure as the SAFOD borehole was being drilled in the vicinity of the San Andreas fault zone, confirming that the mud weights predicted by the wellbore stability analysis were essentially correct. A moderate degree of hole enlargement is seen in the deeper part of the interval logged with LWD (3630–3700 m). (b) Five weeks later, six-arm wireline calipers show deterioration of the borehole with time. In the inset, centralized six-arm caliper pads are plotted in a borehole coordinate system. The borehole diameter is highly enlarged at the top (the blue points indicate the approximate center of the logging tool). From Paul and Zoback (2006). (*For colour version see plate section.*)

This type of problem is best addressed through detailed numerical analysis which is beyond the scope of this book. This type of sophisticated analysis can incorporate the post-failure behavior of a formation which is essential for calculating the volume of produced sand at a given drawdown (*i.e.* production rate) and degree of formation depletion. This said, there are several principles that can be illustrated using the types of analytical wellbore failure models discussed above.

Figure 10.21a illustrates the relative stability of wells drilled at different azimuths and deviations in the Cook Inlet in Alaska (Moos, Zoback *et al.* 1999). The question addressed in that study was whether during production it might be possible to leave multi-lateral wells uncased near the join between the multi-lateral and the main hole. When an uncased well is put into production, the pressure in the wellbore is lower than the pore pressure. Hence, the well is more unstable during production than it is during drilling, somewhat analogous to underbalanced drilling.

The geomechanical model developed for this field predicted that highly deviated wells drilled at an azimuth of ∼N30°W (or S30°E) are most stable whereas those drilled to the ENE or WSW are most unstable. What gave this prediction added credibility is that the drilling history of two near-horizontal wells (shown in Figure 10.21b and pre-dating the geomechanical analysis) was revealed only after developing the geomechanical model upon which Figure 10.21a was developed. The well drilled to the NNW was drilled without wellbore stability problems, whereas that drilled to the NE had severe wellbore stability problems. As such results were consistent with the conclusions derived from the geomechanical analysis, it provided still more evidence that the geomechanical model was correct.

Once the most stable direction for drilling was established, the next step to address was identification of the depths at which the strongest rocks were found. These intervals are preferred as the kick-off depths for the multi-laterals. This was accomplished through utilizing log-based strength estimates calibrated by laboratory tests on core. In fact, equation (5) in Table 4.1 was derived in this study. Once the most stable drilling direction and depths were identified, the key operational question to address was whether or not the well would remain stable as production and depletion occur over time.

The results presented in Figure 10.22a,b address the question of stability of the uncased multi laterals for the case of wells drilled only at the most stable depths and in the most stable direction. Figure 10.22a shows the amount of drawdown in the region around the well associated with a modest rate of production (∼500 psi). A finite element analysis was used to do this calculation. After calculating the effect of the pore pressure change on stresses around the wellbore, the stability of the producing wellbore was calculated assuming a uniaxial compressive strength of 10,000 psi (typical of the stronger intervals in the well). As shown in Figure 10.22b, only a modest degree of wellbore failure is expected to result from the change in pore pressure and stress around the wellbore associated with production. However, Figure 10.22c shows the

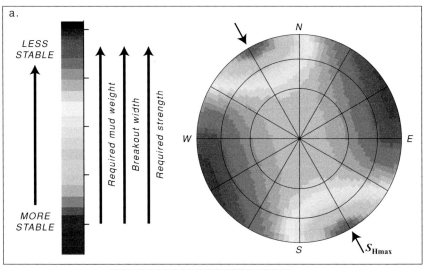

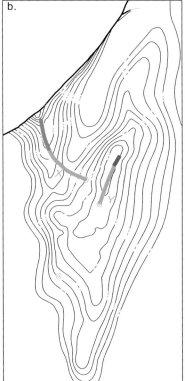

Figure 10.21. (a) Relative stability of multi-lateral wells drilled at various orientations in the Cook Inlet (modified from Moos, Zoback *et al.* 1999). Note that highly deviated wells drilled to the NW and SE are expected to be stable whereas those drilled to the NE and SW are not. (b) Following development of the analysis shown in (a) it was learned that well X (drilled to the NW) was drilled without difficulty whereas well Y (drilled to the NE) had severe problems with wellbore stability. © *1999 Society Petroleum Engineers. (For colour version see plate section.)*

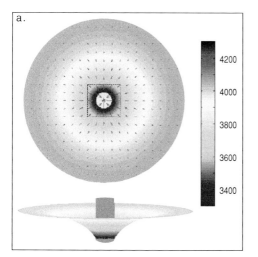
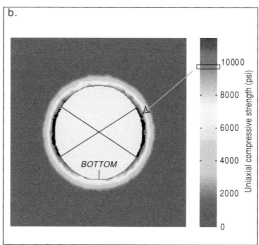
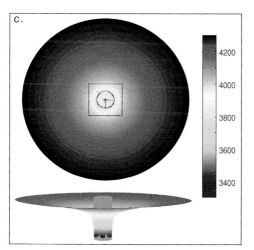
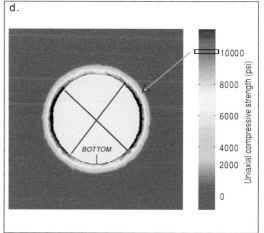

Figure 10.22. Pressure drawdown and predicted width of the failure zone for the Cook Inlet multilateral study shown in Figure 10.21 (modified from Moos, Zoback *et al.* 1999). (a) and (b) Calculations are for the case where a 500 psi drawdown is achieved slowly. Note that the zone of rock failure (breakout width) is limited to about 60°. (c) and (d) Calculations for a very rapid drawdown of 1000 psi. Note that the region of failure around the well is much more severe. © *1999 Society Petroleum Engineers. (For colour version see plate section.)*

same calculation as Figure 10.22a, but for a greater rate of production. The pressure drawdown around the well is much more severe in this case. The resultant effect on the stability of the well is shown in Figure 10.22d. In this case, more than half the wellbore circumference goes immediately into failure. As this situation only gets worse as the depletion occurs over time, it was decided not to leave the multi-laterals uncased because overall, the requirements of maintaining a stable uncased multilateral are too restrictive. It requires kicking off at only the depths where sufficient rock strength is

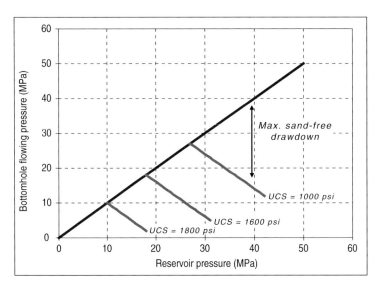

Figure 10.23. The likelihood of sand production in an uncased well as a function of depletion (expressed as reservoir pressure) and drawdown, or production rate (expressed as bottom hole flowing pressure) for formations of variable strength (courtesy M. Brudy). For weak formations with a uniaxial compressive strength of 1000 psi, the maximum drawdown without sand production is over 20 MPa when the reservoir pressure is 40 MPa, but less than 6 MPa when the reservoir pressure is 30 MPa. Hence, reducing production rate can only limit sand production in weak formations prior to significant depletion.

found, drilling only in the most stable direction and limiting the production rate – altogether too restrictive for efficient operations.

Figure 10.23 (courtesy M. Brudy) shows the way in which sand production in an uncased well is related to depletion (expressed as formation pressure) and production rate (expressed as bottomhole flowing pressure) for formations of variable strength. For the stress conditions appropriate to this case, it is clear that in relatively weak formations, the maximum drawdown without sand production is over 20 MPa when the reservoir pressure is 40 MPa, but less than 6 MPa when the reservoir pressure is 30 MPa. Stronger formations can experience appreciably more drawdown at either reservoir pressure. It is also clear in Figure 10.23 that reducing production rates can limit sand production in weak formations prior to depletion.

Using a finite element model of a cased, cemented and perforated well, it is possible to consider the use of perforation orientation to prevent sand production as discussed by Morita and McLeod (1995). Intuitively, one can see that in weak formations, perforating at the azimuth of the minimum compressive stress would not be advisable as one would be perforating at the azimuth where breakouts form and where the formation might already be subjected to very high compressive stress. Figure 10.24a (courtesy M. Brudy)

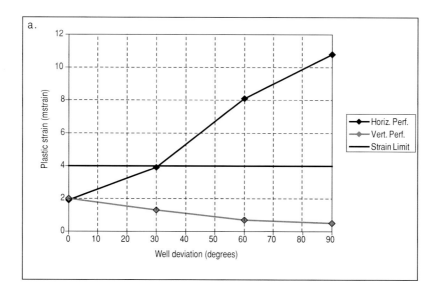

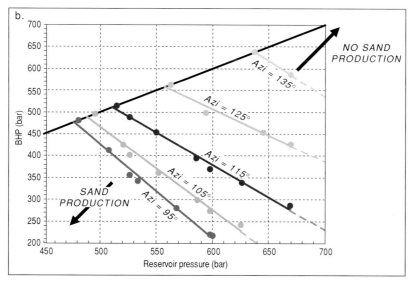

Figure 10.24. (a) Example showing the influence of the orientations of perforations on sand production in wells of varying deviation. Sand production is expected at a plastic strain of approximately 0.4%. As seen in the figure, vertical perforations are stable in wells of all deviations whereas horizontal perforations are unstable in wells deviated more than 30°. (b) Maximum bottom hole flowing pressure for a cased well deviated 60° with horizontal perforations. By varying well azimuth it is possible to alter the degree of sand production. Wells drilled at an azimuth of 130° are most unstable (sand production occurs with only minor depletion or drawdown) whereas wells drilled at an azimuth of 90° are much more stable. (Courtesy M. Brudy.)

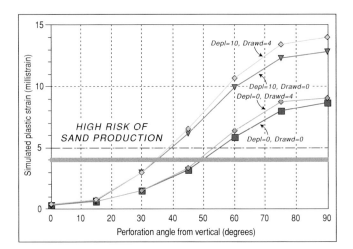

Figure 10.25. An illustration of the relation between perforation angle as a function of drawdown and depletion in a horizontal well in a normal faulting environment (courtesy M. Brudy). Near-vertical perforations are stable at all values of drawdown and depletion. As the perforations are oriented more horizontally, they become unstable. The angle at which sand production occurs depends on the degree of drawdown and depletion.

shows the influence of the orientation of perforations on sand production in wells of varying deviation. It is known that in the formation in question, sand production can be expected at a plastic strain of approximately 0.5%. As seen in the figure, perforations at the top and bottom of the well (regardless of deviation) are stable, whereas horizontal perforations are unstable in wells deviated more than ~40°.

To consider still another case study, Figure 10.24b is similar to what is shown in Figure 10.23 except that it is for a cased well deviated 60° from vertical with horizontal perforations. This figure shows the effect of varying well azimuth on sand production. In this case, wells drilled at an azimuth of 135° are most unstable (sand production occurs with only minor depletion or drawdown) and wells drilled at an azimuth of 90° are most stable as sand production does not occur unless there is appreciable drawdown or depletion. As was the case in the example shown in Figure 10.23, the reason for this is that the perforations in the region of highest stress concentration around the well are most likely to produce sand.

A final example of the value of oriented perforations is presented in Figure 10.25 (courtesy M. Brudy) which shows the combined effects of perforation angle, depletion and drawdown in a horizontal well in a normal faulting environment where the vertical stress is the maximum principal stress. Note that sand production is predicted (*i.e.* plastic strains exceed the expected limit of ~0.5%) for subhorizontal perforations, regardless of the depletion and drawdown. For zero depletion and drawdown, sand production is expected for perforations that are 50° from vertical. As depletion and drawdown increase, sand production is expected for perforations only $35°$ from vertical.

Overall, the examples shown in Figures 10.23–10.25 illustrate the value of numerical modeling in the context of a comprehensive geomechanical model that includes knowledge of the magnitude and orientation of the principal stresses, the strength of the formation the reservoir pressure and the likely changes in pressure associated with both drawdown surrounding the well and long-term depletion of the reservoir.

11 Critically stressed faults and fluid flow

In this chapter I consider three topics related to fluid flow in fractured and faulted oil, gas and geothermal reservoirs. First, I consider the influence of fractures and faults on reservoir permeability with the associated implications for permeability anisotropy within a fractured and faulted reservoir. Second, I consider geomechanical controls on fault sealing and leakage in fault bounded reservoirs. Finally, I consider dynamic constraints on hydrocarbon column heights and reservoir pressures, again in fault bounded reservoirs. A common element of the discussions of these three subjects is fluid flow along critically stressed faults. As discussed in Chapters 4 and 9, the fact that the state of stress in many reservoirs is found to be controlled by the frictional strength of optimally oriented faults has important implications for fluid flow at a variety of scales.

Each of the topics considered in this chapter is of considerable practical interest during the development of oil, gas and geothermal reservoirs. In reservoirs with low matrix permeability, there might be no significant fluid flow in the absence of permeable fractures and faults. Moreover, it is frequently the case that relatively few fractures and faults serve as the primary conduits for flow (Long *et al.* 1996). The question, of course, is to know which of the fractures and faults that might be present in a reservoir are most likely to be hydraulically conductive and why. Accordingly, in the first sections of this chapter I discuss how critically stressed fractures affect permeability in rocks with relatively low matrix porosity and present several case studies that illustrate the principle. I then go on to present a way to use observations of wellbore breakouts and tensile fractures with image logs to identify whether there might be active faults in a reservoir and discuss the use of intentionally induced seismicity to enhance permeability in very low-permeability reservoirs.

There are a number of questions related to how reservoirs become compartmentalized (as discussed in Chapter 2) that are generally considered as a question of *fault seal*. Why do some faults isolate pressure and flow from adjacent portions of a reservoir but other faults do not? Note that the fault that isolates fault block A from fault block B in Figure 2.7 is obviously a sealing fault, whereas the fault which separates fault block B from fault block C is not. The degree to which such behavior could be predicted prior to field development (and answering such questions only after drilling and/or depletion) would appreciably benefit the efficient exploitation of many oil and gas reservoirs.

The subject of dynamic constraints on hydrocarbon column heights arises from the recognition that there appear to be many reservoirs around the world in which the volume of hydrocarbons is inconsistent with that expected from conventional interpretations based on such factors as structural closure, stratigraphic pinch-outs or cross-fault flow. As shown below, dynamic mechanisms not only have the potential for understanding such cases, but may shed light on enigmatic processes such as the accumulation of appreciable volumes of hydrocarbons in geologically young reservoirs that are isolated from possible source rocks by significant thicknesses of essentially impermeable shale (see Finkbeiner, Zoback *et al.* 2001).

Fractured reservoirs and permeability anisotropy

In this section I revisit some of the arguments introduced in Chapter 5 related to the importance of faults (as opposed to Mode I fractures) on fluid flow at depth. Because rock may have several different sets of fractures and faults that have been introduced at different times during its geologic history (in potentially different stress fields), it is important to have a criterion that allows us to determine which of the faults are hydrologically conductive today. Figure 11.1a schematically illustrates a hypothesis introduced by Barton, Zoback *et al.* (1995) that we will refer to in the context of flow through faulted and fractured rock as the *critically-stressed-fault* hypothesis. Briefly, in a formation with faults at a variety of angles to the current stress field (illustrated by the light and dark lines in the cartoon map), the faults that are hydrologically conductive today are those that are critically stressed in the current stress field. We generally assume coefficients of friction of 0.6–1.0 as measured for a wide variety of crustal rocks (Byerlee 1978) for reasons that were discussed previously in Chapters 4 and 9. In other words, because the gray faults in Figure 11.1a are at the appropriate angle to the current stress field as to be mechanically active (as shown on the right side of the Mohr diagram in Figure 11.1a) they are expected to be hydraulically active. The dark faults in Figure 11.1a were, of course, active at some time in the past, but because they are not active today, they are not currently hydraulically conductive. In short, the *critically-stressed-fault* hypothesis posits that *faults that are mechanically alive are hydraulically alive and faults that are mechanically dead are hydraulically dead.*

The fault and fracture data shown in Figure 11.1b are derived from the fault and fracture data shown previously in Figure 5.10 obtained in the highly fractured granitic rock of the Cajon Pass scientific borehole near the San Andreas fault in Southern California. By using detailed temperature logs, Barton, Zoback *et al.* (1995) separated the fault population into permeable faults and impermeable faults. By plotting each set of faults separately in normalized 3D Mohr diagrams (normalized by the vertical stress) in Figure 11.1b, it is clear that the great majority of permeable faults are critically

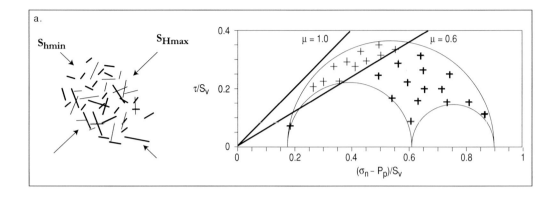

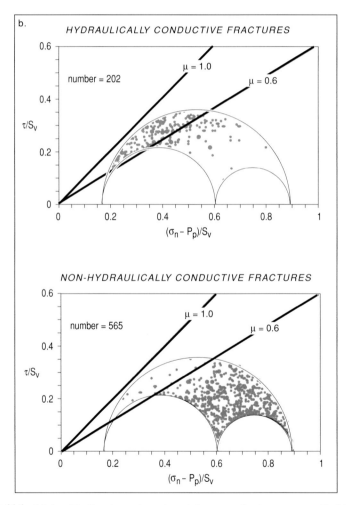

Figure 11.1. (a) A critically stressed crust contains many fractures, some of which are active in the current stress field (light line faults in cartoon on left and light + marks in the normalized Mohr diagram) and some of which are not (heavy line faults and heavy + marks). (b) In the context of the *critically-stressed-fault* hypothesis, hydraulically conductive faults are critically stressed faults (upper diagram) and faults that are not hydraulically conductive are not critically stressed. After Barton, Zoback *et al.* (1995). (*For colour version see plate section.*)

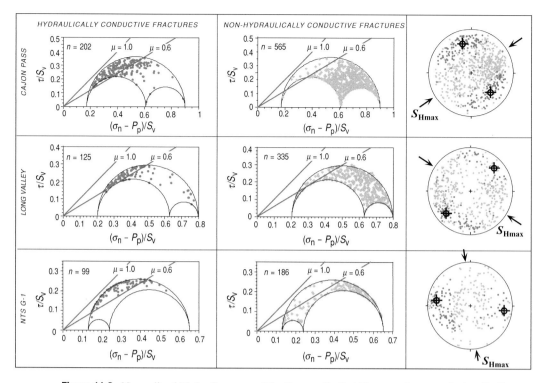

Figure 11.2. Normalized Mohr diagrams of the three wells that illustrate that most hydraulically conductive faults are critically stressed faults (left column) and faults that are not hydraulically conductive are not critically stressed (center column) along with stereonets that show the orientations of the respective fracture sets. The first row shows data from the Cajon Pass well (same Mohr diagrams as in Figure 11.1b), the second from the Long Valley Exploration Well and the third from well G-1 at the Nevada Test Site. After Barton, Zoback *et al.* (1995). (*For colour version see plate section.*)

stressed (upper Mohr diagram), whereas those that are impermeable are not critically stressed (lower Mohr diagram).

Figure 11.2 shows the three sets of data originally presented by Barton, Zoback *et al.* (1995), for the Cajon Pass borehole (fractured granitic rock), the Long Valley Exploratory Well in eastern California (fractured metamorphic rock) and test well USW G-1 at the Nevada Test Site (tuffaceous rock). In the left column, the normalized 3D Mohr diagram is shown with the hydraulically conductive faults (blue dots) and the middle column shows the hydraulically dead fractures (green dots). In all three wells, it is clear that the great majority of hydraulically conductive faults are critically stressed whereas those that are not conductive are not critically stressed.

The stereonets on the right side of Figure 11.2 show the orientations of the fault populations plotted in the corresponding Mohr diagrams. The symbols show the orientation of wells that would intersect the greatest number of critically stressed faults (subparallel to the blue fracture poles).

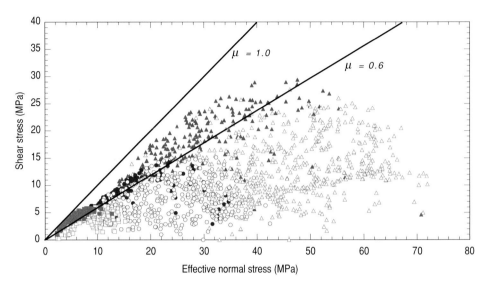

Figure 11.3. Shear and normal stresses on fractures identified with borehole imaging in Cajon Pass (triangles), Long Valley (circles), and Nevada Test Site (squares) boreholes. Filled symbols represent hydraulically conductive fractures and faults, and open symbols represent non-conductive fractures. From Townend and Zoback (2000) based on original data in Barton, Zoback *et al.* (1995). (*For colour version see plate section.*)

The stereonets shown in the right column of Figure 11.2 introduce several basic principles that will be illustrated further in the examples considered below:
• Faults and fractures are often observed at many orientations, implying that they formed at various times during the geologic history of the formation.
• Simple conjugate sets of faults are not usually identifiable.
• The subsets of permeable faults have orientations controlled by the current stress state (blue dots). This is normal/strike-slip for Cajon Pass and normal for Long Valley and the Nevada Test Site. The subset of permeable faults in any given well is not the same as the most significant concentrations. The last point is seen most dramatically in the Cajon Pass data set.

Figure 11.3 presents the data shown in the normalized Mohr diagrams in Figure 11.2 in a single Mohr diagram that is not normalized by the vertical stress (after Townend and Zoback 2000). Each data point indicates the shear and effective normal stress acting on a given fault. Colors and symbol shape distinguish the data from the three wells. Filled symbols indicate hydraulically conductive faults and open symbols indicate hydraulically dead faults. Note that independent of the effective normal stress acting on a given fault, the tendency for a fault to be hydraulically conductive depends on the ratio of shear to effective normal stress, with the majority of conductive faults having a ratio of shear to effective normal stress consistent with coefficients of friction between $\sim$0.6 and $\sim$0.9. The reason for this, we believe, is that in fractured and faulted siliciclastic rocks, most geologic processes, such as precipitation, cementation and alteration of feldspars to clays cause faults (and the secondary fractures and faults in the damage zones adjacent to them) to seal over time. Mechanical processes associated with active faulting such as brecciation counteract these tendencies and help maintain permeability within the faults and in the damage zones adjacent to them. Townend and Zoback (2000) argue that it is the presence of critically stressed faults deep within the brittle crust that keeps the bulk permeability of the crust about four orders of magnitude greater than intact rock samples subjected to appropriate confining pressures.

It is well known that laboratory studies show that the permeability of faults and fractures is a strong function of effective normal stress (Kranz, Frankel *et al.* 1979; Brace 1980; Brown and Scholz 1985). The data shown in Figure 11.3 would seem to contradict this as the tendency for faults to be permeable appears to be independent of normal stress. However, while the critically-stressed-fault hypothesis may help us understand which faults are likely to be permeable, a number of other factors determine what the actual permeability is likely to be. For example, for a given permeable fault, a number of geologic factors control its permeability such as the degree of alteration and cementation of the brecciated rock within the fracture and its diagenetic history (Fisher and Knipe 1998; Fisher, Casey *et al.* 2003), as well as the current effective normal stress.

It is probably useful to discuss briefly why the critically-stressed-fault hypothesis works, and what might be its extent of usefulness (some of which will be discussed below in the context of specific case studies). In the context of the critically-stressed-fault hypothesis, the increase in permeability associated with critically stressed faults results from brecciation during shearing (Figure 5.2) and formation of a damage zone adjacent to the faults (*e.g.* Chester and Logan 1986; Antonellini and Aydin 1994; Davatzes and Aydin 2003, and many others). In formations like the diagenetically immature shales of the Gulf of Mexico (the so-called *gumbo* shales) or diagenetically immature siliceous rocks in which silica is in the form of Opal A (a form of SiO_2 that deforms ductily), shearing would not cause brecciation so slip on active faults may not contribute significantly to formation permeability. The effects of fractures and faults in both of these lithologies are discussed in case studies below. In carbonates, both dissolution and precipitation influence permeability, but there still might be an important role for faulting to contribute to permeability. For example, faulting and brecciation may occur along planes originally formed through dissolution processes and there is no reason to reject out-of-hand the critically-stressed-fault hypothesis for carbonate rocks. The importance of brittle faulting in contributing to bulk permeability in chalk reservoirs of the North Sea is discussed in Chapter 12.

Another geologic setting in which critically stressed faults may not directly contribute significantly to formation permeability is in the case of porous, poorly cemented sandstones and diatomites. In such lithologies, either compaction bands, planes of reduced porosity but little shearing (see Mollema and Antonellini 1996 and Sternlof, Rudnicki *et al.* 2005), or shear bands, planar bands of reduced porosity that form in association with faults (Antonellini and Aydin 1994), may be planes of lower permeability than the matrix rock. As such phenomena are associated with formations of high porosity and permeability formations, the degree to which compaction bands and shear bands affect reservoir permeability is unclear. Sternlof, Karimi-Fard *et al.* (2006) discuss the effects of compaction bands on reservoir permeability. Faults and shear bands in highly porous sands can cause permeability reduction due to the communition and porosity reduction associated with shearing, but the damage zone adjacent to faults and shear bands is an area of enhanced permeability (Antonellini and Aydin 1994). This can result in a situation where cross-fault flow is impeded but flow parallel to the fault plane is enhanced in a *permeability halo* surrounding the fault. This *halo* would result from a damage zone adjacent to the fault that consists of numerous, relatively small critically stressed faults. A similar phenomenon is likely associated with large displacement faults that are frequently associated with a relatively impermeable fault core, consisting of ultra-fine grained cataclasite (Chester, Chester *et al.* 2005) that result from pervasive shearing. In this case too, the fault may be relatively impermeable to cross-fault flow but flow parallel to the fault may be appreciably enhanced in the damage zone surrounding it.

For the cases in which the critically-stressed-fault hypothesis is most likely to be applicable (brittle rocks with low matrix permeability), it is worth considering the implications for permeability anisotropy in a highly fractured medium. As noted in Chapter 5, the implications of the critically-stressed-fault hypothesis for permeability anisotropy in a fractured reservoir are markedly different from those that arise assuming that maximum permeability is parallel to S_{Hmax} because it is controlled by Mode I fractures. This is easily seen in the idealized view of the orientations of critically stressed faults in different tectonic regimes presented in Figure 5.1. As can be seen in that figure, in a normal faulting environment (row 2), if there are conjugate sets of normal faults present (as theoretically expected), the direction of maximum permeability will be subparallel to S_{Hmax} (similar to what would be seen with Mode I fractures, row 1) but the dip of the normal faults will also have an effect on permeability anisotropy. In a strike-slip faulting environment (row 3), conjugate faults would cause flow to be greatest at directions approximately $30°$ to the direction of S_{Hmax}, significantly different from what is expected for Mode I fractures. In reverse faulting environments (row 4) flow along critically stressed conjugate faults would be maximum parallel to the direction of S_{hmin}, orthogonal to the direction of S_{Hmax}.

While the idealized cases shown in Figure 5.1 are helpful in a general sense, there are many places around the world characterized by normal/strike-slip faulting stress states ($S_{\mathrm{V}} \sim S_{\mathrm{Hmax}} > S_{\mathrm{hmin}}$) or reverse/strike-slip faulting ($S_{\mathrm{Hmax}} \sim S_{\mathrm{hmin}} > S_{\mathrm{V}}$) which make the idealized cases shown in Figure 5.1 overly simplified because multiple fault sets are likely to be active. In other words in a normal/strike-slip stress state, one might observe any number of the sets of active faults that are shown in the idealized stereonets for normal and strike-slip faulting in Figures 5.1b,c. An analogous situation is true for a reverse/strike-slip faulting environment (Figures 5.1c,d). An example of such a case will be considered below.

Also, it is important to remember that it is relatively rare to see equally well-developed sets of conjugate faults. This is illustrated by the case shown in Figure 11.4. The tadpole plots show that the great majority of fractures and faults encountered in this particular well are steeply dipping. The stereonets in the figure are colored to indicate how close a given fault is to failure in terms of the Coulomb Failure Function, CFF (defined by equation 4.40), or by the pore pressure required to induce slip on a given fault. As indicated, the distribution of faults and fractures seen in the image logs principally define two families of faults striking to the NE. The majority of the faults and fractures dip steeply to the WNW, but a significant number also dip to the ESE. Figure 11.4 compares the distribution and orientation of critically stressed faults for the case of an assumed normal faulting stress state (Figure 11.4a) and a reverse faulting stress state (Figure 11.4b). In both cases it is assumed that S_{Hmax} strikes N10°E and the vertical stress and pore pressure are the same. As can be seen in Figure 11.4a, in a normal faulting environment, many of these features are critically stressed (and cluster at several specific depths) and thus would be expected to be permeable. Note that the fracture set dipping to the WNW far outnumbers the conjugate set dipping to the ESE. These critically stressed faults would induce strong permeability anisotropy in the NNE direction. However, if this same distribution of fractures were encountered in a well located in a reverse faulting environment (Figure 11.4b), very few faults would be critically stressed. As illustrated in Figure 5.1d, critically stressed faults in a reverse faulting environment are expected to strike in the direction of S_{hmin} (in this case ~N80°W). As the observed distribution of faults in this well appears to define conjugate sets of normal faults (see Figure 5.1a), if the current stress field is characterized by reverse faulting, the steep dip of most of the faults would result in very few of them being active today, regardless of the orientation of S_{Hmax}. The case studies presented below will address these types of issues by presenting data from fractured and faulted reservoirs in a variety of rock types and stress states.

In the case studies presented in Figures 11.2 and 11.3, the subset of permeable faults was identified by comparing faults seen in image logs with high-precision temperature logs that reveal locally anomalous temperatures associated with small amounts of flow in or out of the well. In the cases presented below, we present cases in which temperature logs, packer tests and spinner flow meters were used (sometimes in combination) to identify permeable faults and fractures. Table 11.1 summarizes the various techniques used to identify fractures and faults in wells. Each of the techniques has advantages and disadvantages. While packer tests are the only way to measure the value of permeability of a fault or fracture quantitatively, it is very time consuming (and therefore expensive) to do many tests in any given well and setting packers in an open hole can involve appreciable risk. Spinner flow meters measure how much flow occurs out of a given fault, but they are only sensitive to relatively large flow rates. It has been

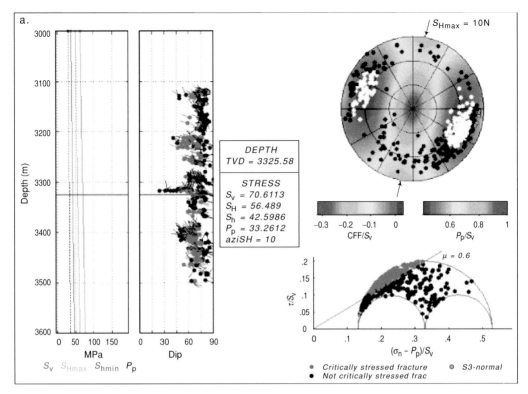

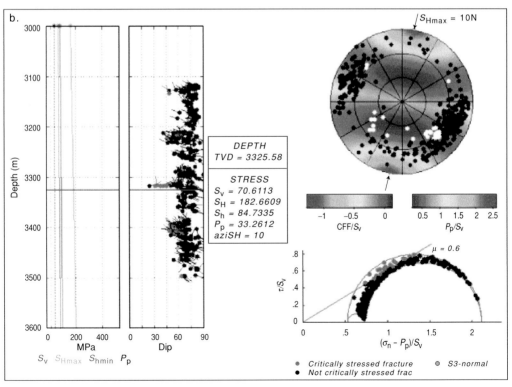

argued that Stonely wave analysis in full waveform sonic data and attribute analysis of electrical image data can discriminate permeable from impermeable faults in logging data but are not without drawbacks (Cheng, Jinzhong *et al.* 1987; Tang and Cheng 1996).

Some representative case studies

One formation where the critically stressed fault hypothesis seems to be applicable is the Monterey formation of central California. The Monterey is a Miocene age siliceous shale with very low matrix permeability. The photograph in Figure 5.2 shows oil concentrated in brecciated fault zones of the Monterey formation (Dholakia, Aydin *et al.* 1998). The map of coastal California in the Point Arguello area (Figure 5.8) shows the location of four wells drilled in the Monterey formation (A, B, C and D) as well as stereonets showing the orientations of faults and fractures (and azimuth of S_{Hmax}) in each well. The stereonets from Figure 5.8 are shown again in Figure 11.5 for wells A, B and C with the tendency for fault slip to occur indicated by color. As in Figure 11.4, red indicates that the Coulomb Failure Function (CFF) is close to zero, hence slip is likely (equation 4.40), or equivalently that a small increase in pore pressure would be sufficient to induce slip. Poles to faults that are critically stressed in the reverse/strike-slip faulting stress state that characterizes this region today are shown in white. These are the same faults shown in red in the 3D Mohr diagrams that correspond to each stereonet.

Note that while the stress orientations are similar for each well (shown with the respective stereonets) and the stress magnitudes are assumed to be the same (corresponding to a reverse/strike-slip stress state), the fact that the fault distribution is so different in each well results in markedly different critically stressed fault orientations in each well and hence directions of enhanced permeability. For example, in well A (left figures) the numerous steeply dipping fractures striking to the northwest (dipping to the northeast) are not critically stressed. The critically stressed faults are principally the moderate dipping, conjugate reverse faults. These faults strike NW–SE, implying that this is the direction of maximum permeability. This direction is orthogonal to the maximum horizontal compression (analogous to what is shown schematically in Figure 5.1d). In contrast, in well B (center figures) the steeply dipping strike-slip faults

Figure 11.4. Illustration of the relationship between critically stressed fault orientations and absolute stress magnitudes (left column). In both cases, the direction of S_{Hmax} is N10°E. (a) For a normal faulting stress state, a large fraction of the fault population is critically stressed as they are well-oriented for slip in a normal faulting stress field (i.e. they strike NNE–SSW and dip relatively steeply). (b) In a reverse faulting stress state, very few of the faults are critically stressed because of the steep dip of the faults. (*For colour version see plate section.*)

Table 11.1. *Detection of permeable faults and fractures in wells*

Technique	Basis	Depth of investigation	Benefits	Drawbacks
Packer tests	Isolation of specific faults and fractures using packers allows the transmissivity (permeability times thickness) to be measured directly	Fault permeability in region surrounding the wellbore.	Determines absolute permeability	Very time consuming and costly to test numerous intervals
Thermal anomalies	Measures flow-induced thermal anomalies	Near wellbore	Easy to acquire and process data	Difficult to use if temperature log is noisy or if there are so many closely spaced fractures and faults that it is difficult to interpret
Electrical images	Quantifies electrical conductivity of fractures with respect to host rock	Near wellbore	Easy to acquire image data and identify fractures	Assumes fluid flow and electrical properties are related at the wellbore wall
Stoneley-wave analysis	Permeable fractures attenuate Stoneley waves	Near wellbore	Straightforward to implement and carry out waveform analysis	Relatively insensitive. Stoneley wave attenuation can be caused by various factors
"Spinner" flowmeter logs	Measures variation of flow rate with depth as the logging tool is lowered, or raised, in the well	Formation surrounding the wellbore	Directly measures fluid flow	Requires high flow rates

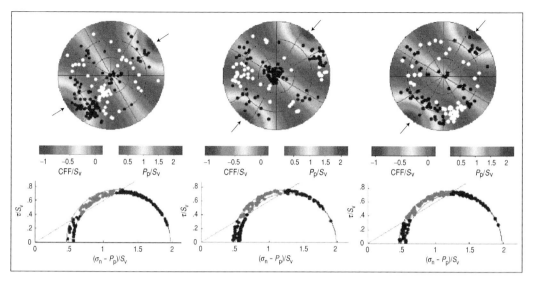

Figure 11.5. Faults identified in image logs from wells A, B, C in the Monterey formation of western California shown previously in Figure 5.8. As in Figure 11.4, the color of the stereonets indicates the tendency for fault slip to occur for a fault of given orientation, in terms of either the CFF or pore pressure needed to induce fault slip. Critically stressed faults are shown in white on the stereonet and red in the Mohr diagram. Note that although each well is considered to be in the same reverse/strike-slip stress state, the distribution of critically stressed faults in each well is quite different because of the distribution of faults that happen to be present in the three respective wells. (*For colour version see plate section.*)

that strike ∼N–S (dipping principally to the east) define the direction of maximum permeability (similar to some of the faults shown in Figure 5.1c). Finally, in well C, the majority of critically stressed faults are steeply dipping strike-slip faults that strike to the NE. Needless to say, the highly varied orientation of faults that might be encountered at a given site makes it difficult to generalize about permeability anisotropy on the basis of knowledge of the stress state alone. In other words, the general relationships shown in Figures 5.1b,c,d are valid in establishing a framework for understanding permeability anisotropy induced by critically stressed faults. This framework requires knowledge of both stress magnitudes and orientations as well as the distribution of faults in the location of interest.

As mentioned above, it is important to keep in mind that critically stressed faults are permeable because of the brecciation that accompanies fault slip. Hence, in well-cemented, brittle rocks with low matrix permeability, the brecciation that accompanies shear deformation on faults should clearly enhance permeability whereas in materials that shear without brecciation, critically stressed faults would not be expected to enhance permeability of the host rock. One example of this is presented below in the context of dynamic constraints on fluid flow along faults in the Gulf of Mexico. Another is illustrated in Figure 11.6 for measurements of permeability in the Monterey formation

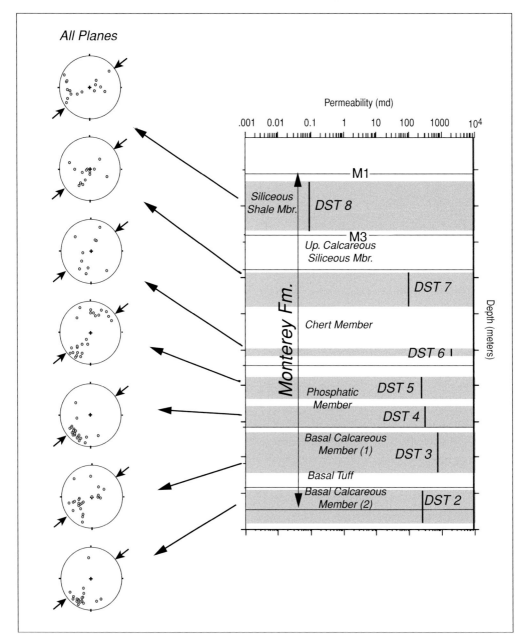

Figure 11.6. Permeability measurements from drill stem tests in well A (Figures 5.8 and 11.5). Note that the presence of critically stressed faults results in significant permeability increases (matrix permeability is negligible) except for the case of DST 8, where the silica is in the form of Opal A. From Finkbeiner, Barton *et al.* (1997). *AAPG© 1997 reprinted by permission of the AAPG whose permission is required for futher use.*

derived from drill-stem tests (DSTs) in well A of Figure 11.5 (and Figure 5.8) from Finkbeiner, Barton *et al.* (1997). Note that in DSTs 2–7 (all intervals tested except the shallowest) the bulk permeability of the tested interval is on the order of several hundred md, many orders of magnitude greater than the almost immeasurably small matrix permeability of the formations. However, in DST 8, the bulk permeability is only about 0.1 md, four orders of magnitude smaller than the deeper intervals. The reason for this is that in the formation in which DST 8 was conducted, silica is in the form of Opal A, an amorphous and weak mineral that smears out ductily with shear deformation. Hence, shearing on critically stressed faults did not contribute appreciably to matrix permeability. In the formations in which DST's 2–7 were conducted, the silica is in the form of Opal CT and quartz, which are both brittle minerals.

Figure 11.7a presents data from the Sellafield project in Great Britain (Rogers 2002), a site where research related to radioactive waste disposal is being conducted. On the left is a rose diagram showing the strikes of all faults and fractures in the test interval. The data show a broad distribution with concentrations of fractures with a strike of ∼N55°E and N65°W. The rose diagram on the right shows the orientations of permeable faults as determined from over 90 transmissivity tests. As this region is characterized by a strike-slip faulting regime with the direction of maximum horizontal stress is N30°W, the orientation of permeable faults is exactly what as predicted by the Critically-Stressed-Fault hypothesis for conjugate shear faults in a strike-slip faulting regime (Figure 5.1c). The majority of these tests were carried out in volcanic rock.

Figure 11.7b shows a seismic attribute analysis of a depth slice from a 3D seismic survey in the Mediterranean Sea in which the faults associated with fluid migration pathways are highlighted yellow (from Ligtenberg 2005). Known sealing faults strike in the orthogonal direction. Note that for the ENE local orientation of S_{Hmax}, the NNE striking permeable faults are exactly what would be expected for a strike-slip faulting regime. There is no conjugate fault set that is observable in the seismic data, but there may be subseismic faults visible in image log data with a NNW orientation. It should be noted that Ligtenberg (2005) argued that the highlighted faults are permeable because they are subparallel to the direction of S_{Hmax}, implicitly suggesting that the permeable faults are Mode I fractures.

Figure 11.8 shows a map of stress orientations and a schematic cross-section of the Dixie Valley, Nevada geothermal field. The map (after Hickman, Barton *et al.* 1997) shows the direction of the minimum horizontal principal stress, S_{hmin}, in geothermal wells drilled through alluvium and strata in Dixie Valley and into the Stillwater fault. This fault is the basin-bounding normal fault responsible for ∼3 km of uplift of the Stillwater range with respect to Dixie Valley over the past ∼10 million years (see cross-section). The geothermal reservoir is the fault and fracture system associated with the Stillwater fault at depth. Note that, as expected, the direction of least principal stress determined from wellbore breakouts and drilling-induced tensile fractures (the formation of which was substantially aided by cooling-induced stresses) is essentially orthogonal to the strike of the normal fault as expected (Figure 5.1b). Prior to the

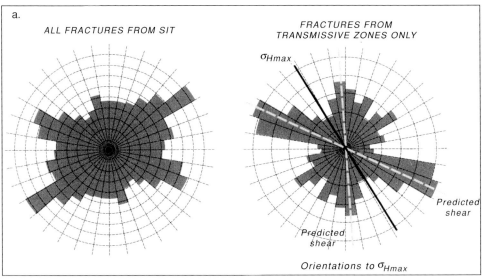

a.

ALL FRACTURES FROM SIT

FRACTURES FROM
TRANSMISSIVE ZONES ONLY

σ_{Hmax}

Predicted
shear

Predicted
shear

Orientations to σ_{Hmax}

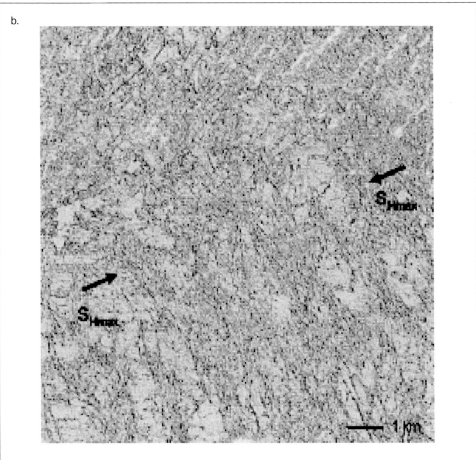

b.

S_{Hmax}

S_{Hmax}

1 km

Figure 11.7. (a) Rose diagrams showing the strike direction of all faults in a test well (Engelder and Leftwich) and only the strike direction of faults shown to be permeable in numerous packer tests (right). The orientation of permeable faults is consistent with the critically-stressed-fault hypothesis for strike-slip faulting (from Rogers 2002). (b) An attribute analysis of a depth slice from a 3D seismic survey in the Mediterranean Sea (from Ligtenberg 2005). Faults that are fluid migration pathways (yellow) are at the appropriate orientation to S_{Hmax} for a strike-slip faulting regime. (*For colour version see plate section.*)

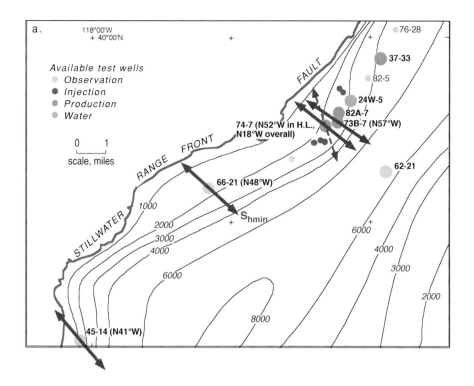

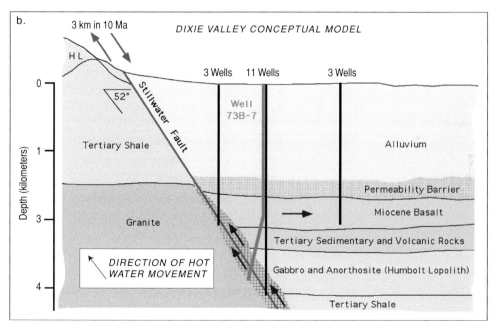

Figure 11.8. (a) Location map of the Dixie Valley geothermal area in central Nevada. Contours indicate depth to basement. Arrows indicate the direction of S_{hmin} from observations of wellbore breakouts and drilling-induced tensile fractures in the wells indicated. (b) Schematic cross-section of the Dixie Valley system showing that hot water comes into the fault zone reservoir at ~4 km depth. Prior to exploitation of geothermal energy (hot water that flashes to steam in producing wells), fluid flowed out of the fault zone and into fractured basalts beneath the valley. After Hickman, Barton *et al.* (1997). (*For colour version see plate section.*)

exploitation of the geothermal reservoir, the natural hydrologic system was characterized by fluid moving up the fault zone from ~4 km depth and then out into fractured basalt in the valley at 2–3 km depth (as shown in the cross-section).

The permeability structure of the Dixie Valley geothermal system appears to be the result of two competing processes: the creation of porosity and permeability by brecciation associated with slip along the Stillwater fault (and subsidiary faults) and the loss of permeability caused by precipitation of quartz in the fault zone. Prior to production of hot water for geothermal energy, the water coming into the fault zone at 3 km depth is saturated with silica at a temperature of ~250 °C (~450 ppm). The water leaving the fault at ~2.3 km depth at a temperature of ~220 °C also saturated with silica (~320 ppm) (Hickman, Barton *et al.* 1997). The drop in the amount of silica in water as it rises within the fault zone is the result of precipitation in the fault zone – in other words, the fault would gradually seal up, becoming a quartz vein – except for action of the fault slip events (earthquakes) which cause brecciation and create permeable pathways for fluid flow up the fault. The fact that geothermal systems such as the one in Dixie Valley are fairly rare in the basin and range province simply means that in most areas, fault sealing processes through quartz precipitation dominate the earthquake/brecciation process. Because of this, one of the critical questions the Dixie Valley study was designed to address is whether brecciation (in areas where the Stillwater fault is critically stressed) could be the reason why some parts of the fault zone are permeable (and can host an active geothermal system) and others not.

The Mohr diagrams in Figure 11.9a indicate that many of the faults encountered in well 73B-7 controlling fluid flow in the geothermal field at Dixie Valley are critically stressed (i.e. the Stillwater fault, subparallel faults and conjugate normal faults) in a manner similar to that illustrated in Figure 11.1. The great majority of faults that are not hydraulically conductive are not critically stressed. Note that in this case, all of the major flow anomalies (detected with spinner flowmeter logs) are associated with critically stressed faults (the + symbols in the Mohr diagram) as is the Stillwater fault itself (large dot). One of the hydraulically conductive normal faults is shown in the borehole televiewer image log in Figure 11.9b along with the associated temperature anomaly indicating significant fluid flow along this fault.

The chalk reservoirs of the central North Sea (such as Valhall and Ekofisk) are another example of the importance of critically stressed faults in controlling formation permeability. As shown in Figure 9.6 the stress state on the crest of the Valhall anticline reservoir prior to production was characterized by active normal faulting. The presence of permeable fractures and faults in Ekofisk raise the bulk permeability of the reservoir from the 0.1–1 md of the matrix to a mean effective permeability of ~50 md (Toublanc, Renaud *et al.* 2005). As depletion occurred during production (but prior to fluid injection to offset compaction and subsidence) normal faulting appears to have continued on the crest (Figure 9.6) and spread out to the flanks of the structures (see Chapter 12) due to poroelastic stress changes. In this highly compressible chalk, one would have

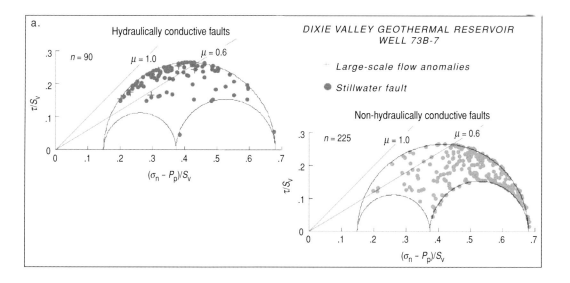

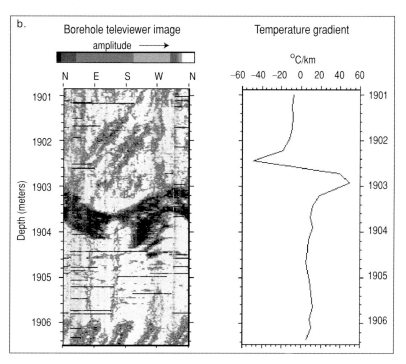

Figure 11.9. (a) Normalized Mohr diagrams for hydraulically conductive and non-conductive faults encountered in well 73B-7. Note that the great majority of the hydraulically conductive faults are critically stressed, especially the faults associated with the major flow zones. (b) An example of a significant temperature anomaly associated with a large normal fault in one of the Dixie Valley wells. After Hickman, Barton *et al.* (1997). *Reprinted with permission of Elsevier.*

expected significant decreases in permeability to accompany the increase in mean effective stress associated with depletion (Teufel 1992). However, Sulak (1991) pointed out that the permeability did not decrease as depletion and compaction occurred and that productivity remained high. We argue that the movement on active normal faults within the reservoir appears to have maintained formation permeability despite the compaction that was occurring. Toublanc, Renaud *et al.* (2005) argue that *tectonic* fractures and faults are critical in controlling fluid flow and use history matching to demonstrate the importance of permeable fractures and faults on reservoir performance. That said, they argue that two nearly orthogonal tectonic fault sets are present. If this is true, it would be inconsistent with active normal faulting as the latter would imply a uniform strike for permeable fault sets, essentially parallel to the direction of maximum horizontal compression (Figure 5.1b).

Identification of critically stressed faults and breakout rotations

In the sections above it was argued that the movement on critically stressed faults in reservoirs with low matrix permeability results in enhanced bulk permeability of the formation. Slip on faults causes the stress field surrounding the fault to be perturbed (see, for example, Pollard and Segall 1987). Hence, one way to readily determine if wells are being drilled in formations containing active faults is that fluctuations of stress orientation should be detected using observations of wellbore failure, such as stress-induced breakouts. Examples of this are illustrated in Figure 11.10. On the left, the profile of breakout orientations from 3000–6800 m depth in the KTB research hole in Germany is shown (Brudy, Zoback *et al.* 1997). Note that perturbations of stress orientation are seen at a wide variety of scales, superimposed on an average stress orientation with depth that is consistent with regional stress indicators. Breakouts also seem to stop abruptly at certain depths, only to resume at slightly greater depths. Borehole image data from the Cajon Pass research hole in southern California show similar fluctuations of breakout orientation over a 16 m long interval and 4 m interval (the middle and right panels), respectively (Shamir and Zoback 1992). Again, it can be seen that the fluctuations of breakout orientation occur at a variety of scales.

Shamir and Zoback (1992) considered a number of mechanisms that might be responsible for fluctuations of breakout orientation seen in the Cajon Pass well and concluded that the stress perturbations were due to fault slip. Not only was there a spatial correlation between breakout rotations and terminations, but using a dislocation model of slip on faults (after Okada 1992), they could explain the general nature of the breakout rotations observed as well as the cessation of breakouts often observed to be coincident with the location of the apparently causative faults (Figure 11.11). This was subsequently confirmed through more detailed modeling by Barton and Zoback (1994). In addition, through spectral analysis of the profile of breakout orientations in the Cajon Pass

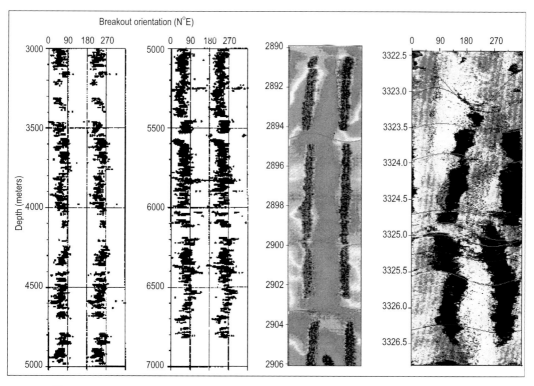

Figure 11.10. Examples of breakout orientations at a variety of scales. The two left panels show breakout orientations from 3000–6800 m depth in the KTB research hole in Germany (Brudy, Zoback *et al.* 1997). The middle and right panels (from the Cajon Pass research hole in southern California) show fluctuations of breakout orientation in image logs over a 16 m and 4 m interval, respectively (Shamir and Zoback 1992).

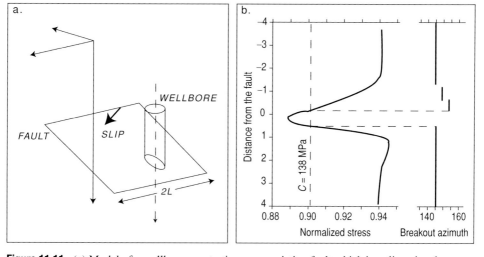

Figure 11.11. (a) Model of a wellbore penetrating a pre-existing fault which has slipped and perturbed the stress field in the surrounding rock mass. (b) The observed stress rotations (as well as the abrupt termination of breakouts) can be explained by the stress perturbations associated with fault slip. After Shamir and Zoback (1992).

hole, Shamir and Zoback (1992) showed that the number of stress fluctuations at various scales is quite similar to that of earthquake magnitude scaling. In other words, in a manner similar to the well-known Gutenberg–Richter relation (in a given region, there are approximately 10 times as many magnitude 3 earthquakes as magnitude 4, 10 times as many magnitude 2 events as magnitude 3, etc.), the frequency of stress orientation fluctuations is inversely related to the scale at which they are observed along the wellbore.

It should also be noted that anomalous stress orientations can be seen at much larger scales in regions affected by large earthquakes. An example of this at the field scale is seen in Figure 6.10 for an oil field in California. Many of the rotations occurring within the various structures of the field are undoubtedly associated with slip on the many active faults in the field. Another California example at the basin scale is seen in Figure 1.7, a stress map of the southern San Joaquin valley based on analysis of stress-induced breakouts in oil wells (Castillo and Zoback 1995). The rotation of the direction of maximum horizontal compressive stress rotates slowly from NE–SW in central California to ~N–S as the San Andreas fault (and subparallel fold and thrust belt) rotates to a more easterly strike. This is illustrated by the states of stress in the Yowlumne North, Paloma, Yowlumne and Rio Viejo fields. The anomalous NE–SW compressive stress field seen in the San Emidio, Los Lobos, Plieto, Wheeler Ridge and North Tejon fields have been modeled by Castillo and Zoback (1995) as having resulted from the stress perturbation caused by slip on the NE–SW trending reverse fault responsible for the 1952, M7+ Kern County earthquake. The southernmost fields shown in Figure 1.9 (characterized by NE–SW compressive stress direction) are located in the hanging wall of the thrust fault and were developed in the 1960s and 1970s. Hence, at the time these oil fields were developed, the stress state was already perturbed by the occurrence of the Kern County earthquake.

Intentionally induced microseismicity to enhance permeability

As discussed in Chapter 4, the triggering of induced seismicity by fluid injection in the Rocky Mountain arsenal and Rangely oil field in Colorado (Figure 4.22) demonstrated that lowering the effective normal stress on pre-existing fault planes can induce slip on otherwise stable faults. It is obvious that inducing seismicity and fault slip might be detrimental in many oil and gas fields (see Chapter 12). However, when permeability is very low (such as in tight gas shales), intentionally induced microseismicity by fluid injection can be used to stimulate formation permeability (Rutledge, Phillips *et al.* 2004; Maxwell, Urbancic *et al.* 2002). Operationally, this is sometimes called *slick-water frac'ing* because the technique that is used to induce fault slip is to induce a relatively large-scale hydraulic fracture with low-viscosity water rather than gel with proppant. Water is used in order to promote fluid penetration into pre-existing faults in the country rock adjacent to the plane of the hydrofrac. Hence, the microseismicity tends

Figure 11.12. Perspective view of four wells in the Yufutsu gas field, some of the larger faults in the reservoir and the cloud of microseismicity induced by injection of 5000 m^3 of water over 7 days (after Tezuka 2006). The cloud of seismicity is elongated along the direction of the vertical plane of a hydrofrac at the azimuth of S_{Hmax}.

to occur in a cloud around the fracture plane. When the state of stress is in frictional failure equilibrium (as is often the case, as demonstrated in Chapter 9) and there are pre-existing faults present at various orientations, fault slip on well oriented planes can be induced at an injection pressure essentially equivalent to the least principal stress. Similar operations have been carried out in hot-dry-rock geothermal systems where hydraulic fracturing and induced microseismicity are used to increase permeability and area of contact for heat exchange (Pine and Batchelor 1984; Baria, Baumgaertner *et al.* 1999; Fehler, Jupe *et al.* 2001).

An example of permeability stimulation through induced microseismicity in the Yufutsu gas field in Japan is shown in Figure 11.12 (from Tezuka 2006). Injection of 5000 m^3 of water over 7 days produced the cloud of seismicity shown in the figure, parallel to the NNE direction of S_{Hmax} in the field. The reservoir is in fractured granitic rock with both large and small scale faults in the reservoir. A strike-slip faulting stress state (in frictional equilibrium) exists in the reservoir. An injectivity test carried out after inducing microseismicity resulted in a factor of 7 increase. Interestingly, modeling of flow in the reservoir also confirms the applicability of the critically-stressed-fault as fluid flow seems to be dominated by critically stressed faults in the reservoir (Tezuka 2006). The same interpretation was reached following slick-water frac operations in the Cotton Valley formation of east Texas (Rutledge, Phillips *et al.* 2004).

Fault seal/blown trap

In this and the following section I consider the critically-stressed-fault hypothesis in the context of faults that cut through reservoirs. If a fault cutting through a reservoir is a sealing fault, it could potentially compartmentalize a reservoir (as discussed in Chapter 2) which is an issue of appreciable importance as a field is being developed. In fact, it is important to know which faults in a reservoir might be sealing faults and capable of compartmentalizing a reservoir and which are not. A second reason flow along critically stressed faults may be important to the question of fault seal relates to the understanding of why some exploration targets show evidence of potentially commercial quantities of hydrocarbons in the past which are not present today. This is sometimes known as the *blown trap* problem. To the degree that the reasons for this can be understood, some exploration failures might be avoided.

While this section (and the one that follows) focuses on natural process linking fault slip and hydrocarbon accumulations, anthropogenic processes such as water flooding, steam injection or hydraulic fracturing can potentially induce fault slip and potentially induce hydrocarbon leakage. Although this topic is not discussed explicitly here, the principles being discussed can easily be applied to the case of injection-related pressure increases.

To address the problem of slip on reservoir bounding faults that may not be planar, breaking up the fault into small, planar fault sections allows us to assess shear and normal stress on each section, as we did for small faults within a reservoir earlier in this chapter. We now express the likelihood of fault slip in terms of the amount of excess pressure required to cause fault slip resulting from a trapped, buoyant column of hydrocarbons on one side of the fault. The likelihood of fault slip directly affects the potential sealing capacity of a given reservoir-bounding fault. As discussed in Chapter 2, when relatively buoyant hydrocarbons accumulate in a permeable reservoir bounded by a sealing fault, the pore pressure at the fault–reservoir interface increases because the pore-pressure gradient in the hydrocarbon column is considerably less than the hydrostatic gradient owing to its low density (*e.g.* Figure 2.8). As the height of the hydrocarbon column increases, at some point the pore pressure will be sufficient to induce fault slip, providing a mechanism of increasing fault permeability and allowing leakage from the reservoir.

Figure 11.13 revisits the Visund field in the northern North Sea which was previously discussed in the context of using drilling-induced tensile fractures to determine stress orientation (Figure 6.7) and stress magnitude (Figure 7.13). In this case, we consider the question that originally motivated our study: *why has there been gas leakage along the southern portion of the A-central fault?* The Visund field is located in offshore Norway in the easternmost major fault block of the Tampen spur (Færseth, Sjøblom *et al.* 1995) along the western edge of the Viking graben. The reservoir is divided into several oil and gas compartments, some of which are separated by the A-Central fault (map in Figure 11.13). Hydrocarbon columns were detected in the Brent group, which is the primary reservoir, as well as in the Statfjord and Amundsen formations. As shown in Figure 11.13a, low seismic reflectivity along the southern part of the A-Central fault at the top of the Brent reservoir horizon is interpreted to be the result of gas leakage from the reservoir (indicated by the area enclosed by the dashed white line). The data in this region are of very high quality and there are no notable changes in lithology that might account for the change in seismic reflectivity.

The contour map in Figure 11.13a is the top Brent reservoir horizon (red lines), with the faults, lateral extent of gas leakage (dashed black line), and outline of the map area shown to the left (blue rectangle) superimposed on the structural contours. Exploration wells that yielded stress and pore pressure data are shown with black circles. The Brent reservoir consists of a ridge running northeast–southwest with a saddle crossing perpendicular to the ridge between wells B and C. The ridge is trapping gas along most of its length except for the portion of the ridge defined by the dashed low-reflectivity area. The southern boundary of the Brent reservoir plunges steeply into the Viking graben. This is the result of a large northeast–southwest trending graben-bounding fault that intersects the southern end of the A-Central fault. The effect of the graben-bounding fault can be seen as a sharp transition from high to low reflectivity in the southern portion of the seismic map.

Figure 11.13b shows a generalized geologic cross-section running approximately east–west through well D and the A-Central fault. The A-Central fault developed during the Jurassic as a normal fault with a $\sim 60°$ dip (Færseth, Sjøblom *et al.* 1995) and as much as 300 m of normal throw. Since that time, the fault appears to have rotated and now dips between 30° and 45° with the result that the A-Central fault is well oriented for being reactivated in a reverse sense in the current stress field (Figures 1.10, regional stress field; 6.7, Visund stress orientations; and 7.13, Visund stress magnitudes). Recall that in Chapter 1, we discussed the strike-slip/reverse state of stress in the northern North Sea (see also Lindholm, Bungum *et al.* 1995) in terms of the compressive stresses induced by deglaciation in the past $\sim 10,000$–$15,000$ years. Hence, the conditions leading to potential reactivation of long-dormant normal faults in the current compressive stress field are, in geologic terms, very recent, and represent the potential for their being *blown traps*. In other words, previously inactive faults that had been capable of sealing hydrocarbons (such as the bounding fault of the footwall reservoir in the Brent shown in the cross-section in Figure 11.13b) could have lost their sealing capacity if they have

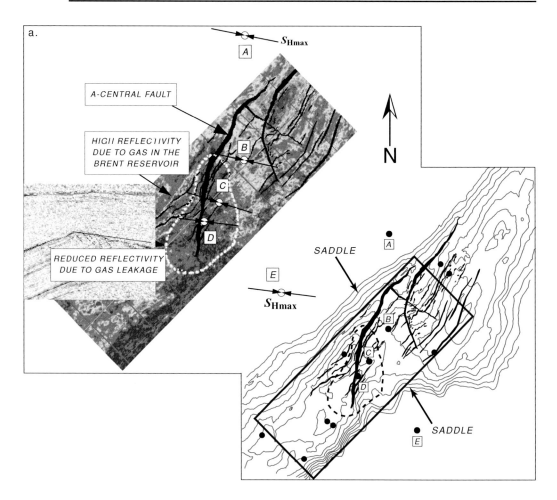

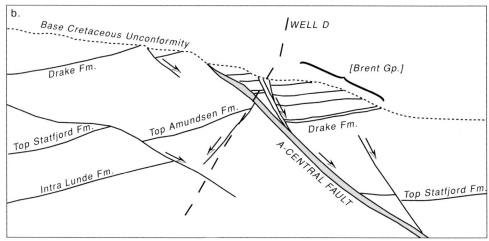

been reactivated by a combination of recently increased compressive stress and fluid pressure at the top of the reservoir.

To evaluate the hypothesis that the southern portion of the A-central fault has been reactivated and is thus the cause of localized leakage, Wiprut and Zoback (2000) resolved the Visund stress orientations and stress magnitudes (after Wiprut, Zoback *et al.* 2000) onto ~100 m × 100 m triangular elements of individual fault planes to calculate the shear and normal stress on each part of the fault. We use Coulomb frictional failure to determine which fault element is expected to slip. We rearrange the terms in equation (4.39) to determine the pore pressure at which a fault element will begin to slip (equation 11.1), and refer to this pore pressure as the critical pore pressure, $P_\mathrm{p}^\mathrm{crit}$,

$$P_\mathrm{p}^\mathrm{crit} = S_\mathrm{n} - \tau/\mu \qquad (11.1)$$

In order to calculate the shear and normal stress we determine the orientation of the unit normal to the fault element in a coordinate system defined by the stress field. The inset of Figure 11.14 shows a fault element defined in the principal stress coordinate system, S_1, S_2, and S_3. Points a, b, and c are the vertices of the fault element, n̂ is the unit normal to the fault element, and **t** is the traction acting on the surface of the fault element. The unit normal to the fault element is defined by the cross product in equation (11.2):

$$\hat{n} = \frac{f \times g}{|f||g|} \qquad (11.2)$$

where f and g are any two vectors defined by the points a, b, and c. The traction acting on the fault plane is the product of the stress tensor and unit normal vector (equation 11.3)

$$t = \mathbf{S}\hat{n} \qquad (11.3)$$

(see inset of Figure 11.14). Because the stresses do not vary significantly between the study wells in individual fields, we define one stress tensor for each field using a single one-dimensional model that varies with depth. The stress tensor is defined in

Figure 11.13. (a) Seismic reflectivity map of the top of the Brent formation (Engelder and Leftwich 1997) and a structure contour map (right). The dashed lines indicate the region of apparent gas leakage from along the southern part of the A-central fault. The stress orientations in Visund wells were shown previously in Figure 6.7. The inset is a portion of a seismic section showing an apparent gas chimney in the overburden above the leakage point. (b) A generalized geologic cross-section showing the manner in which well D penetrates a splay of the A-central fault. Note that hydrocarbons in the Brent formation between the A central fault and the splay fault would have trapped the hydrocarbons in a footwall reservoir (after Wiprut and Zoback 2002). *Reprinted with permission of Elsevier. (For colour version see plate section.)*

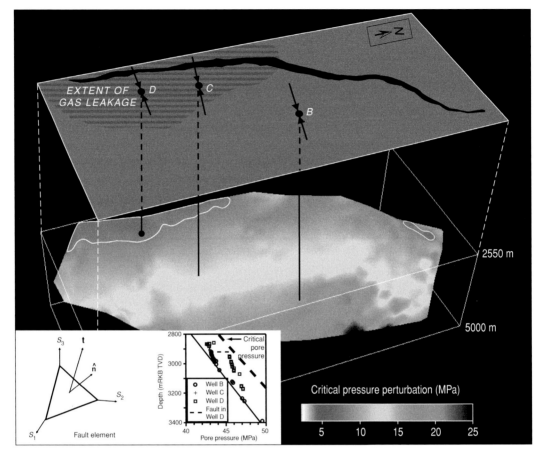

Figure 11.14. Perspective view of the A-central fault and area of fault leakage shown in Figure 11.13a. The stress orientations associated with wells B, C and D were previously shown in Figure 6.7. The color on the fault plane (dipping toward the viewer) indicates the critical pore pressure perturbation (or pressure above ambient pore pressure at which fault slip is likely to occur). The inset illustrates how stress is resolved on individual segments of the fault and the fact that the calculated critical pore pressure (dashed line) is within one MPa of measured pressure in well D on the foot all side of the fault. After Wiprut and Zoback (2000). (*For colour version see plate section.*)

equation (11.4).

$$\mathbf{S} = \begin{bmatrix} S_1 & 0 & 0 \\ 0 & S_2 & 0 \\ 0 & 0 & S_3 \end{bmatrix} = \begin{bmatrix} S_{\mathrm{Hmax}} & 0 & 0 \\ 0 & S_v & 0 \\ 0 & 0 & S_{\mathrm{hmin}} \end{bmatrix} \tag{11.4}$$

Taking the dot product of the unit normal vector and the traction vector gives the magnitude of the normal stress (equation 11.5). The magnitude of the shear stress is determined simply using the Pythagorean theorem (equation 11.6).

$$S_n = \hat{n} \cdot t \tag{11.5}$$

$$\tau^2 = t^2 - S_n^2 \tag{11.6}$$

We calculate the critical pore pressure at each fault element assuming a coefficient of sliding friction of 0.6. The difference between the critical pore pressure and the reference pore pressure is called the critical pressure perturbation. This value shows how close the fault element is to slipping given the reference pore pressure determined for the field, and hence is a measure of the leakage potential.

Figure 11.14 shows a perspective view of the A-Central fault as determined from three-dimensional seismic reflection data (modified from Wiprut and Zoback 2000). In the upper part of the figure a simplified map view of the fault is shown along with the orientation of the maximum horizontal stress in the three wells closest to the fault. The shaded area shows the lateral extent of gas leakage (simplified from Figure 11.13). In the lower part of Figure 11.14, a perspective view of the approximately east-dipping fault surface is shown. A dark circle on the fault plane indicates the point where well D penetrates the A-Central fault. The fault plane is colored to indicate the leakage potential based on the orientation of the fault, the stress, and the pore pressure. The color shows the difference between the critical pore pressure we calculate and the reference pore pressure line shown in Figure 7.13. This difference is the critical pressure perturbation (defined previously). Hot colors indicate that a relatively small increase in pore pressure is enough to bring the fault to failure. Cool colors indicate that the pore pressure must rise significantly (>20 MPa) before those parts of the fault will begin to slip in the current stress field. Note that the largest part of the fault that is most likely to slip (indicated by the white outline) is located along the same part of the fault where leakage seems to be occurring. Note also that this portion of the fault is coincident with a change in the fault plane strike. Thus, there appears to be a correlation between the critically-stressed-fault criterion and the places along the fault where leakage appears to be occurring.

Well D was deviated to penetrate the A-Central fault at 2933 m true vertical depth (the cross-section in Figure 11.13 and Figure 11.14). Because well D penetrates the fault in this area, we can evaluate the correlation between the gas leakage and our prediction of leakage more quantitatively. Pore pressures in Visund are significantly above hydrostatic throughout the reservoir (Figure 7.13). The inset of Figure 11.14 shows a detailed view of the pore-pressure measurements in the three wells closest to the A-Central fault. The steep pressure gradient in well D is the result of light oil rather than free gas. As shown in the inset of Figure 11.14, the pressure below the fault (indicated by the position of the dashed horizontal line) is within ~ 1 MPa of the theoretical critical pore pressure for fault slippage (the thick dashed line). Above the fault, pore pressures are significantly reduced, indicating that there is pore-pressure communication along the fault, but not across the fault.

Figure 11.15a shows a perspective view, looking down and toward the north, of all the major faults in the Visund field with colors indicating the potential for hydrocarbon leakage as in Figure 11.15 (Wiprut and Zoback 2002). The perspective view in this figure creates distortions such that the scales are approximate. Many of the faults indicate a high potential for leakage for the same reason that the A-central fault did. They strike nearly orthogonal to the direction of S_{Hmax} (Figure 6.7) and because they dip at relatively

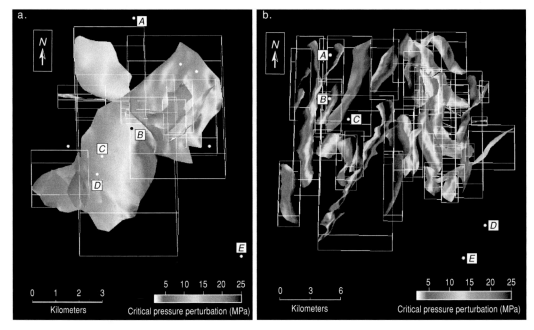

Figure 11.15. (a) Leakage potential map for the faults of the Visund field (similar to that shown for the A-central fault, the large fault in the center dipping to the east) in Figure 11.14. Note that most of the faults in this field have high leakage potential because they strike almost normal to S_{Hmax} (see Figure 6.7) and have shallow dips making them relatively easy to reactivate as reverse faults. The letters refer to well locations. It should be noted that because of the perspective view, the scale is only approximate. (b) A leakage potential map for a field that is relatively near Visund with a very similar stress state. Because these faults have steeper dips, S_{Hmax} tends to resolve high normal stress on these faults making them unlikely to be reactivated. After Wiprut and Zoback (2002). *Reprinted with permission of Elsevier.* (*For colour version see plate section.*)

low angle, the faults can be reactivated as reverse faults in the strike-slip/reverse faulting stress state that exists in field (Figure 7.13).

It is important to recognize that this *leakage map* only indicates the potential for hydrocarbon leakage along a given fault in terms of the pore pressure required to cause fault slip. It does not imply any fault with red colors is currently leaking as there must be hydrocarbons present to leak and the pore pressure must be elevated to the level shown in order to reactivate the fault in order for the leakage to take place.

Figure 11.15b is similar to that on the left for a relatively nearby field in the northern North Sea where the stress state is essentially the same as in Visund (Wiprut and Zoback 2002). Note that most of the faults do not show any significant potential for leakage. This is primarily the result of the steep dip of the faults, which makes them poorly oriented for frictional failure in the current stress field, unlike the faults in the Visund field that have rotated to shallower dip over time. This prediction is consistent with the absence of hydrocarbon leakage and migration in the field shown in Figure 11.15b. According to the analysis of Wiprut and Zoback (2002), the major faults in the center

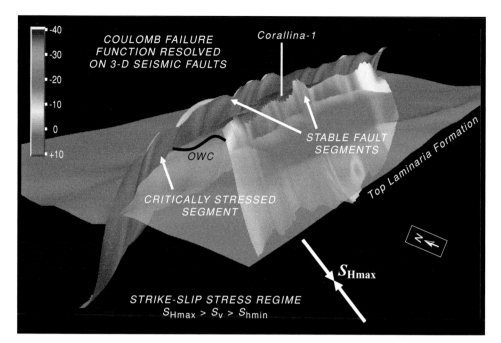

Figure 11.16. Leakage potential map for the Corallina field in the Timor Sea. In this strike-slip faulting regime, there is significant leakage potential (which correlates with the oil–water contact) where the fault has the appropriate strike to the direction of S_{Hmax}. After Castillo, Bishop *et al.* (2000). (*For colour version see plate section.*)

of the field can potentially maintain up to 15–17 MPa pore pressure difference across its surface at the weakest points. Observations of pore pressure from wells on opposite sides of the faults in this field show a pressure difference of approximately 15 MPa.

An additional example of how the critically-stressed-fault hypothesis can help distinguish between leaking and sealing faults is shown in Figure 11.16 (modified from Castillo, Bishop *et al.* 2000) for the Corallina field in the Timor sea. The fault in question is steeply dipping and bends from a NW–SE orientation to a more easterly direction. As this is a strike-slip faulting area in which S_{Hmax} trends to the NNE, the NW–SE trending section of the fault is critically stressed while the more easterly striking section is not. Hence, the position of the oil–water contact (OWC) is consistent with the transition from critically stressed to stable fault segments.

Dynamic constraints on hydrocarbon migration

In this section we consider whether dynamic mechanisms play an important role in hydrocarbon migration in some sedimentary basins and whether dynamic processes play an important role in determining how much oil and gas may accumulate in a given reservoir. Figure 11.17 illustrates in cartoon form how hydrocarbons may accumulate in an anticlinal reservoir bound on one side of a sealing fault. Classical structural

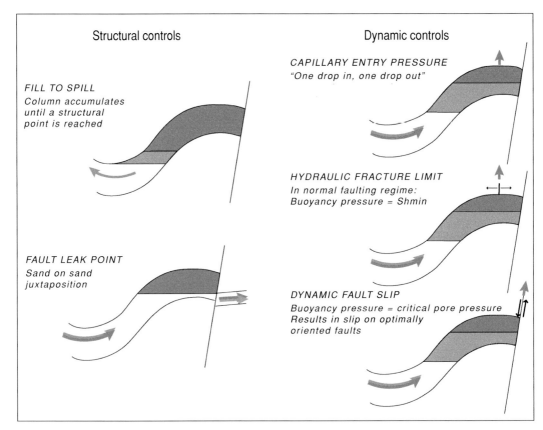

Figure 11.17. The cartoons on the left illustrate classical structural controls on the accumulation of hydrocarbons in a fault-bounded anticlinal sand reservoir (dark gray represents gas, light gray represents oil). These mechanisms include *fill-to-spill* (once the structural closure has filled with hydrocarbons) and *cross-fault* flow if the fault juxtaposes impermeable shale against the reservoir in some places but permeable sands in others. Three dynamic mechanisms are associated with the buoyancy pressure at the top of a reservoir. These include capillary entry pressure into the caprock, hydraulic fracturing of the caprock and fault slip induced by reservoir pressure.

controls on hydrocarbon accumulation involve either the geometry of the structure or the lithologic juxtaposition across the fault. When the anticlinal structure controls the amount of hydrocarbons, as in the example shown, the reservoir is said to be *filled-to-spill* such that adding any additional oil or gas would cause it to flow up dip, or off to the left in the cartoon. When permeable sands are juxtaposed across the fault, *cross-fault* flow can control hydrocarbon accumulation. In this case, the fault could be a sealing fault when the reservoir is juxtaposed against low-permeability shale but a leaking fault when the reservoir is juxtaposed against a permeable sand. In this case, the amount of hydrocarbons that accumulate in the reservoir is controlled by the structural position of the sand, as shown in Figure 11.17.

Dynamic mechanisms may also be important in controlling the accumulation of hydrocarbons in a reservoir. The accumulation of hydrocarbons increases the pressure at the top of the structure through buoyancy (as previously discussed in Chapter 2). Three dynamic mechanisms for hydrocarbon leakage are illustrated on the right side of Figure 11.17. Capillary entry pressure simply means that the buoyant pressure of the oil or gas column at the top of a reservoir exceeds the force required to flow through the caprock (Schowalter 1979). The *capillary entry* pressure can be thought of as the pressure at which a substantial volume of a *non-wetting* phase such as oil or gas might penetrate a cap rock saturated with water which is the *wetting* phase, in direct surface contact with the minerals surrounding the pores. Because rocks have a distribution of pore throat sizes, as more pressure is applied to the non-wetting phase, increasingly smaller pore openings are invaded. This results in a capillary pressure curve which represents saturation as a function of pressure. Capillary pressure is typically measured in the laboratory using air and mercury (or air/water and oil/water) and rescaled to fit field conditions for oil/water or gas/water systems. As it is beyond the scope of this book, a more complete discussion of capillary pressure can be found in Dullien (1992). It should be noted, however, that there are cases in which it is difficult to distinguish between mechanisms acting to limit the amount of hydrocarbons in a reservoir. In a case in which a reservoir is not filled-to-spill, in the absence of other data, it might lead one to suspect that the buoyancy pressure at the top of the reservoir has reached the capillary entry pressure of the cap rock when, in fact, another dynamic mechanism is limiting hydrocarbon column heights.

The other two dynamic mechanisms controlling hydrocarbon column heights in a reservoir that are illustrated in Figure 11.17 are hydraulic fracturing of the cap rock when the pressure in the reservoir exceeds the magnitude of the least principal stress in the caprock. In this case, the reservoir is said to be *at leak-off*. Any additional hydrocarbon added to a reservoir would cause vertical fracture growth and leakage through the caprock. An example of this process is described by Seldon and Flemings (2005) for the Popeye-Genesis deepwater mini-basin of the Gulf of Mexico. The third mechanism, dynamic fault slip, implies that buoyancy pressure at the top of the structure exceeds the frictional strength of the bounding fault. In other words, the fluid pressure in the reservoir, in part resulting from hydrocarbon buoyancy, is sufficient to induce fault slip on reservoir-bounding faults. An example of this process is described below for the South Eugene Island Field, also in the Gulf of Mexico.

It should be noted that with all three of the dynamic mechanisms illustrated in Figure 11.17, if the initial water phase pore pressure (prior to the introduction of hydrocarbons) is quite high, even a relatively small accumulation of hydrocarbons could result in very high pressure on the seal, in fact reaching a critical pressure at which the capacity to seal hydrocarbons is exceeded. This is illustrated in Figure 11.18. Regardless of which of the three dynamic mechanisms illustrated in Figure 11.18 might be operative, the higher the initial water phase pore pressure, the lower the dynamic

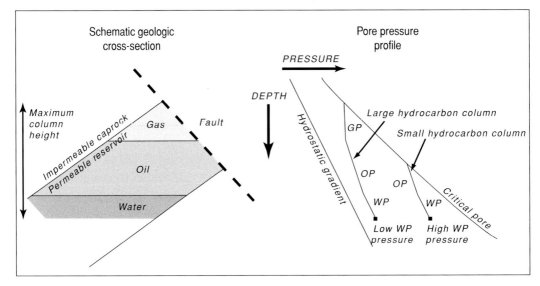

Figure 11.18. Illustration of how the accumulation of hydrocarbons in a footwall reservoir induces buoyancy pressure on the seal (in this case a reservoir bounding fault). For dynamic leakage mechanisms, when the water phase pore pressure is relatively high, the quantity of hydrocarbons that can accumulate in a reservoir before the dynamic limit is reached is smaller. After Wiprut and Zoback (2002). *Reprinted with permission of Elsevier.*

sealing capacity and the less the total accumulation of hydrocarbons is likely to be (Finkbeiner, Zoback *et al.* 2001).

In many ways, the case where buoyancy pressure induces fault slip is analogous to the case described for the A-central fault (and other potentially leaking faults) at Visund. However, there is an important difference between that case and what may happen in areas of young, uncemented and ductile formations such as those found in the Gulf of Mexico. In a brittle rock (such as encountered in Visund) fault slip is expected to cause brecciation and a permanent loss of fault seal capacity. In an uncemented and ductile shale in the Gulf of Mexico, fluid flow along a fault may only occur while the fault is slipping. Hence, following a slip event, the fault surfaces in the ductile shale may easily deform plastically and seal up after slip has ended such that the fluid flow up the fault plane is episodic, something akin to the fault valve model of Sibson (1992). An example of this will be discussed below.

In Figure 11.19 (from Finkbeiner, Zoback *et al.* 2001) we revisit fluid pressures in a dipping sand reservoir surrounded by relatively impermeable shale in the context of the centroid concept introduced in Figure 2.12. We do so to further examine the dynamic controls on column heights illustrated in Figure 11.19. In Figure 11.19a we show the conditions under which hydraulic fracturing of the caprock will occur. Above the centroid, the pore pressure in the reservoir is higher than that in the shale caprock and eventually reaches the least principal stress in the shale. Note that the Mohr diagrams in

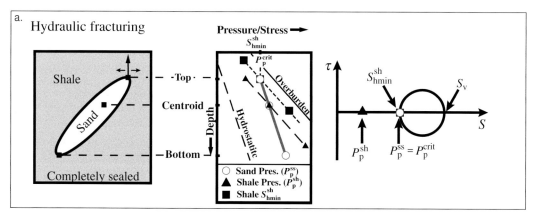

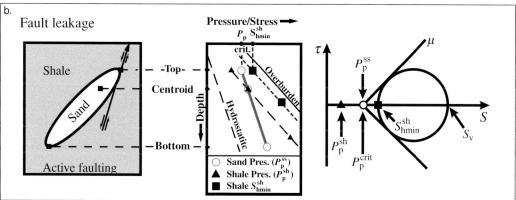

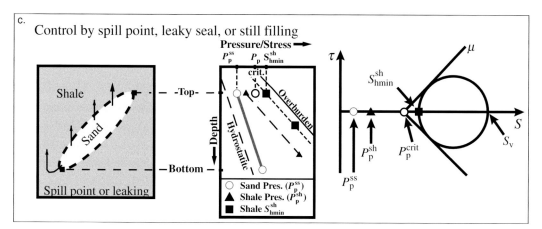

Figure 11.19. Illustration of the pressure and stress conditions in a dipping sand reservoir surrounded by shale under which (a) hydraulic fracturing of the caprock, (b) pressure-induced slip on a reservoir bounding fault occurs or (c) conventional limits on column occur. Note that for both hydraulic fracturing of the caprock and pressure-induced fault slip, pore pressure at the top of sand reservoir is higher than that in adjacent shales. After Finkbeiner, Zoback *et al.* (2001). *AAPG© 2001 reprinted by permission of the AAPG whose permission is required for futher use.*

Figure 11.19 are somewhat different than those introduced previously. In this case, the abscissa represents total stress, rather than effective stress, such that one can plot pore pressure, as well as the stress magnitudes on the diagram. Hence, the conditions under which the caprock hydrofracs is one in which the pore pressure in the sand reservoir (P_p^{ss}) reaches the value of the least principal stress in the shale (left edge of Mohr circle). Note that the pore pressure in the sand, P_p^{ss}, exceeds P_p^{sh}, the pore pressure in the shale.

The conditions under which dynamic fault slip occurs is shown in Figure 11.19b. Note that as in the hydrofrac case, the pore pressure in the reservoir at the top of the structure exceeds that in the shale but does not reach the value of the least principal stress. Rather, because of the presence of the reservoir-bounding fault at the top of the structure, the pore pressure induces slip at a pressure in the reservoir, P_p^{ss}, at which the Mohr circle touches the failure line. Hence, slip on the reservoir bounding fault occurs at a lower pressure than that required for hydraulic fracturing. In other words, breach of the sealing faulting and fluid migration may occur at an earlier stage as is often assumed.

Under conventional structural controls on reservoir column heights (or capillary leakage), the pore pressure in the sand is below that in the shale (such that there is no centroid) and below that at which either hydraulic fracturing or fault slip occurs (Figure 11.19c).

We apply these concepts to the South Eugene Island 330 field located 160 km offshore of Lousiana in the Gulf of Mexico following the study of Finkbeiner, Zoback *et al.* (2001). South Eugene Island 330 is a Pliocene-Pleistocene salt-withdrawal minibasin bounded by the north and east by a down to the south growth fault system (Alexander and Flemings 1995). A cross-section of the field is shown in Figure 2.6a and a map of the OI sand, one of the major producers in the area, is shown in Figure 2.7. There were several questions that motivated the Finkbeiner, Zoback *et al.* (2001) study, including why there are such different hydrocarbon columns in adjacent compartments (as illustrated for fault blocks A and B in the OI sand in Figure 2.7). Note that while there is an oil column of several hundred feet in fault blocks A, D and E there is a very large gas column (and much smaller oil columns) in fault blocks B and C. While fault blocks B and C (and D and E) appear to be in communication across the faults that separate them, fault blocks A and B (and C and D) are clearly separated. As there appears to be an ample source of hydrocarbons to fill these reservoirs (S. Hippler, personal communication), *why are fault blocks A, D and E not filled-to-spill?* A still more fundamental question about this oil field is of how such a large volumes of hydrocarbons could have filled these extremely young sand reservoirs separated by large thicknesses of essentially impermeable shale (Figure 2.6). The South Eugene Island field is one of the largest Plio-Pleistocene oil and gas reservoirs in the world and yet the manner in which the reservoirs have been filled is not clear (Anderson, Flemings *et al.* 1994).

To examine these questions, Finkbeiner, Zoback *et al.* (2001) made a detailed examination of pressures in various reservoirs. As shown in Figure 11.20a, in South Eugene Island fault block A, the JD, KE, LF, NH and OI sands all indicate clear centroid effects with the gas (or oil) pressure at the top of the reservoirs exceeding the shale pore pressure at equivalent depths (Figure 2.8b). However, only in the OI sand does there appear

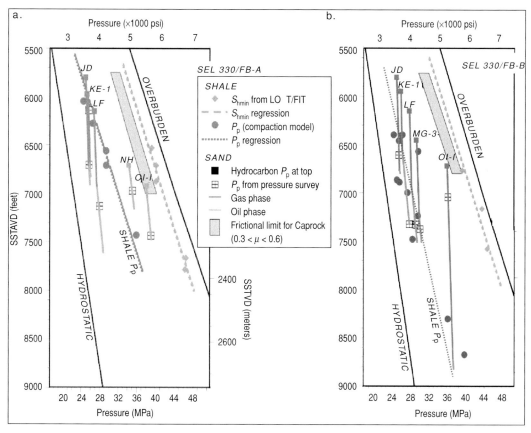

Figure 11.20. Pressure in various reservoirs in South Eugene 330 Field in (a) fault block A and (b) fault block B. The map of the OI sand in Figure 2.7 identifies the location of fault blocks A and B. The geologic cross-section shown in Figure 2.6a identifies the various reservoirs. The square-with-cross symbol indicates the measurement point with the pressures extrapolated to greater and lesser depth from knowledge of the hydrocarbon column heights and fluid densities. Note that only the pressure at the top of the OI sand columns are near the dynamic limit for inducing slip on reservoir bounding faults. After Finkbeiner, Zoback *et al.* (2001). *AAPG© 1994 reprinted by permission of the AAPG whose permission is required for futher use. (For colour version see plate section.)*

to be a pressure at the top of the reservoir equivalent to one of the pressures associated with the dynamic leakage mechanisms discussed above. In this case, the pressure at the top of the reservoir is equivalent to the pressure at which the bounding fault is expected to slip (indicated in yellow for coefficients of friction that range between 0.3 and 0.6). Hence, the several hundred feet high oil column in the OI sand appears to be at its dynamic limit, with any additional oil (or gas) added to the reservoir causing fault slip and hydrocarbon migration up the fault (Finkbeiner, Zoback *et al.* 2001). The same thing appears to be true in fault blocks D and E where relatively short oil columns are seen (see Figure 2.7). The situation in fault block B (Figure 11.20b) is similar in that in fault block A as centroid effects are clearly seen for all of the reservoirs. However,

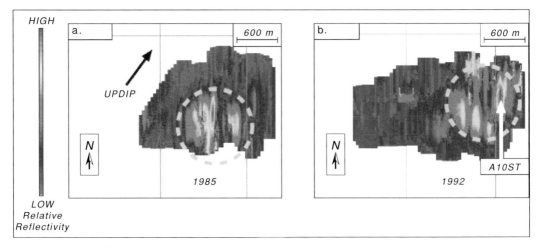

Figure 11.21. Maps of fault plane reflectivity in the SEI-330 field in (a) 1985 and (b) 1992 that appears to show a pocket of hydrocarbons moving updip along the fault plane. After Haney, Snieder *et al.* (2005). The location of the fault along which this *burp* of hydrocarbons is moving (and position of well A10ST along the fault) can be deduced by comparing the maps in Figures 8.11b and 2.7. (*For colour version see plate section.*)

because the water phase pore pressure in fault block B is lower than in fault block A (Figure 2.8a), a large gas column (see Figure 2.7) can be supported before the pressure at the top approaches that associated with fault slip. Hence, the initial pore pressure in the respective reservoirs, prior to them being filled with hydrocarbons, established the condition that controlled how much hydrocarbon would eventually fill the reservoir. It should be noted that while the pressure at the top of the OI sand in fault block B is near the dynamic limit, the column is also close to the spill point. While it is not clear which mechanism is responsible for the resultant hydrocarbon column, it would not have been possible to support the column shown unless water phase pore pressure in fault block B is appreciably lower than in fault block A.

Because of the young age of the sediments in the South Eugene Island field, the sands and shales are uncemented and the shales deform ductily. Because of this, fault slip is not expected to cause brecciation and the fault would be expected to *heal* and show little, if any, increase in permeability after a slip event. Hence, it is presumed that fluid flow along the faults would be episodic in nature, occurring at the time that fault is actually slipping (Anderson, Flemings *et al.* 1994; Losh, Eglinton *et al.* 1999). Interestingly, a seismic image of a pulse (*burp?*) of hydrocarbons moving up the NNW trending growth fault bounding fault block B to the north has been captured through 4D seismic profiling (Haney, Snieder *et al.* 2005). As shown in Figure 11.21, between 1985 and 1992 a *bright spot* in the plane of the fault, presumably due to the presence of gas (or gas dissolved in oil), is seen at two different places in the fault plane, ∼1 km apart.

Finally, Figure 11.22 shows the seismic section previously shown in Figure 2.6b, with a summary of the mechanisms controlling hydrocarbon accumulations at various

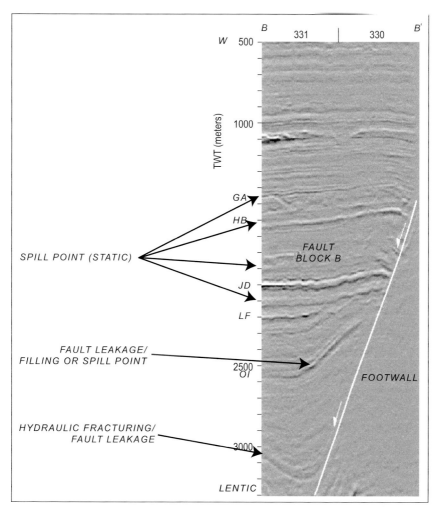

Figure 11.22. The seismic section from the South Eugene Island 330 field previously shown in Figure 2.6b with a summary of the mechanisms controlling hydrocarbon accumulations at various depths (Finkbeiner, Zoback *et al.* 2001). *AAPG© 2001 reprinted by permission of the AAPG whose permission is required for futher use.*

depths in the South Eugene Island 330 field (Finkbeiner, Zoback *et al.* 2001). The relatively shallow reservoirs (GA, HB, JD and LF) are all *filled-to-spill.* Dynamic fault slip appears to be the mechanism controlling the accumulation of oil and gas in the OI sand, as discussed above. Finkbeiner, Zoback *et al.* (2001) argued that the small hydrocarbon columns in the deep Lentic sand which is very severely overpressured (but is not shown here), appears to be controlled by a dynamic mechanism (it is not *filled-to-spill* nor does there appear to be evidence for cross-fault flow). However, because the overpressure is so high, uncertainties in the magnitude of pore pressure and the least principal stress make it unclear if the dynamic mechanism limiting the column height is hydraulic fracturing of the caprock or dynamic fault slip.

12 Effects of reservoir depletion

Addressing problems associated with the deformation and changes of stress within and surrounding depleting reservoirs is important for many reasons. Most well known are the problems associated with casing collapse and surface subsidence that create substantial difficulties in some oil and gas reservoirs due to compaction in weak formations. The significant stress changes that occur in highly depleted reservoirs (*e.g.* Figure 2.10a) can make drilling a new well to deeper targets quite problematic due to the need to lower mud weights in depleted formations to avoid lost circulation. Depletion also has the potential to induce faulting, both within and outside reservoirs in some geologic environments. While these problems can be formidable in some reservoirs, depletion can also have beneficial impact on reservoir performance. For example, hydraulic fracturing can be more effective in depleted reservoirs than in the same reservoirs prior to depletion. In some weak reservoirs, compaction drive is an effective mechanism for enhancing the total amount of hydrocarbons recovered, especially if the permeability changes accompanying compaction are not severe.

To address these issues in a comprehensive manner, this chapter is organized in three sections. In the first section I consider processes accompanying depletion *within* reservoirs and focus initially on the stress changes associated with depletion. We begin by discussing reservoir *stress paths*, the reduction of horizontal stress magnitude within the reservoir resulting from the decrease in pore pressure associated with depletion. The next topic considered is that of depletion-induced faulting *within* reservoirs which may seem counter-intuitive because it is well known that raising pore pressure through injection can induce seismicity by decreasing the effective normal stress on pre-existing faults (as discussed in Chapter 4). I then discuss the conditions under which depletion can induce rotations of horizontal principal stress directions. Under most circumstances, horizontal principal stress magnitudes are expected to change the same amount during depletion such that there is no rotation of horizontal principal stress directions. I demonstrate below that when an impermeable fault bounds a reservoir, depletion on one side of the fault makes it possible to induce rotations of horizontal principal stress directions. In such cases, hydraulic fractures induced after depletion would be expected propagate in a different direction than those made prior to depletion – potentially having a particularly beneficial effect on production. Finally, I address the problems associated with drilling through depleted intervals to deeper reservoirs. In such cases, the reduction of the least principal stress would seem to require lower mud weights to be used (to prevent unintentional hydraulic fracturing and lost circulation) and, in some cases, drilling with the mud weight required in the depleted reservoir would not appear to be feasible in the context of that required to achieve wellbore stability (or offset formation pressure) in adjacent formations.

In the second section of this chapter I discuss deformation within a depleting reservoir. After discussing compaction in general terms, we introduce *end-cap* models of deformation (introduced in Chapter 4) and a formalism we call DARS (Deformation Analysis in Reservoir Space) that considers irrecoverable (plastic) compaction and the potential for production-induced faulting in terms of parameters measured easily in the laboratory and parameters frequently measured in reservoirs. Because uncemented reservoirs can experience viscoplastic, or time-dependent, irreversible compaction, I briefly discuss the constitutive law for long-term reservoir compaction from laboratory measurements previously described in Chapter 3 that addresses this phenomenon. I also discuss how compaction can induce significant permeability loss in weak sediments and show that by using DARS (or a viscoplastic constitutive law) to estimate the total porosity loss in a weak sand reservoir, one can estimate the associated permeability changes. I present a case study in which we compare predicted permeability changes with those measured in a depleting oil field in the Gulf of Mexico. Using a simple reservoir model, I briefly illustrate how compaction drive in a weak sand reservoir improves recovery while evaluating the importance of permeability loss on the rate of total recovery.

In the final section of this chapter we consider the stress changes and deformation that occurs in formations surrounding a depleting reservoir. These are illustrated in Figure 12.1 (after Segall 1989). Surface subsidence above compacting reservoirs is a well-known phenomenon. As shown below, the amount of surface subsidence depends on the reservoir depth, lateral extent and the amount of compaction. I will use DARS to predict the amount of compaction occurring within a reservoir, then utilize an analytical method for evaluating the degree of surface subsidence. We also investigate production-induced faulting outside of reservoirs utilizing both analytical and numerical approaches. There are now numerous documented examples of sheared well casings outside of depleting reservoirs (as well as seismicity and surface offsets along pre-existing faults) such that induced slip on faults outside of a depleted reservoir is an important topic to consider. As illustrated qualitatively in the upper part of Figure 12.1, reverse faulting is promoted by reservoir depletion immediately above and below the reservoir whereas normal faulting is promoted near its edges. This is discussed more quantitatively later in this chapter.

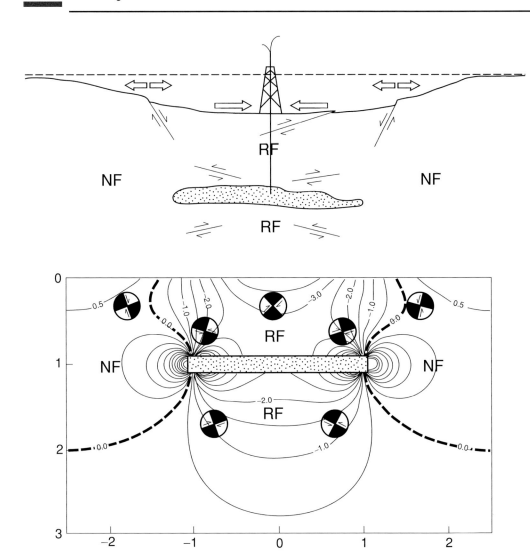

Figure 12.1. Schematic diagram illustrating the effects of reservoir depletion on deformation surrounding the reservoir and contours of the stress changes resulting from depletion in a reservoir at unit depth and radius (from Segall 1989). Note that in compressional tectonic settings, reverse faulting is promoted above and below the reservoir whereas in extensional tectonic settings, normal faulting is promoted around the edges of the reservoir.

Stress changes in depleting reservoirs

Reservoir stress paths

Poroelastic theory can be used to predict the magnitude of stress changes with depletion. In an isotropic, porous and elastic reservoir that is infinite in extent, if the only source of horizontal compressive stress is instantaneously applied gravitational loading, the relationship between vertical effective stress and the corresponding horizontal effective stress (assuming no lateral strain) is given by:

$$S_{\text{Hor}} = \left(\frac{\nu}{1-\nu}\right)(S_{\text{v}}) + \alpha P\left(1 - \frac{\nu}{1-\nu}\right) \tag{12.1}$$

where S_{Hor} corresponds to both S_{Hmax} and S_{hmin} (Lorenz, Teufel *et al.* 1991), α is Biot's coefficient and ν is Poisson's ratio. Taking the derivative of both sides with respect to pore pressure and simplifying yields

$$\Delta S_{\text{Hor}} = \alpha \frac{(1-2\nu)}{(1-\nu)}\Delta P_{\text{p}} \tag{12.2}$$

(see Brown, Bekins *et al.* 1994), assuming that α is constant with respect to changes in P_{p}. To get a sense of the magnitude of stress changes with depletion, note that if $\nu = 0.25$ and $\alpha = 1$, the change in horizontal stress corresponding to a change in pore pressure is

$$\Delta S_{\text{Hor}} \sim \frac{2}{3}\Delta P_{\text{p}}$$

Rearranging equation (12.2), it is possible to define a *stress path* of a reservoir that corresponds to the change in horizontal stress with changes in production, A, as

$$A = \alpha \frac{(1-2\nu)}{(1-\nu)} = \frac{\Delta S_{\text{Hor}}}{\Delta P_{\text{p}}} \tag{12.3}$$

It is obvious that because equation (12.2) has been derived for an infinite, horizontal reservoir of finite thickness, there is no change in the vertical stress, S_{v}. Using an elliptical inclusion model of a compacting reservoir, Segall and Fitzgerald (1996) have shown that once the ratio of lateral extent to thickness of a reservoir is greater than 10:1 (which is almost always the case), equation (12.2) is nearly exactly correct, despite the assumption that the reservoir is infinite in extent. Hence, in reservoirs that are laterally extensive with respect to their thickness, the horizontal stresses will decrease with depletion but the vertical stress remains essentially constant. In a more equi-dimensional reservoir, this must be modified. Utilizing the poroelastic theory of Rudnicki (1999). Holt, Flornes *et al.* (2004) discusses depletion and stress path effects in reservoirs as a function of the aspect ratio of the reservoir and the stiffness of the reservoir with respect to the surrounding medium.

Both theoretical and observed poroelastic stress paths are shown as a function of α and ν in Figure 12.2 (after Chan and Zoback 2002). As can be seen, for reasonable values of ν and α, the theoretical change in horizontal stress with depletion will generally be in the range 0.5–0.7, which corresponds to observed values for many reservoirs. Note that the stress paths associated with the fields listed in italics on the side of Figure 12.2 may only represent *apparent* stress paths, as it was not clear from the data presented in the original reference that the reported stress changes were accompanying pore pressure changes through time in the same part of the reservoir. The line labeled *normal faulting*

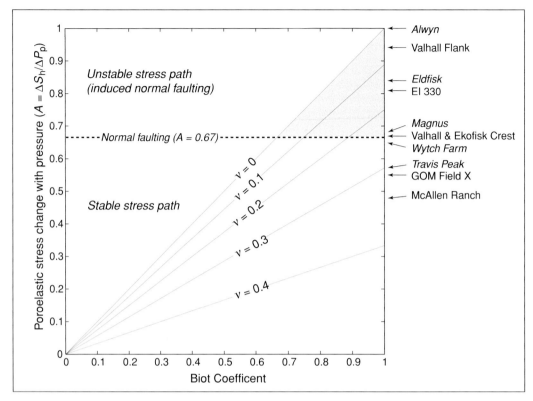

Figure 12.2. Variation of stress change with pressure as a function of Biot coefficient, α, and Poisson's ratio, v (after Chan and Zoback 2002). The normal faulting line ($A = 0.67$) is explained in the text. The gray area represents the possible combination of α and v such that stress path leads to production-induced normal faulting. Observed stress paths in different reservoirs are shown on the right-hand side of the diagram. For the fields listed in *italics*, it is not clear whether the reported *stress path* indicates a change of stress with depletion or variation of stress with pore pressure in different part of the fields. © *2002 Society Petroleum Engineers*

in Figure 12.2 corresponds to values of the stress path above which depletion would ultimately lead to movement on pre-existing normal faults as explained below. This is derived below.

    Depletion data from several wells studied in a Gulf of Mexico oil field (Field X) are shown in Figure 12.3 (from Chan and Zoback 2002). All of the wells in this field (indicated by the different symbols) deplete along the same path, indicating that where the reservoir is penetrated by wells, it is interconnected and not sub-compartmentalized. The least principal stress, measured during *fracpac* completions, indicates a stress path of ~0.55 (Figure 12.2), easily explained by reasonable combinations of α and v. Data from this field will be considered at greater length below. Note that considerable depletion occurred, from an initial pore pressure of ~80 MPa prior to significant production in 1985 to ~25 MPa in 2001, after ~15 years of production.

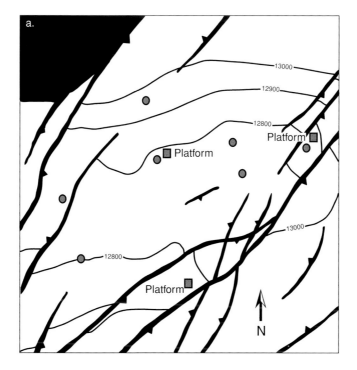

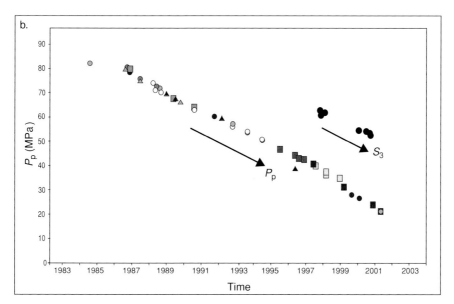

Figure 12.3. (a) Map of Field X in the Gulf of Mexico. (b) Pore pressure history of Field X (after Chan and Zoback 2002). The symbols represent measurements made in different wells. The magnitude of the pore pressure is then adjusted to the datum. Notice that the continuous decline of pore pressure in the different wells implies that there is no sub compartmentalization within the reservoir. © 2002 *Society Petroleum Engineers*

There are several important qualifications about equation (12.2) that should be noted. It is most important to note that while equation (12.2) has some applicability for predicting stress changes in a reservoir, it should *not* be used for predicting actual stress values at depth (see Chapter 9). Unfortunately, this distinction has been confused by some authors. As already alluded to, this equation is derived for a homogenous, isotropic, linear poroelastic formation. If deformation in the reservoir is inelastic, the analysis above will not accurately predict the stress path. As indicated in Figure 12.2 and discussed below, observed stress paths on the flanks of the Valhall and Ekofisk fields of the North Sea are appreciably greater than what is theoretically reasonable (Zoback and Zinke 2002). As the chalks in these fields are rather weak and compliant, this could be due to inelastic effects. Another reason for questioning the applicability of equation (12.2) is the assumption of no horizontal strain (Teufel, Rhett *et al.* 1991).

Production induced faulting in normal faulting areas

There have been a number of case studies in which both fluid withdrawal and fluid injection appear to have induced active faulting in oil and gas reservoirs (see the review by Grasso 1992). Since the classic studies of injection-induced earthquakes at the Rocky Mountain Arsenal (Healy, Rubey *et al.* 1968) and the Rangely oil field in Colorado (Raleigh, Healy *et al.* 1972) discussed in Chapter 4, induced faulting in oil and gas reservoirs is usually thought to be associated with pore pressure increases due to water-flooding or hydraulic fracturing. The cause of the induced fault slip is obvious in these cases as increases in pore pressure cause a reduction of the effective normal stress on the fault plane (as discussed in Chapter 4). In this section, we follow the analysis of Zoback and Zinke (2002) and discuss a mechanism by which normal faulting within reservoirs might be induced by poroelastic stress changes associated with production if the stress path is sufficiently large. We demonstrate that the initial stress path (*i.e.* prior to fluid injection for pressure maintenance and subsidence control) associated with reservoir production in the Valhall field of the North Sea (as well as the Ekofisk field) is such that normal faulting was promoted by depletion. Similar processes may be active in other oil and gas fields where normal faulting within reservoirs appears to have been induced by hydrocarbon production (Doser, Baker *et al.* 1991). Faulting that occurs outside of reservoirs in response to reservoir depletion (Figure 12.1) is discussed below.

Fault slip within reservoirs induced by decreases in pore pressure seems counter-intuitive in light of the conventional relationship between pore pressure, effective normal stress and shear failure. In light of the fact that S_v is expected to remain essentially constant during depletion of laterally extensive reservoirs, the reduction of S_{hmin} and P_p can induce normal faulting within a reservoir if the depletion stress path exceeds a critical value. This value can be calculated from the Coulomb failure condition for normal faulting. Modification of equation (4.45) for depletion in a laterally extensive reservoir yields:

$$\frac{[S_V - (P_p - \Delta P_p)]}{[(S_{hmin} - \Delta S_{hmin}) - (P_p - \Delta P_p)]} = f(\mu) \tag{12.4}$$

where

$$f(\mu) = (\sqrt{\mu^2 + 1} + \mu)^2$$

Simplifying this results in:

$$\frac{S_V - P_p}{S_{hmin} - P_p} = \left[1 - \frac{\Delta S_{hmin} - \Delta P_p}{S_{hmin} - P_p}\right] f(\mu) - \frac{\Delta P_p}{S_{hmin} - P_p} \tag{12.5}$$

In areas where normal faults are in frictional equilibrium, the left-hand side of equation (12.5) is equivalent to $f(\mu)$ such that,

$$f(\mu) = f(\mu) - \frac{\Delta S_{hmin} - \Delta P_p}{S_{hmin} - P_p} f(\mu) - \frac{\Delta P_p}{S_{hmin} - P_p}$$

$$\frac{\Delta S_{hmin} - \Delta P_p}{S_{hmin} - P_p} f(\mu) = -\frac{\Delta P_p}{S_{hmin} - P_p}$$

$$\frac{\Delta S_{hmin} - \Delta P_p}{\Delta P_p} = -\frac{1}{f(\mu)}$$

Substituting $A = \Delta S_{hmin}/\Delta P_p$ yields the stress path, A^*, which if exceeded, can lead to production-induced normal faulting:

$$A^* = 1 - \frac{1}{(\sqrt{\mu^2 + 1} + \mu)^2} \tag{12.6}$$

For $\mu = 0.6$, the theoretical stress path corresponding to normal faulting will be roughly equal to 0.67. This corresponds to the horizontal dashed line in Figure 12.2 such that stress paths in excess of 0.67 will be unstable. In other words, sufficient depletion will eventually result in production induced faulting. Stress paths less than 0.67 never lead to faulting, regardless of the amount of depletion. Hence, the fields shown above the horizontal dashed line in Figure 12.2 are located in a normal faulting environments, where sufficient depletion could eventually induce slip on pre-existing normal faults in the reservoir.

This is illustrated in Figure 12.4a in a type of plot we refer to as *reservoir space* because it expresses the evolution of the state of stress in a reservoir in terms of pore pressure and the least principal stress. The vertical stress is assumed to remain constant with depletion (for the reasons stated above). The value of the least principal stress that corresponds to the vertical stress defines the horizontal line labeled S_V. In a normal faulting environment, we can assess the potential normal faulting on pre-existing faults simply by defining a failure line that corresponds to equation (4.45). The diagonal line

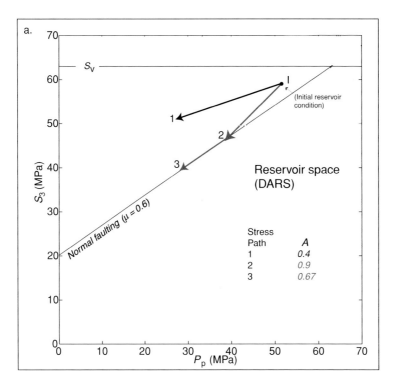

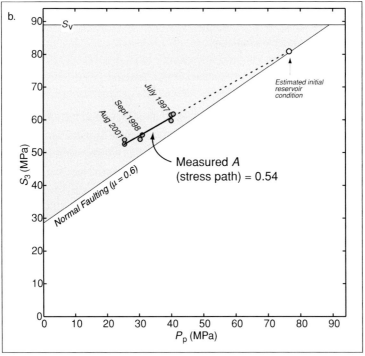

shown labeled $\mu = 0.6$, expresses the magnitude of the S_{hmin} at which normal faulting occurs as a function of pore pressure. As noted above, the slope of this line is 0.67. As depletion occurs, the least principal stress decreases along a line with slope A. As we track the evolution of the least principal stress with depletion, Figure 12.4a shows that when $A < 0.67$, the state of stress moves away from the normal failure line. The line from the initial stress state to point 1 has a slope of 0.4, slightly lower than that observed at McAllen ranch (Figure 12.2) and similar to the expected stress path for a poroelastic rock with a Biot coefficient of ~0.6 and a Poisson's ratio of ~0.25 (Figure 12.2). This represents a stable stress path, as depletion would not be expected to induce fault slip. In contrast, the line from the initial stress state to point 2 in Figure 12.4a has a slope of 0.9, similar to that observed on the flanks of the Valhall reservoir (as discussed in more detail below). When $A > 0.67$ (as observed in a number of fields in Figure 12.2) depletion will eventually cause normal faulting to occur. Once the stress path hits the failure line, the state of stress evolves along the failure line because, for a given value of the vertical stress and pore pressure, S_{hmin} cannot be lower than the value predicted by equation (4.45). Such a situation would be equivalent to a Mohr circle exceeding the failure line, as illustrated in Figure 4.27.

For such a stress path, regardless of how much depletion has occurred, normal faulting becomes less likely. Figure 12.4b (from Chan and Zoback 2002) shows that the stress path for GOM Field X is 0.54, such that depletion would not be expected to result in production-induced faulting. Interestingly, extrapolation of stress changes back to initial pore pressure conditions indicates that the reservoir was initially in a state of frictional failure equilibrium (although data on the magnitude of S_{hmin} are not available) and depletion moved the pre-existing normal faults away from the failure state.

While schematic, Figure 12.4a was drawn to track the evolution of stress and pore pressure in the Valhall field in the central graben in the southern part of the Norwegian North Sea (Zoback and Zinke 2002). The Valhall structure trends NW–SE and is an elongated anticline with normal faults across the crest of the structure. The reservoir is at a depth of approximately 2400 m subsea and consists of two late Cretaceous oil bearing formations: the Tor formation and the underlaying Hod formation which are overlain by Paleocene and Eocene age shale cap rock. Both formations are soft chalk facies with a primary porosity that varies between 36 and 50%. There has been significant concern

Figure 12.4. (a) Schematic stress paths in *reservoir space* illustrate how the magnitude of the least principal stress evolves with depletion within a reservoir. The horizontal line corresponds to the vertical stress, which is not expected to change with depletion. Relatively steep stress paths (such as 2) lead to production-induced faulting in normal faulting environments, whereas relatively shallow stress paths (such as 1) do not. If the failure line is intersected by the stress path, further decreases in pore pressure will cause the stress path to follow the normal faulting failure line (such as 3). (b) Stress and pore pressure measurements throughout the lifetime of the Field X reservoir (after Chan and Zoback 2002). The initial reservoir condition is extrapolated from the measured stress path to initial pore pressure conditions. © *2002 Society Petroleum Engineers*

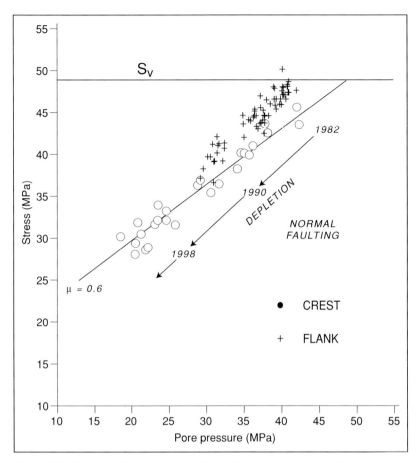

Figure 12.5. The evolution of the least principal stresses with decreasing pore pressure in the Tor reservoir of the Valhall field in the North Sea (after Zoback and Zinke 2002). As in Figure 9.6, the horizontal line denotes an average value of the vertical stress in the field and the inclined straight line corresponds to the Coulomb failure criterion for a coefficient of friction of 0.6. The + s denote values obtained for the flanks, and circles show values obtained for the crest. Note that initially, there was nearly an isotropic stress state on the flanks of the reservoir whereas the crest was in frictional failure equilibrium. As depletion progressed, pore pressure and stress dropped dramatically such that normal faulting would be expected both on the flanks of the reservoir and on the crest.

about active faulting. There have been numerous occurrences of casing failures in both fields which have been interpreted as being the result of shear along active faults and there is appreciable gas leakage through the shale cap rocks which may be exacerbated by flow through faults (Munns 1985).

Figure 12.5 presents least principal stress data from the reservoir section of the Valhall field (principally from the Tor formation) to evaluate the magnitude of the least principal stress with respect to pore pressure and position within the reservoir (from Zoback and Zinke 2002). The vertical stress shown was derived by integration of density logs. Leak-off tests (LOT's) and mini-fracs were used to constrain the least principal stress and wireline sampler data were used to constrain pore pressures. We separate the measurements with respect to their position in the crest or flank. The approximate time that a given observation was made is illustrated.

Several trends are apparent in these data. First, there is a clear reduction of pore pressure and S_3 with time that shows the overall effects of depletion. It is clear that both on the crest of the structure and on the flanks, pore pressures and least principal stresses in the reservoir were quite high in the early 1980s. As production occurred, pressure and stress dropped dramatically. On the crest of the structure, the initial stress and pore pressure were in a normal faulting stress state and the change of the least principal stress with production follows the normal faulting line (as illustrated in Figure 12.4a). As depletion occurred, the stress path was such that the crest remained in a normal faulting stress state even though pore pressure was decreasing with time. The natural state of stress was active normal faulting and as production occurred, the stress path was sufficiently steep that normal faulting continued (unlike Field X). The intermediate principal stress, S_2, in normal faulting areas corresponds to the maximum horizontal stress, S_{Hmax}. It is expected that both horizontal stresses were affected more or less equally by depletion. While we have no direct estimates of S_{Hmax} magnitudes at Valhall, utilization of a variety of techniques indicates very little difference between the two horizontal stresses (Kristiansen 1998).

Figure 12.5 also demonstrates that stress magnitudes on the flanks of the Valhall structure were initially appreciably higher than on the crest. The least principal stress values were initially not far below S_v, a stress state that does not favor normal faulting. In fact, as the maximum horizontal stress is intermediate in magnitude between the least horizontal stress and vertical stress in normal faulting areas (as defined by the state of stress on the crest of the structure), the initial stress state on the flanks was almost isotropic. What is interesting about the evolution of stress with depletion on the flank is that despite the nearly isotropic initial stress state, the stress path accompanying production on the flanks of the reservoir is so steep ($A \sim 0.9$) that once depletion has reduced pore pressure to about 30 MPa, a normal faulting stress state is reached. Thus, depletion of the Tor formation appears to have induced normal faulting on the flanks of the reservoir. As normal faulting had already been occurring on the crest, it appears that as production and depletion occurred, normal faulting spread outward from the crest of the structure onto the flanks. At Valhall, an array of six, three-component seismometers was deployed between June 1 and July 27, 1998 in a vertical section of one of the wells near the crest of the structure. The seismic array was deployed about 300 m above the reservoir. The majority of events occurred about 200 meters to the west of the monitoring well and were occurring either at the very top of the reservoir or in the Paleocene shale cap rocks that overlay the Tor reservoir (Maxwell 2000). Zoback and Zinke (2002) show that the focal mechanism of these events indicate normal faulting on a $\sim$NE-trending plane.

Zoback and Zinke (2002) argued that a similar process appears to have occurred in the Ekofisk field based on data from Teufel, Rhett *et al.* (1991) which came from the time period prior to the initiation of water flooding operations. The crest of the structure was initially in a state of normal faulting but the flank and outer flank were not. As reservoir pressure decreased throughout, normal faulting was expected to spread out onto the flanks of the reservoir. The outer flank of the reservoir was almost in a state of normal faulting at the time of the last set of measurements. Teufel, Rhett *et al.* (1991) discussed the change in stress and pore pressure at Ekofisk in the context of faulting by relating the stress and pore pressure measurements to strength measurements made on Ekofisk core. However, the Zoback and Zinke (2002) interpretation of incipient normal faulting in the crest of the structure and induced normal faulting on the flanks due to the poroelastic stress path expands upon their interpretation.

With respect to reservoir permeability, it is clear that active faulting in the reservoirs is capable of increasing matrix permeability, as discussed in Chapter 11. In this light, it is quite interesting that despite the reservoir compaction accompanying depletion at Ekofisk prior to pressure maintenance, reservoir productivity remained steady, or slightly increased, despite appreciable depletion (Sulak 1991). From this observation, it is reasonable to assume that active shear faults may be enhancing the low matrix permeability (Brown 1987) and counteracting the permeability reductions accompanying compaction.

Stress rotations associated with depletion

As argued above, depletion in a laterally extensive reservoir would cause S_{Hmax} and S_{hmin} to decrease by the same amount, assuming that the medium is homogeneous and isotropic. In such cases, no stress rotation is expected to accompany depletion. However, on the basis of wellbore stress orientation measurements in several fields, it has been argued that there are stress orientation changes near faults. Note that the stress orientations in the Scott field of the North Sea (Figure 8.14) appear to follow the strike of local faults (Yale, Rodriguez *et al.* 1994). Wright and Conant (1995) and Wright, Stewart *et al.* (1994) argue for the re-orientation of stress in depleted reservoirs.

In this section, we address depletion in an idealized finite reservoir, where a reservoir is bounded by an impermeable fault. Because the fault acts as a barrier to fluid flow, the stress change is not expected to be isotropic when the reservoir is depleted. Figure 12.6 schematically shows the mechanism we are considering. We consider there to be a reservoir-bounding (impermeable) fault at an angle to the current direction of maximum horizontal compression (Figure 12.6a). We seek to investigate whether depletion could cause the direction of S_{Hmax} to rotate so as to become more parallel to the local strike of the fault (Figure 12.6b), resulting in a relationship between stress orientation and fault strike similar to that shown in Figure 8.14.

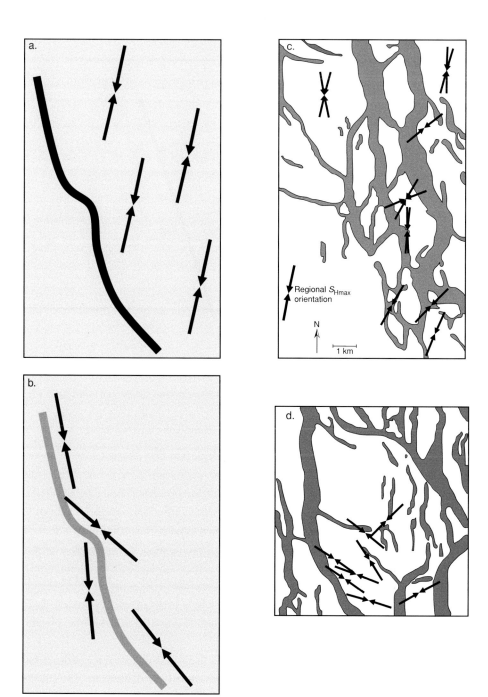

Figure 12.6. (a) Cartoon of a hypothetical oil field in which the current direction of maximum horizontal stress is at an oblique angle to the strike of inactive normal faults in the region. (b) Following depletion, the direction of maximum horizontal stress appears to follow the local strike of the faults in the region. (c) A portion of the Arcabuz–Culebra field in Mexico in which the current direction of maximum horizontal stress is at an oblique angle to the predominant trend of normal faults in the region. (d) A depleted section of the Arcabuz–Culebra field (near that shown in c) where the direction of maximum horizontal stress appears to follow the strike of the normal faults in the area. (c) and (d) are modified from Wolhart, Berumen *et al.* (2000). © *2000 Society Petroleum Engineers*

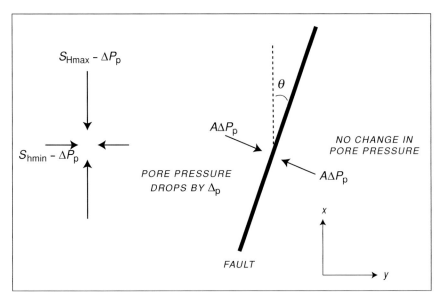

Figure 12.7. Schematic diagram of depletion (ΔP_p) on one side of a sealing fault at angle θ to the direction of S_Hmax. The result of depletion results in an increase of normal stress across the fault of $A\Delta P_\mathrm{p}$, where A is the stress path.

There are two reasons why we suspect depletion may be the cause of apparent stress rotations near faults. First is a case where the orientation of maximum horizontal stress in a reservoir appears to change with time. Wright, Stewart *et al.* (1994) utilized tilt meters to determine the azimuth of hydraulic fracture in the San Joaquin basin of California before and after significant depletion. They report that orientation of the hydraulic fractures changes with time. Second is a case of adjacent sections of the Arcabuz–Culebra gas field in northern Mexico. Here, stress orientations in an undepleted section of the field are subparallel to the regional direction of S_Hmax of ~N22°E (Figure 12.6c). In a nearby section of the field that is depleted, stress orientations seem to follow the local strike of the reservoir bounding faults (Figure 12.6d, modified from Wolhart, Berumen *et al.* 2000).

Figure 12.7 illustrates the state of stress near an impermeable, vertical fault within a reservoir which, during production, experienced a reduction in pore pressure by an amount ΔP_p. θ is the acute angle between the fault strike and the regional maximum horizontal stress direction. If permeability in the reservoir is homogeneous and isotropic (that is, as pore pressure drops it decreases the same amount in every direction), then the depletion decreases the magnitudes of the horizontal stresses by $A\Delta P_\mathrm{p}$, where A, the stress path, defined above, is known. In addition, the traction induced to balance the strain on either side of the fault also reduces the fault-normal stress by $A\Delta P_\mathrm{p}$. Continuity of the normal stress across the fault means that this change affects both sides of the fault and decays rapidly with distance from the fault.

Following the approach used by Sonder (1990) and Zoback (1992), the new stress state in the reservoir and *near the fault* can be found by superimposing the uniaxial fault-normal stress perturbation onto the background stress state. The resultant stress components in the original coordinate system (Figure 12.7) are

$$S_x = (S_{Hmax} - A\Delta P_p) - \frac{A\Delta P_p}{2}(1 - \cos 2\theta)$$ (12.7)

$$S_y = (S_{hmin} - A\Delta P_p) - \frac{A\Delta P_p}{2}(1 + \cos 2\theta)$$ (12.8)

$$\tau_{xy} = \frac{A\Delta P_p}{2}\sin 2\theta$$ (12.9)

The rotation, γ, of the new maximum, principal horizontal stress near the fault relative to the original S_{Hmax} azimuth can be found by

$$\gamma = \frac{1}{2}\tan^{-1}\left[\frac{2\tau_{xy}}{S_x - S_y}\right] = \frac{1}{2}\tan^{-1}\left[\frac{A\Delta P_p \sin 2\theta}{(S_{Hmax} - S_{hmin}) + A\Delta P_p \cos 2\theta}\right]$$ (12.10)

The sign of γ is the same as the sign of θ. If we define q as the ratio of the pore pressure change (positive for depletion) to the original, horizontal differential stress,

$$q = \frac{\Delta P_p}{(S_{Hmax} - S_{hmin})}$$ (12.11)

following Zoback, Day-Lewis *et al.* (2007) we can express the stress rotation simply as a function of q, the stress path (A), and the fault orientation (θ):

$$\gamma = \frac{1}{2}\tan^{-1}\left[\frac{Aq \sin 2\theta}{1 + Aq \cos 2\theta}\right]$$ (12.12)

The resulting stress rotation actually occurs on both sides of the fault, because the perturbation provides the only contribution to shear (τ_{xy}) in the *x-y* coordinate system, and $S - S_y$ is the same on either side.

As shown in Figure 12.8, depletion can induce appreciable stress rotations when both q (the ratio of the change in pore pressure to the original difference between the horizontal principal stresses) and θ (the difference between the azimuth of S_{Hmax} and the strike of the sealing fault) are large.

In Figure 12.9 we revisit the depleted section of the Arcabuz–Culebra field illustrated in Figure 12.6d to see if the stress rotation model derived above can explain the observed orientations of S_{Hmax}. The value of q in this field is estimated to be between 0.5 and 2, although little is actually known about the magnitude of S_{Hmax} and there is a possibility that there might be significant local variations in ΔP_p. For $A = 2/3$ and $q = 0.5$, the maximum possible rotation of S_{Hmax} is approximately 10° (not illustrated), which clearly does not explain the observations. Using $q = 2$, however, we show in Figure 12.9 the fault orientation needed to match the observed rotation at each well, and we find that the majority of the observed stress orientations can be explained by the presence of nearby faults (even if the theoretically influential fault is not always the closest or

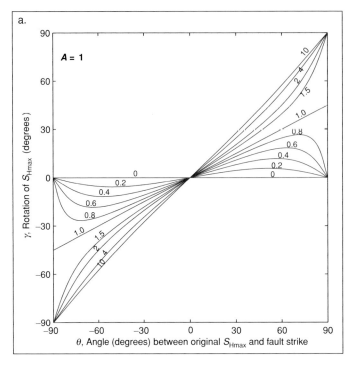

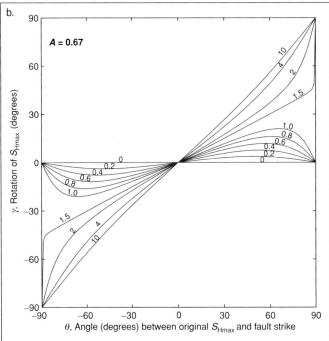

Figure 12.8. Rotation, γ, of the direction of S_{Hmax} as a function of the original angle between S_{Hmax} and the fault trend, θ, and the normalized depletion, q. Note that very large rotations can occur when the depletion is on the order of, or exceeds, the initial difference of the horizontal principal stress magnitudes. After Zoback, Day-Lewis *et al.* (2007)

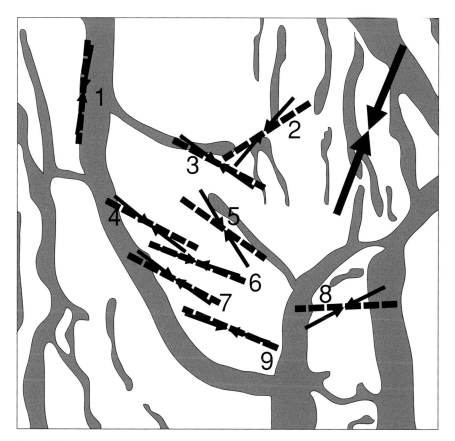

Figure 12.9. Comparison of the calculations presented in Figure 12.8 with data from the field presented in Figure 12.6d. The observed stress orientation is shown by the inward arrows. The dashed lines show the orientation of a sealing fault required to explain the observed rotation. Because there is little specific information available on the amount of depletion or the magnitude of S_{Hmax}, the calculations were done assuming $q = 2$. After Zoback, Day-Lewis *et al.* (2007)

largest fault). The clustering of the three observations that are not near a fault of appropriate orientation (wells 4, 6, and 7) and are similar in their S_{Hmax} orientations implies that they are affected by a sealing fault in this part of the field similar in orientation to the fault seen near well 5.

Drilling and hydraulic fracturing in depleted reservoirs

As noted at the beginning of this chapter, drilling and hydraulic fracturing are affected by the poroelastic stress changes accompanying depletion. When there is a need to drill through depleted reservoirs to reach deeper formations, a variety of drilling problems could occur. Unless relatively low mud weights are used, there could be unintentional hydraulic fracturing and lost circulation in the depleted reservoir due to the decrease of the least principal stress in the depleted zone (but not adjacent formations). There can also be differential pipe sticking due to the difference between the mud weight and pore pressure in the depleted formations and there could be considerable invasion and formation damage in the depleted formation if further production was planned for the depleted reservoir. If much lower mud weights are used to avoid these problems in the depleted zone, wellbore instability could be a significant problem above and below it.

As discussed by van Oort, Gradisher *et al.* (2003) there are a variety of techniques that can be used to address the problem of drilling through depleted formations. Some of these are related to the use of water-based mud and lost circulation additives discussed in the section of Chapter 10 addressing drilling with mud weights above the fracture gradient. As was discussed in Chapter 10, it could also be advantageous to drill in optimal directions to avoid hydraulic fracturing near the wellbore and lost circulation when drilling with mud weights above the least principal stress. Other techniques include the use of additives to prevent mud penetration into the formation (see also Reid and Santos 2003) and the use of formation "strengthening" additives which, in effect, cements the grains of the formation out in front of an advancing wellbore thus making drilling easier (see also Eoff, Funkhauser *et al.* 1999 and Webb, Anderson *et al.* 2001).

While drilling through depleted reservoirs can be considerably more problematic than drilling through the same reservoirs prior to depletion, hydraulic fracturing in depleted reservoirs can be easier than prior to depletion (and surprisingly effective if stress rotation has accompanied depletion). The various papers presented in the compilation of Economides and Nolte (2000) discuss many different aspects of reservoir stimulation using hydraulic fracturing. It is worth briefly discussing the advantages of repeating hydraulic fracturing operations (or re-fracturing) depleted reservoirs, two topics not considered by the papers in that compilation.

One type of reservoir that would be particularly advantageous to consider hydraulic fracturing after depletion is those in which significant rotation of principal stress directions occurs. The conditions under which such cases are likely to occur were discussed in the previous section. Clearly, if hydraulic fracturing was used when wells were initially drilled in a tight reservoir, rotation of principal stress directions during depletion would cause any new fracture to propagate at a new azimuth, possibly accessing previously undrained parts of the reservoir.

A second advantage of hydraulic fracturing a depleted reservoir is illustrated in Figure 12.10. In cases in which there is only a small contrast in the magnitude of the least principal stress between the reservoir and caprock prior to depletion (Figure 12.10a, modified after Wolhart, Berumen *et al.* 2000), it is difficult to extend a hydraulic fracture far from a well without the potential for vertical hydraulic fracture growth. This is illustrated by the fracture growth simulation in Figure 12.10c. It is important to avoid vertical fracture growth because of the potential of connecting to water-bearing strata. Hence, reservoirs in which there is only a small contrast in the magnitude of the least principal stress between the reservoir and adjacent formations may be poor candidates for hydraulic fracturing, or at least limit the degree to which hydraulic fracturing can

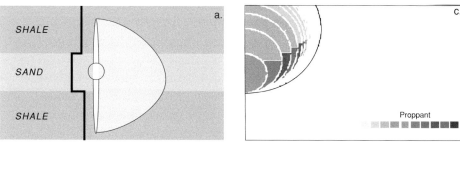

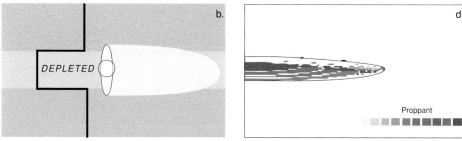

Figure 12.10. (a) Cartoon illustrating the difficulty of propagating a hydraulic fracture in a sand reservoir when there is only a small contrast in the magnitude of the least principal stress between the reservoir and adjacent units. The magnitude of S_{hmin} is indicated by the straight lines. Because of the low stress contrast there would be a tendency for vertical fracture growth as the pressure required to exceed the least principal stress in the reservoir might also exceed that in the adjacent units as well. (b) Calculations using a commercial hydraulic fracture growth simulator that illustrates vertical fracture growth when the stress contrast between the reservoir and surrounding units is small. The shading indicates the distribution of proppant in the fracture. (c) Following depletion, a larger stress contrast between the sand and adjacent units exists. (d) Simulation of fracture growth (and distribution of proppant) in a depleted reservoir indicates effective propagation in the reservoir as well as placement of the proppant. Modified from Wolhart, Berumen *et al.* (2000). © *2000 Society Petroleum Engineers*

be used to stimulate productivity. In a depleted reservoir, however, poroelastic effects amplify the contrast in the magnitude of the least principal stress with adjacent formations over that which existed initially. This has the potential for dramatically improving the efficacy of hydraulic fracturing by reducing the potential for vertical fracture growth considerably, as illustrated by the fracture simulation shown in Figure 12.10d (Wolhart, Berumen *et al.* 2000).

Deformation in depleting reservoirs

In this section we consider deformation within depleting reservoirs. We focus on very weakly cemented sands because depletion effects in such reservoirs are considerable and there are many such reservoirs around the world. We begin by discussing laboratory experiments of porosity loss with hydrostatic pressure, simulating the increase in effective stress in a reservoir caused by reduction in reservoir pore pressure over time. Next, we revisit *end-cap* models (shear-enhanced compaction) first introduced in Chapter 4 (see Figures 4.19 and 4.20). This allows us to consider more fully how porosity will evolve in a reservoir as depletion occurs and horizontal stresses decrease due to poroelastic effects. This discussion leads to a formalism for representation of porosity loss in depleting weak sediments using DARS (Chan and Zoback 2002). This formalism enables one to evaluate the potential for porosity loss (and the possibility that depletion-induced faulting might occur) as pore pressure decline and stresses change along whatever stress path is characteristic of the reservoir in question.

Fundamentally, DARS is a technique for predicting irreversible porosity loss with depletion (and the possibility of production-induced faulting within a reservoir) using parameters normally available in oil and gas fields. We can further utilize this information to estimate the possible range of permeability changes in depleting weak sands. We compare theoretical predictions of permeability loss against observations made in a depleting reservoir in the Gulf of Mexico. In the final part of this section, we utilize reservoir simulation in an idealized reservoir to briefly consider how depletion, porosity loss and permeability loss affect productivity.

Compaction with increased confining pressure

In this section, we consider data in Field X in the Gulf of Mexico discussed above to evaluate how porosity would be expected to evolve with depletion (Chan and Zoback 2002). Laboratory measurements of the porosity loss with increasing effective stress on seven samples from Field X are presented in Figure 12.11. Initial porosities of these samples varied from 22% to 32%. Note that the three samples marked *low-porosity samples* in Figure 12.11 were not loaded beyond 48 MPa. This is noteworthy because the first set of tests (on the samples marked *high porosity samples*) experienced an abrupt loss in porosity when loaded beyond 55 MPa. The pressure where the abrupt porosity loss occurs is close to the pre-consolidation pressure the samples experienced *in situ*, estimated assuming the sediments were buried under hydrostatic pressure and whatever overpressure exists in the reservoir prior to depletion developed during burial. For the seven samples tested, the pre-consolidation pressure is estimated to be about 48 MPa. It is reasonable to assume the abrupt change in porosity at ~55 MPa is related to exceeding the pre-consolidation pressure in the laboratory experiments. In other words, the laboratory samples were compacting along the reloading path until they reach the pre-consolidation pressure at which point they begin to compact following a steeper compaction curve.

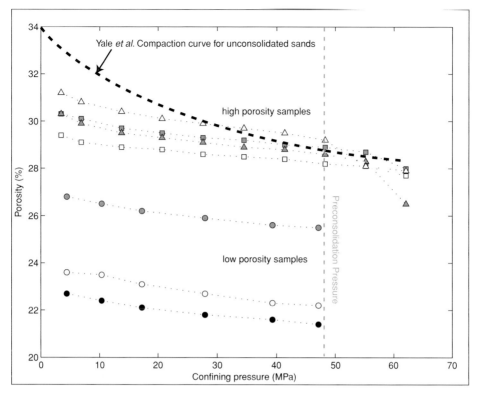

Figure 12.11. Pressure dependence of porosity for seven samples collected from the producing formation of GOM Field X. The three circles with different shades are experiments conducted in one study while the squares and triangles are samples tested in another study. There is a marked change in the rate of porosity and permeability reduction at about 50 MPa, close to the apparent preconsolidation pressure. Note that the porosity changes for all samples are very similar even though the initial porosities are quite different. The experimental compaction curve of Yale, Nabor *et al.* (1993) for unconsolidated sand reservoirs is shown for comparison. Modified from Chan and Zoback (2002). © *2002 Society Petroleum Engineers*

The dashed line in Figure 12.11 is the generalized compaction curve defined by Yale, Nabor *et al.* (1993) for Gulf of Mexico sands. The form of their compaction curve is

$$C_f = A(\sigma_{lab} - B)^C + D \tag{12.13}$$

where C_f is the formation compressibility and σ_{lab} is the laboratory stress. A, B, C and D are constants derived from laboratory experiments and, in the case of poorly sorted unconsolidated, they have the values of -2.8×10^{-5}, 300, 0.14, and 1.18×10^{-4}, respectively. Given that $C_f = \Delta\phi/\Delta p$, by rearranging equation (12.13), the porosity as a function of increasing confining pressure is plotted. Note that the curve seems to overestimate the amount of compaction for the samples, probably because Yale's curve is for virgin compaction but the laboratory tests are reloading samples to the pre-consolidation pressure.

End-cap models and DARS

In this section we discuss the DARS (Deformation Analysis in Reservoir Space) formalism introduced by Chan and Zoback (2002). The principal idea of DARS is to bridge simple laboratory compaction measurements with *in situ* stress measurements to predict reservoir deformation associated with depletion. While end-cap models described in Chapter 4 are widely used in soil mechanics and straightforwardly defined using laboratory experiments, it is not obvious how changes in p and q (defined in equations 4.35 and 4.36) may be readily applicable to a producing reservoir during day-to-day operations. While, in theory, *in situ* stress measurements could be conducted in reservoirs through time and yield knowledge of the magnitudes of the three principal stresses and pore pressure (allowing one to compute p and q), it is much more likely to know only pore pressure and the magnitude of the least principal stress (*i.e.* the initial state of the reservoir and stress path). As a result, transforming the end-caps from the laboratory $p{:}q$ space into least principal stress-pore pressure space (which we refer to as *reservoir space*) simplifies the use of end-cap models appreciably. Combining and rearranging equations (4.35)–(4.37) as a function of the three principal stresses and pore pressure (*i.e.* S_{Hmax}, S_{hmin}, S_{V}, P_{p}, p^* and M) results in:

$$9P_{\mathrm{p}}^2 + \left(1 + \frac{9}{M^2}\right)\left(S_{\mathrm{V}}^2 + S_{\mathrm{Hmax}}^2 + S_{\mathrm{hmin}}^2\right)$$
$$+ \left(2 - \frac{9}{M^2}\right)\left(S_{\mathrm{V}}S_{\mathrm{Hmax}} + S_{\mathrm{V}}S_{\mathrm{hmin}} + S_{\mathrm{Hmax}}S_{\mathrm{hmin}}\right)$$
$$+ 9P_{\mathrm{p}}p^* - 3(2P_{\mathrm{p}} + p^*)(S_{\mathrm{V}} + S_{H\mathrm{max}} + S_{\mathrm{hmin}}) = 0 \tag{12.14}$$

M defines the failure line in $p{:}q$ space and can be defined in terms of μ from the Mohr–Coulomb failure criterion assuming the cohesion C_0 is negligible (combining equations 4.1–4.3) to yield:

$$M = \frac{6\mu}{3\sqrt{\mu^2 + 1} - \mu} \tag{12.15}$$

For $\mu = 0.6$, M is roughly equal to 1.24.

Rearranging equation (12.14) yields a relationship between the *in situ* reservoir stress measurements and the pre-consolidation pressure (substituting S_{H} and S_{h} for S_{Hmax} and S_{hmin} for simplicity)

$$p^* = \frac{1}{3(S_{\mathrm{V}}S_{\mathrm{H}} + S_{\mathrm{h}}) - 9P_{\mathrm{p}}}\left\{9P_{\mathrm{p}}^2 + \left(1 + \frac{9}{M^2}\right)\left(S_{\mathrm{V}}^2 + S_{\mathrm{H}}^2 + S_{\mathrm{h}}^2\right)\right.$$
$$\left. + \left(2 - \frac{9}{M^2}\right)(S_{\mathrm{V}}S_{\mathrm{H}} + S_{\mathrm{V}}S_{\mathrm{h}} + S_{\mathrm{H}}S_{\mathrm{h}}) - 6P_{\mathrm{p}}(S_{\mathrm{V}} + S_{\mathrm{H}} + S_{\mathrm{h}})\right\} \tag{12.16}$$

Rock properties measured in the laboratory such as p^* (*e.g.* the porosity at which irreversible plastic deformation occurs) can then be transformed into the reservoir domain of the least principal stress and pore pressure, as S_V is expected to remain constant with depletion for a laterally extensive reservoir. In this way, the two-dimensional end-caps in $p{:}q$ space are transformed into three-dimensional end-cap ellipsoids in the reservoir domain (S_{Hmax}, S_{hmin} and P_p). Thus, we combine the shear (Coulomb) failure envelope with the transformed end-cap ellipsoids to project the ellipsoids onto the $S_{hmin}{:}P_p$ domain assuming S_{Hmax} has only a minor effect on deformation because it corresponds to the intermediate principal stress in normal faulting regimes. This yields a new composite diagram that can be created for analyzing the degree of shear and compaction deformations that are associated with reservoir depletion in $S_{hmin}{:}P_p$ (*i.e.* reservoir) space Hence, end-caps associated with the Cam-Clay model, such as those shown in Figure 4.19 are transformed into equivalent end-caps in reservoir space in presentations such as Figure 12.12. The evolution of the end-caps of any given reservoir rock at different porosities can now be used as an indicator of the deformation induced by the increase of the effective stresses due to the decrease in pore pressure during production. In this way, it is possible to examine directly from the initial state of a reservoir and the stress path accompanying depletion how porosity will evolve and whether faulting is likely to occur.

To implement DARS for a producing reservoir, the value of the vertical stress is derived from density logs while M and p^* are determined from relatively simple laboratory experiments. As M is based on the frictional strength of faulted rock, it is reasonable to assume it has a value of about 1.2. p^* is straightforwardly determined from measuring porosity change with increased confining pressure, going beyond the pre-consolidation pressure. For this discussion, we limit ourselves to normal faulting regions where S_{hmin} is the least principal stress, which can be obtained from LOTs and mini-fractures, and the initial S_{Hmax} is intermediate in value between S_{hmin} and S_V.

In summary, there are three steps in the DARS formalism:
(1) The initial stress state and pore pressure in the reservoir are determined (if possible).
(2) The reservoir depletion stress path must be measured, or estimated from poroelastic theory.
(3) Laboratory measurements of porosity reduction as a function of pressures are needed. If only hydrostatic experiments are available, the theoretical plasticity model can be utilized to extrapolate these data into $p{:}q$ space and then into reservoir space. A Cam-Clay model is used in this study because of its simplicity. These laboratory end-caps are then transformed into reservoir space, the $S_{hmin}{:}P_p$ domain.

We apply DARS to Field X using the laboratory data introduced in Figure 12.11 and the pore pressure and stress data shown in Figure 12.4b. While initial depletion caused relatively minor porosity changes, once pore pressure dropped below ~50 MPa, porosity reduction was much more rapid as the end caps were being encountered. As noted above, the stress path in Field X is such that depletion-induced faulting is not

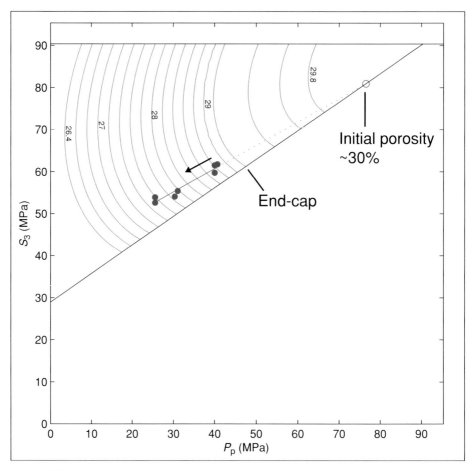

Figure 12.12. Using the DARS formalism, end-caps computed with a Cam Clay model (such as shown in Figure 4.19) are transformed into reservoir space (S_{hmin}:P_{p}). DARS predictions of porosity change (or compaction) for Field X are based on the laboratory experiments shown in Figure 12.11. The data points represent the *in situ* stress and pore pressure measurements shown in Figure 12.4b. The closely spaced contours correspond to the onset of plastic deformation once the pre-consolidation pressure is reached. After Chan and Zoback (2002). © *2002 Society Petroleum Engineers*

expected to occur. However, the overall porosity loss accompanying depletion is quite significant, from about 30% to 27%. The DARS methodology was used by Chan and Zoback (2002) to predict porosity decreases (and production-induced faulting) in the weak chalk reservoir of the Valhall field in the North Sea.

Depletion-induced permeability loss

Large reductions in the pore volume with depletion can assist in the expulsion of oil (compaction drive). Before one can assess the overall significance of compaction drive on the production of a reservoir, it is necessary to estimate the degree to which the porosity decrease affects reservoir permeability. Based on experimental data of samples from five different reservoirs and sixteen outcrops, Schutjens, Hanssen *et al.* (2001) illustrate that the change in axial permeability is independent of how the sample is loaded when the applied stress deforms the sample within the elastic domain (*i.e.* loading within the end cap or at pressures less than the pre-consolidation pressure). However, a significant change in permeability is triggered by the onset of shear-enhanced compaction once the sample is loaded beyond the elastic domain into the plastic deformation domain in the reservoir stress space (*i.e.* when the stress state exceeded the pre-consolidation pressure causing an expansion of the end-cap). Clearly, the effects of plastic deformation and permeability alteration can be significant in reservoir simulations of a highly compressible formation. Using coupled simulations, Yale (2002) showed that the initial stress state and plasticity significantly increases the compressibility of the formation and the compaction drive energy of the reservoir; modeling the changes in permeability with plastic deformation shows an extremely large effect on near wellbore pressure drawdown and deformation over conventional simulations in which only elasticity is assumed. Crawford and Yale (2002) use an elastoplastic model (also referred to as a critical state model) to study the relationship between deformation and the corresponding permeability loss. They show that an elastoplastic model captures the main characteristics of experimental results for permeability changes as a function of both stress and strain, following a constitutive model similar to that for deformation of weak and unconsolidated sand samples.

The samples from Gulf of Mexico Field X shown in Figure 12.11 were used for determining the way in which permeability changes with changes in porosity. The range of confining pressure used in these experiments (0–60 MPa) represents the possible range of depletion that might occur in the field. The initial permeability of the samples varies between 80 mD and 1050 mD and reflects the variation of initial porosity, grain packing, clay abundance (and distribution) and cementation. To examine the relationship between compaction and permeability loss, normalized permeability is shown as a function of normalized porosity (Figure 12.13a after Chan, Hagin *et al.* 2004). The data for Field X, shown in Figure 12.13a with triangles and squares follow two general trends we label as the upper and lower bounds of permeability change with porosity change that are probably related to the initial porosity of the samples. The low-porosity samples appear to follow the lower bound while the high-porosity samples have a more drastic change in permeability as a result of porosity loss. To examine the generality of these empirical trends, results from published experimental data on 22 deep-water turbidites from different fields within the Gulf of Mexico (Ostermeier 2001) are used for comparison. By re-plotting Ostermeier's experimental data (+ signs in Figure 12.13a) in terms of normalized permeability vs. normalized porosity and superimposing them on the plot of experimental data from Field X, we see that 95% of the Ostermeier's data fall within the upper and lower bounds defined on the basis of Field X samples, regardless of the initial porosity of the samples and the location and depth at which the

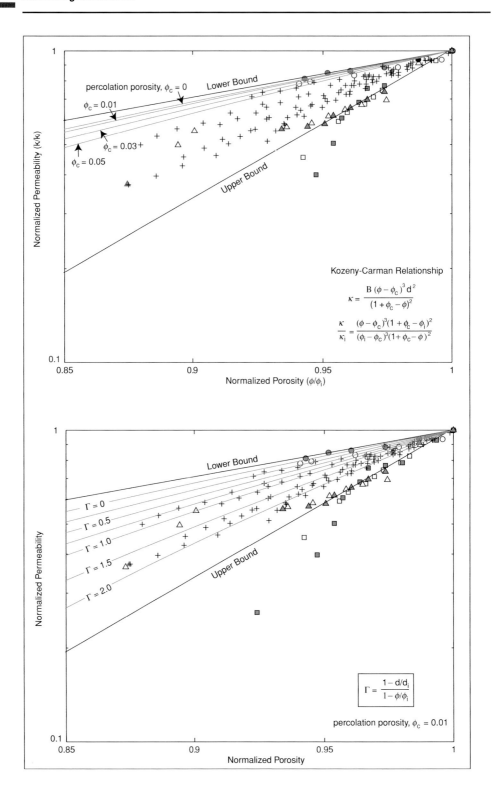

samples were collected. In other words, the two empirical trends presented here can be used as a general estimation on how porosity reduction will affect the permeability of the samples for turbidite sands from the Gulf of Mexico. Note that the reduction of permeability can be as high as 70% of the original permeability for a 10% change in porosity. This drastic variation in permeability change as a result of production-induced compaction could greatly affect reservoir simulation if ignored.

To put the permeability change in a theoretical context, we consider the predictions of the Kozeny–Carman relationship to examine the physical implications of the two empirical permeability trends described in the previous section. This relationship is a widely used method to determine the permeability of a porous formation in terms of generalized parameters such as porosity (Carman 1961; Mavko *et al.* 1998). To estimate fluid flow in a porous medium, the Kozeny–Carman relationship idealizes the medium as a twisted circular pipe of known dimensions. Applying Darcy's law for laminar flow through the circular pipe, the Kozeny–Carman relationship states that

$$\kappa = \frac{B\phi^3}{\tau^2 S^2} = B\phi^3 \frac{d^2}{\tau} \qquad (12.17)$$

where k is the permeability, B is a geometric factor, τ is tortuosity and d is the average grain diameter. The porosity, ϕ, and the specific surface area, S, can be expressed by:

$$\phi = \frac{\pi R^2}{A} \quad \text{and} \quad S = \frac{2\pi R}{A} \qquad (12.18)$$

where R and A are the radius and the cross-sectional area of the imaginary pipe.

In general, the Kozeny–Carman relationship implied that permeability is proportional to the porosity cubed. Mavko and Nur (1997) introduce the percolation porosity, ϕ_c, to the Kozeny–Carman relationship. They define the percolation porosity as the limiting porosity at which the existing pores within the formation are disconnected and do not contribute to flow. The modified Kozeny–Carman relationship that includes the percolation porosity becomes:

$$\kappa = B \frac{(\phi - \phi_c)^3}{(1 + \phi_c - \phi)^2} d^2 \qquad (12.19)$$

where ϕ_c ranges from 0 to 0.05 in most cases.

Figure 12.13. (a) Comparing the empirical permeability–porosity relationship derived from laboratory studies with the Kozeny–Carman relationship (after Chan, Hagin *et al.* 2004). The solid lines define an upper and lower bound to observed changes in permeability with changes in porosity. The dotted lines are derived from a modified Kozeny–Carman relationship (Mavko and Nur 1997) that introduces a percolation porosity, ϕ_c, to the Kozeny–Carman relationship. (b) If one assumes that grain-size reduction occurs during compaction, the modified Kozeny–Carman relationship predicts a greater change in permeability with porosity change (Chan, Hagin *et al.* 2004). However, because $\Gamma \geq 1$ is unlikely, grain-size reductions alone do not explain the large reductions in permeability observed in the samples that fall near the upper bound of permeability reduction.

To determine the permeability change as a result of porosity change, Chan, Hagin *et al.* (2004) used the modified Kozeny–Carman relationship and simplify equation (12.19) such that both geometric factors are removed (assuming a constant B for simplicity):

$$\frac{k}{k_i} = \left(\frac{\phi - \phi_c}{\phi_i - \phi_c}\right)^3 \left(\frac{1 + \phi_c - \phi_i}{1 + \phi_c - \phi}\right)^2 \tag{12.20}$$

where k_i and ϕ_i are the initial permeability and initial porosity, respectively. The theoretical values of compaction-induced permeability changes for ϕ_c between 0 and 0.05% using equation (12.19) are then superimposed onto the laboratory data from the GOM core samples (Figure 12.13a). The theoretical permeability changes calculated using the modified Kozeny–Carman relationship are similar to the lower bound estimated from the laboratory data. This similarity might imply that the empirical lower bound represents the lower limit of permeability changes for most Gulf of Mexico sands for which percolation porosity does not exist. In other words, if the producing formation is composed of porous materials in which all pore spaces are well connected, the Kozeny–Carman relationship with $\phi_c = 0$ could be used as a reference for the lower limit of permeability changes as a result of production-induced compaction, although this obviously does not fully capture the significantly large permeability loss due to compaction in weak sediments.

The estimated change in permeability from the normalized Kozeny–Carman relationship (equation 12.20) assumes a constant grain size during compaction. Hence, one possible explanation for the reductions in permeability is that they are caused by grain crushing during compaction. This phenomenon was noted by Zoback and Byerlee (1976) during deformation tests and permeability measurements on crushed granite. A reduction in average grain size, d, change in tortuosity or grain arrangement (resulting in a change in the geometric factor, B) would allow for this to be predicted by the modified Kozeny–Carman relationship. To incorporate the change in average grain size within the modified Kozeny–Carman relation, Γ is introduced such that:

$$\Gamma = \frac{1 - d/d_i}{1 - \phi/\phi_i} \tag{12.21}$$

where d_i and d are the average grain size prior to and after compaction. $\Gamma = 0$ implies the average grain size did not change during porosity reduction. This term is introduced in order to simplify the various responses of grain size reduction as a function of porosity reduction, as illustrated in Figure 12.21b below. Equation (12.21) suggests that a big reduction in porosity is required in order to get a large reduction in average grain size (or a large Γ). Note that $\Gamma > 1$ is only possible for a very small range of porosity changes. Introducing the variable Γ, equation (12.20) becomes:

$$\frac{k}{k_i} = \left(\frac{\phi - \phi_c}{\phi_i - \phi_c}\right)^3 \left(\frac{1 + \phi_c - \phi_i}{1 + \phi_c - \phi}\right)^2 \left[1 - \Gamma\left(1 - \frac{\phi}{\phi_i}\right)\right] \tag{12.22}$$

Figure 12.13b suggests that although grain-size reduction might have some influence on the permeability reduction, a very high grain-size reduction (high Γ) is required to explain the upper bound of permeability loss. As a result, a more complicated model that includes changes in tortuosity and the geometric factors might be needed to fully describe the physical mechanism that causes the large drop in permeability represented by the upper bound.

To evaluate whether the range of predicted permeability changes in Figure 12.13 (*i.e.* the upper and lower bounds) are applicable to reservoirs, Chan, Hagin *et al.* (2004) evaluated Field Z in the Gulf of Mexico. It is a deepwater, over-pressured and uncemented sand reservoir juxtaposed against a large salt dome. The formation is mainly turbidite sands with an average porosity of 30%. The initial horizontal permeability of the sands ranged from 60 to 168 mD for a moderate quality sand and 350 to 540 mD for a good quality sand interval. In addition to measurements of S_{hmin} and P_p for this field, horizontal permeability measurements are also available through time for several wells in the field.

Pore pressure measurements from most wells in the field are compiled and corrected to a datum and a continuous decrease in P_p and the least principal stress was observed. Similar to Field X, there is a relatively low stress path in Field Z, $A = 0.54$, indicating that production-induced normal faulting is unlikely to occur. *In situ* permeability measurements from three different wells in Field Z are shown as a function of depletion in Figure 12.14. Well A is located near the center of the reservoir and wells B and C are located near the edge of the reservoir. Permeability measurements in these wells A and C were collected immediately after production began and the first permeability measurement in well B was collected after about 10 MPa of depletion. Well A has a relatively low permeability and is within the range of permeability for a moderate quality sand interval (the initial permeability for well A is assumed to be about 140 mD); while the initial permeability for wells B and C is estimated to be 470 mD (the average value for good reservoir quality sands). Without measurement of initial permeability in these wells, we assume an average value of the reported permeability from Field Z based on the reservoir quality.

Based on the *in situ* stress and pressure measurements, Chan, Hagin *et al.* (2004) predicted the porosity change for wells A, B and C using DARS. Utilizing the two empirical porosity–permeability relationships shown in Figure 12.13, the possible range of permeability changes associated with depletion for the three wells A, B and C is shown in Figure 12.14. The *in situ* permeability for well A seems to follow the lower bound of the permeability loss while wells B and C appear to agree with the upper bound of permeability loss. Note that the absence of knowledge of the initial permeability for these wells makes it difficult to determine the accuracy of the prediction. As initial permeability for good quality reservoir sand ranges from 350 to 540 mD, measurements from well B can easily be fit to the predicted values if the initial permeability used in the analysis is reduced. However, only the average value is used in this case to show

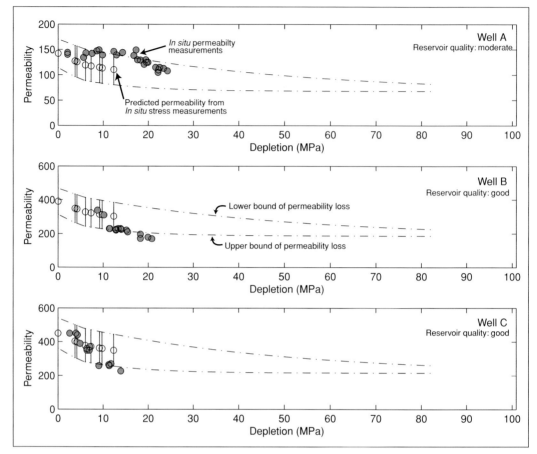

Figure 12.14. Comparison between *in situ* permeability measurements from wells A, B and C in Gulf of Mexico Field Z with the predicted permeability using DARS and the upper and lower bounds of the empirical porosity–permeability relationship shown in Figure 12.13 (after Chan, Hagin *et al.* 2004). The open circles are the predicted average permeability values corresponding to the *in situ* stress measurements. The dashed lines are the lower and upper bounds of permeability loss assuming the reservoir will deplete along the same stress path. Filled circles are *in situ* stress measurements from each of the three wells. Initial permeability measurements from these wells are not available.

that uncertainties associated with *in situ* measurements can also affect the accuracy of the DARS prediction.

A compaction drive exercise

As noted above, while compaction drive is an effective recovery mechanism, the reduction in permeability would tend to reduce recovery. The trade-off between these two phenomena requires detailed modeling of rock compaction during reservoir simulation since the result might affect the prediction of reservoir recovery, production forecast and well placement decisions.

To illustrate the impact of porosity and permeability loss during depletion on reservoir performance, a simple 2D conceptual single-phase flow model based on Field Z was constructed (courtesy Inegbenose Aitokhuehi). The reservoir was assumed to be elliptical with dimensions of 1900 m by 960 m and a thickness of 21 m, a 50 by 50 grid is generated with an average permeability of 350 mD and an initial porosity of 30%. Three scenarios were investigated using a commercial reservoir simulator to demonstrate the effects of compaction and permeability reduction:

1. Constant rock compressibility: The rock compressibility is estimated as an average change in porosity associated with the expected depletion.
2. Compaction drive: Incorporating the DARS formalism, porosity change as a function of depletion and stress reduction is estimated. The predicted change in porosity is input as varying pore volume multipliers in the simulators. In this scenario, no permeability change is assumed to occur during depletion.
3. Compaction drive with permeability loss: By relating the transmissibility multiplier to the pore volume multiplier based on the two empirical bounds of permeability changes, both permeability and porosity loss will contribute to the estimated cumulative production of the conceptual reservoir.

Several assumptions are made to simplify and to shorten the time required for the simulation. The initial production rate is set to be at 10 MSTB/d (thousand surface tank barrels per day) and no water influx or injection. This single-phase simulator is allowed to run until it reaches a minimum bottom hole pressure of 1000 psi ($\sim$7 MPa), an economic limit of 100 STB/d or a maximum time of 8000 days ($\sim$22 years).

Figure 12.15 illustrates the result of the idealized reservoir for the three scenarios outlined above. Using constant compressibility throughout the entire production in scenario 1, the conceptual reservoir will yield about 12 MMSTB cumulative oil over 2500 days ($\sim$7 years). If depletion-induced compaction is considered (as calculated with DARS) for a formation with properties similar to Field Z, the recovery for this reservoir is increased significantly to about 26 MMSTB over 7500 days ($\sim$20.5 years). In other words, compaction drive enhanced the recovery and extended the production life of this conceptual reservoir. When permeability loss associated with compaction is taken into consideration, the times estimated for recovery are extended. The predicted recoveries ranged from 16 to 25 MMSTB over 8000 days ($\sim$22 years) depending on whether the upper-bound or lower-bound permeability change with porosity is used in the simulator. In the last two cases, the production life of the reservoir is extended. As a result, incorporating both depletion-induced compaction and permeability loss into the simulator will significantly affect the anticipated recovery and the production lifetime of a reservoir. In terms of recovery, production-induced compaction provides an additional driving mechanism that increases the recovery estimate; a small reduction in permeability (lower bound) might not have as much of an impact as a large reduction in

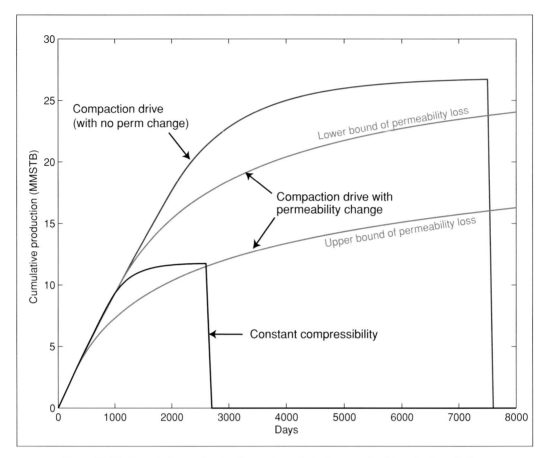

Figure 12.15. Cumulative production from a hypothetical reservoir with a single well. One simulation represents compaction using constant compressibility throughout production. When compaction drive (porosity loss due to depletion) is considered, the cumulative production from this reservoir is increased significantly. However, the loss in permeability associated with a loss in porosity reduces the reservoir productivity to some degree, depending on the severity of compaction-induced permeability loss.

permeability (upper bound) on the estimated recovery. The trade-off between porosity changes and permeability changes has significant implications for the determination of the recovery rate and the overall exploitation scheme for the reservoir and will affect critical decisions such as the need to drill additional wells.

Viscoplastic deformation and dynamic DARS

One of the limitations of traditional end-cap models, such as the modified Cam-Clay described in the previous section, is that the models only describe materials with a static, time-independent yield surface. For materials with significant time-dependent

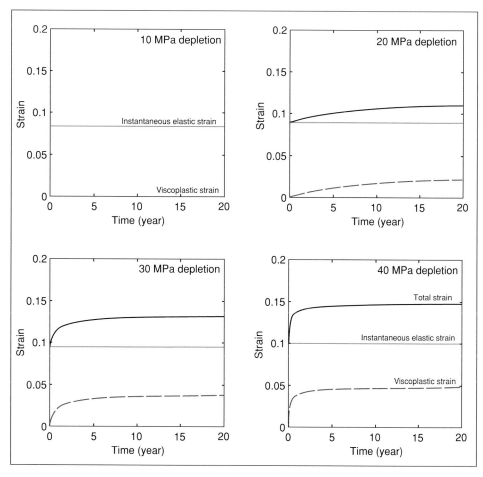

Figure 12.16. Compaction strain for different amounts of depletion for Gulf of Mexico Field Z. Note that there is almost no viscous compaction for 10 MPa of depletion but successively more viscous compaction as depletion increases.

deformation such as the viscoplastic sands described in Chapter 3, such models will generally not predict all of the compaction that is likely to occur unless they fully incorporate viscoplastic deformation. Recall that incremental 5 MPa loading steps in a Wilmington sand produced roughly equal amounts of instantaneous compaction and time-dependent compaction (Figure 3.9). Figure 12.16 shows a series of experiments on GOM Field Z, mentioned earlier in this chapter. Note that there is essentially no viscoplastic compaction with 10 MPa of depletion but successively more viscoplastic strain with higher amounts of depletion. This is because it was necessary to exceed the *in situ* pre-consolidation pressure to see any viscous effects, as previously alluded to in Chapter 3. The importance of viscous compaction in an unconsolidated sand reservoir can be quite appreciable. Note that at 40 MPa depletion, the viscous component of compaction increases total compaction by about 50%.

Hagin and Zoback (2004c) describe a way of modifying traditional end-cap models to include time-dependent plastic deformation, by incorporating viscoplastic effects following the theory of Perzyna (1967). The basic concept behind the inclusion of viscoplasticity into end-cap models is relatively simple; the single static time-independent yield surface is simply replaced with a dynamic yield surface whose position in stress space is dependent on time or rate in addition to porosity or state. In other words, the elastic–plastic constitutive law which defines the static yield surface becomes an elastic–viscoplastic constitutive law. The inclusion of viscoplasticity can be described simply by adding a scaling term to the traditional static end-cap.

Specifically, the theory of viscoplasticity described by Perzyna (1967) can be thought of as a yield stress that depends on some function of strain and follows a power law function of strain rate. Adachi and Oka (1982) used Perzyna viscoplasticity successfully in conjunction with the original Cam-Clay model to describe the deformation of clays and soils. While a detailed discussion of this topic is beyond the scope of this book, it is important to correctly estimate the total amount of compaction in a depleting formation. In an uncemented sand where viscous compaction may be important, such effects need to be considered to correctly estimate the total amount of compaction in the reservoir. Because of this the change in porosity considered in the previous section is a lower bound of that expected to occur *in situ*.

Deformation and stress changes outside of depleting reservoirs

Issues related to depletion so far in this chapter have considered only what happens within a reservoir. As illustrated in Figure 12.1, however, there are a number of important effects of compaction outside a depleting reservoir – namely the potential for surface subsidence and induced faulting.

Compaction and subsidence

For a disk-shaped reservoir of thickness H and radius R at depth D, Geertsma (1973) estimated the effect of compaction on surface subsidence based on a nucleus-of-strain concept in an elastic half-space. Based on poroelastic theory, subsidence due to a uniform pore pressure reduction, ΔP_p, can be treated as the displacement perpendicular to the free surface as a result of the nucleus of strain for a small but finite volume, V, such that:

$$u_z(r, 0) = -\frac{1}{\pi} c_m (1 - v) \frac{D}{(r^2 + D^2)^{3/2}} \Delta P_p V \qquad (12.23)$$

$$u_r(r, 0) = +\frac{1}{\pi} c_m (1 - v) \frac{r}{(r^2 + D^2)^{3/2}} \Delta P_p V \qquad (12.24)$$

where c_m is defined as the formation compaction per unit change in pore-pressure reduction, and Poisson's Ratio, ν. Assuming both c_m and ν are constant throughout the entire half-space, the amount of subsidence caused by a producing disk-shaped reservoir at depth can then be estimated by integrating the nucleus-of-strain solution over the reservoir volume. This yields complex formulae involving integrals of Bessel functions. Introducing the dimensionless parameters $\rho = r/R$ and $\eta = D/R$, the solutions can be simplified to

$$\frac{u_z(r, 0)}{\Delta H} = A(\rho, \eta) \tag{12.25}$$

$$\frac{u_r(r, 0)}{\Delta H} = B(\rho, \eta) \tag{12.26}$$

where A and B are linear combinations of the elliptic integrals of the first and second kind (F_0, E_0) and Heuman's Lambda function (Λ_0)

$$A = \begin{cases} -\dfrac{k\eta}{4\sqrt{\rho}} F_0(m) - \dfrac{1}{2}\Lambda_0(p, k) + 1 & (p < 1) \\[2mm] -\dfrac{k\eta}{4} F_0(m) + \dfrac{1}{2} & (p = 1) \\[2mm] -\dfrac{k\eta}{4\sqrt{\rho}} F_0(m) + \dfrac{1}{2}\Lambda_0(p, k) & (p > 1) \end{cases} \tag{12.27}$$

$$B = = \frac{1}{k\sqrt{\rho}}\left[\left(1 - \frac{1}{2}k^2\right) F_0(m) - E_0(m)\right] \tag{12.28}$$

where

$$m = k^2 = \frac{\rho}{(1 - \rho)^2 + \eta^2} \quad \text{and} \quad p = \frac{k^2[(1 - \rho)^2 + \eta^2]}{(1 - \rho)^2 + k^2}$$

Expressing surface deformation as shown in equations (12.25) and (12.26) allows us to predict surface deformation in terms of ΔH, compaction of the reservoir, and equate the change in thickness of the reservoir with the change in porosity, $\Delta\phi$, assuming that $\Delta\phi = \Delta H/H$ and utilize DARS (as outlined above) to predict the porosity change. As originally formulated by Geertsma (1973), reservoir compaction was calculated assuming linear elasticity and constant compressibility.

To put all of this in practical terms, Figure 12.17 illustrates the amount of surface displacement (normalized by the change in reservoir thickness as expressed in equations 12.27 and 12.28) as a function of normalized distance from the center of the reservoir. Note that for very shallow reservoirs ($D/R \sim 0.2$), the amount of subsidence directly above the reservoir is ~ 0.8 of the total compaction (Figure 12.17a). Hence, cases where there has been substantial subsidence above weak compacting reservoirs are relatively easy to understand. If, for example, $\Delta\phi = 3\%$ (the amount predicted from initial conditions for Field X, Figure 12.12) and the original reservoir thickness is 300 m, $\Delta H \sim 9$ m. If a similar reservoir was sufficiently shallow and broad that

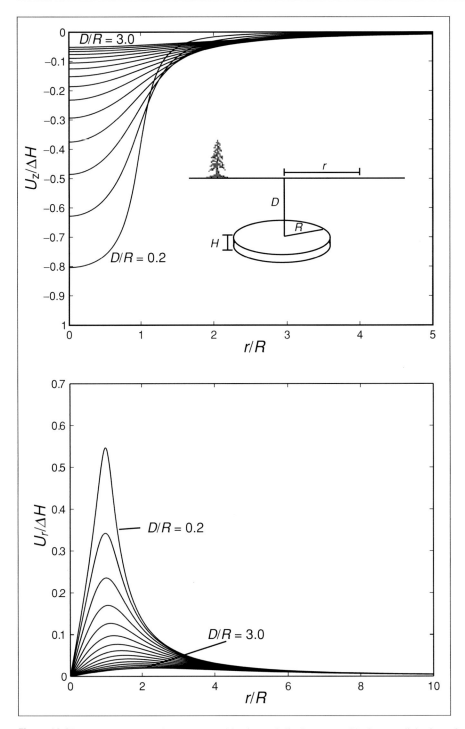

Figure 12.17. Normalized subsidence (a) and horizontal displacement (b) above a disk-shaped reservoir of radius, R, at depth D and initial thickness H, that compacts ΔH following the analytical solution of Geertsma (1973). Note that subsidence is maximum over the center of the reservoir whereas horizontal displacement is maximum at its edge.

$D/R \sim 0.2$, about 7 m of subsidence would be expected. In the Wilmington field in southern California, prior to a fluid injection efforts undertaken to stop subsidence, about 9 m of subsidence occurred. Production was from a shallow ($\sim$1 km deep) laterally extensive (approximately elliptical with long and short axes of 18 km and 5 km, respectively) weak sand reservoir.

The horizontal displacement above a shallow, laterally extensive reservoir is also quite substantial if there has been significant compaction (Figure 12.17b). Note, however, that while subsidence is concentrated directly over the center of a depleting reservoir, the horizontal displacement is concentrated at the boundary of the reservoir. To follow the hypothetical example above, when 7 m of subsidence occurs in a compacting reservoir characterized by $D/R \sim 0.2$, about 5 m of lateral displacement is expected at the edge of the field. That said, Figure 12.17 illustrates that for the hypothetical case being considered, if the reservoir had been deeper and/or less laterally extensive (such that $D/R \sim 1$), there would have been only about 38% as much subsidence over the center of the reservoir and only $\sim$13% as much horizontal displacement at the edges.

We apply the modified Geertsma model for subsidence to two reservoirs in southern Louisiana shown in Figure 12.18a, the Leeville and Lapeyrouse fields. Regional growth faults in the area are shown by the black lines. Leveling lines (indicated by the lines of dots) cross both fields. Information on the structure of the reservoirs and depletion history is also available for both fields. We seek to document the degree of subsidence over each reservoir associated with production and to test the degree to which the modified Geertsma model is capable of predicting the observed subsidence.

The amount of subsidence over the Leeville field between 1982 and 1993 (with respect to a base station approximately 7 km south of the field) is shown in Figure 12.18b along with the associated uncertainties. Over this time period, the maximum subsidence over the field appears to be 6 ± 3 cm. Mallman and Zoback (2007) developed a model of the depleting reservoirs at depth. They utilized DARS to predict the amount of reservoir compaction and equation (12.25) to predict the amount of subsidence. The reservoirs at depth were modeled as multiple circular disks of appropriate thickness and depletion. An elastic–plastic constitutive law (with end-caps) developed for a poorly consolidated sand reservoir in the Gulf of Mexico (Chan and Zoback 2002) was utilized to predict the amount of reservoir compaction in each reservoir. The amount of predicted subsidence associated with depletion in all of the reservoirs is shown by the dashed line in Figure 12.18b. Note that the overall shape of the subsidence bowl is approximately correct, although subsidence seems to be occurring over a somewhat wider area than that predicted. This, along with the observation that the model somewhat underpredicts the observed subsidence, is probably related to the fact that production data from all of the wells in the field were not available. Hence, there is likely depletion occurring in reservoirs not included in the model that are contributing to the observed subsidence.

Chan and Zoback (2006) compared the observed subsidence above the reservoirs of Lapeyrouse field in southern Louisiana with that indicated by leveling data. As illustrated in Figure 2.10, these weak sand reservoirs have undergone significant depletion.

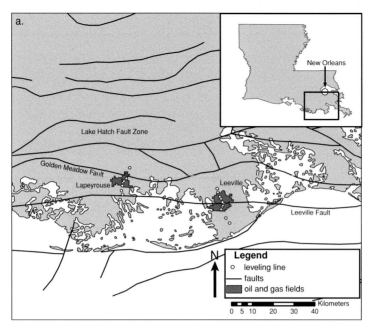

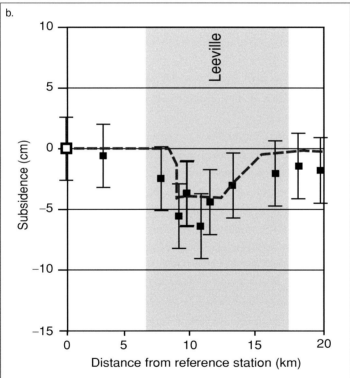

Figure 12.18. (a) Map of southern Louisiana with the regional growth faults shown by black lines. We consider deformation along leveling lines (indicated by the lines of dots) that cross the Leeville and Lapeyrouse oil fields. (b) Relative changes in elevation across the Leeville field (and associated uncertainties) between 1982 and 1993 with respect to a base station approximately 7 km to the south of the field. Predictions of the modified Geertsma model is shown by the dashed line (see text). (after Mallman and Zoback 2007).

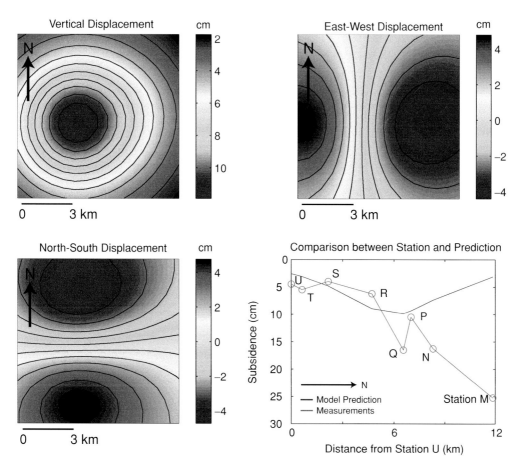

Figure 12.19. Cumulative subsidence and horizontal displacement calculated from superposition of many disk-shaped reservoirs in the Lapeyrouse field using the solution (Geertsma 1973) and DARS (with a viscoplastic rheology) for calculating the total reservoir compaction (after Chan and Zoback 2006). The predicted surface displacements are measured in cm. The predicted subsidence (lower right) is comparable to the measured elevation change from the leveling survey (red line) in the middle of the field but underpredicts the apparent subsidence at the north end of the field near benchmark M. (*For colour version see plate section.*)

However, the reservoirs are relatively deep ($\sim$4.5 km) and thin (individual sands are $\sim$10 m thick), but as many as five reservoirs (ranging in diameter from $\sim$1 to 2.5 km) are stacked upon each other in some places. Hence, D/R is quite large (ranging from about 3 to 9) and the subsidence and surface displacements are expected to be small despite the appreciable depletion. As shown in Figure 12.19, about 12 cm of subsidence is expected over the center of the reservoir and horizontal surface displacements are as much as 5 cm near the edges of the field.

Figure 12.19d compares the predicted cumulative subsidence from all reservoirs (modeled as individual circular reservoirs of constant thickness) with that measured along a ~N–S leveling line going through the field. Note that maximum predicted subsidence (with respect to a benchmark outside the field) is centered on the reservoir and reaches a maximum of about 12 cm. The subsidence indicated by the leveling data appears to be somewhat greater in the center of the reservoir where stations Q and N indicated about 15 cm of subsidence, although station P subsided about 10 cm. Subsidence is much greater near station M, which abuts the Golden Meadow fault. The possibility that slip on this fault contributes to the observed subsidence is discussed in the next section.

Slip on faults outside reservoirs

A second mechanism of induced faulting in the vicinity of oil and gas reservoirs is poroelastic stress changes in the medium surrounding a compacting reservoir. As illustrated in Figure 12.1, faulting is induced by the superposition of the stresses caused by pore pressure decreases in the reservoir with the pre-existing stresses prior to depletion. Hence, in a compressional stress environment, above and below a compacting reservoir we expect to observe further increases of horizontal compressive stresses and potentially, induced reverse faulting. In an extensional area, one would not expect to see induced faulting above or below the reservoir because the stress change due to depletion acts in an opposite direction to the pre-existing stresses. In contrast, near the edges of a reservoir, the stresses induced by depletion would encourage normal faulting to occur. For the sign convention employed by Segall (1989), contours >0 indicate areas where the stress changes due to depletion induce normal faulting in the surrounding medium whereas contours <0 indicate the regions where there is a potential for reverse faulting to be induced. These predicted stress changes are consistent with the occurrence of reverse faulting above the Wilmington field in the Los Angeles area, as well as induced slip on reservoir-bounding normal faults in the Gulf of Mexico region (see below). Feignier and Grasso (1990) applied these ideas to the Lacq field in France and demonstrated a reasonably good correlation between the theoretically expected location of seismicity and that observed, assuming that the pre-existing stress state in the vicinity of the reservoir is characterized by normal faulting with a vertical maximum principal stress. Maury and Zurdo (1996) argued that slip on active faults appears to be the cause of sheared casings of production wells in a number of fields.

   A number of studies of induced seismicity resulting from subsurface fluid injection and withdrawal have been conducted since the 1960s (*e.g.* Healy, Rubey *et al.* 1968; Raleigh, Healy *et al.* 1972; Segall 1985; Mereu, Brunet *et al.* 1986; Pennington, Davis *et al.* 1986; Segall 1989; Grasso and Wittlinger 1990; Grasso 1992; Doser, Baker *et al.* 1991; McGarr 1991; Davis, Nyffenegger *et al.* 1995). Most of these studies demonstrated that the number of seismic events in the proximity of producing oil or gas fields increased significantly after production or injection began. While many of these cases are associated with fluid injection (as discussed in Chapter 4) and result from the increase of pore pressure and lowering effective normal stress, observations and studies of seismic events around some oil and gas fields around the world suggested that depletion will result in a change in stress around the reservoir that may encourage slip on faults outside of the reservoir.

It is important to keep in mind that the magnitudes of stress perturbations outside a depleting reservoir are extremely small. Segall and Fitzgerald (1996) show that in a homogeneous linearly elastic medium, immediately above a flat-lying ellipsoidal reservoir, the stress perturbation is given by

$$\frac{\Delta S_h}{\Delta P_p} = \alpha \left(\left(\frac{1 - 2\nu}{1 - \nu} \right) \left(\frac{\pi}{4} \right) \left(\frac{H}{2R} \right) \right) \tag{12.29}$$

where H is the height of the reservoir and $2R$ is its lateral extent. $H/2R$ is typically very small. To put this in the context of the stress changes occurring within the Valhall reservoir illustrated in Figure 12.5, we note that $H/2R$ is approximately 0.03. Comparing the change in stress immediately above the reservoir (from equation 12.29) with that illustrated in Figure 12.5, we see that the increase in horizontal stress above the reservoir due to production is only about 1% of the decrease in horizontal stress within it. In other words, the increase in horizontal compressive stress would be ~0.2 MPa above the reservoir as opposed to ~20 MPa within it. Thus, the only way that faulting can be induced outside a reservoir (either reverse faulting above and below it or normal faulting near its edges) is that pre-existing faults are already in frictional failure equilibrium. Moreover, for reservoirs located in reverse faulting stress regimes, one would expect the induced faulting to occur only above and below the reservoir during depletion. The sign of the stress perturbation on reverse faults to the sides of the reservoir would not encourage slip. In comparison, in normal faulting regimes, faulting would only be expected to occur on pre-existing critically stressed faults around the edges of the reservoirs.

The impact of the compaction of an irregular shaped reservoir on a non-planar fault surface is best estimated using numerical modeling. Therefore, Chan and Zoback (2006) used a numerical code, Poly3D (Thompson 1993) to examine the impact of hydrocarbon production on the Golden Meadow fault located at the northern edge of the depleting Lapeyrouse reservoirs. They calculated compaction from a DARS analysis of each individual reservoir. Driven by reservoir compaction, Poly3D was used to determine the location and magnitude of slip along the fault surface. They modeled the compacting reservoir as a planar discontinuity surface embedded in an elastic medium. Assuming the fault surface is free of traction and is able to slip in any direction within the fault plane (*i.e.* no opening or closing of the fault), the magnitude and location of slip induced by reservoir compaction were estimated. In reality, fault surfaces are not traction-free, but because growth faults in the coastal area are active and constantly slipping, assuming that they are traction free is equivalent to the assumption that they are incipiently active,

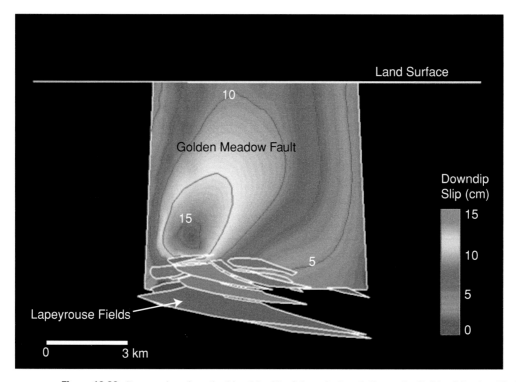

Figure 12.20. Perspective view (looking North) of the calculated slip on the Golden Meadow Fault using Poly3D (Thompson 1993) and a reasonable representation of the actual shape of each of the reservoirs. Note that the highest amount of slip on the fault is just above the shallowest reservoir (after Chan and Zoback 2006). (*For colour version see plate section.*)

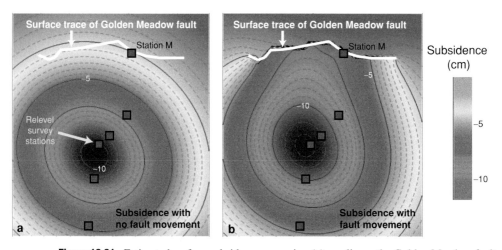

Figure 12.21. Estimated surface subsidence assuming (a) no slip on the Golden Meadow fault and (b) the slip on the fault shown in Figure 12.19 utilizing Poly3D. The shape of the subsidence bowl is altered when the fault is allowed to move freely but the predicted subsidence is still much less than the apparent elevation change at benchmark M (after Chan and Zoback 2006). (*For colour version see plate section.*)

which is the case. As a result, the estimate from Poly3D represents the maximum slip that can occur on the fault plane due to reservoir deformation.

As noted above, the misfit at Station M in Figure 12.19 and the proximity of the survey station to the approximate location of the surface trace of the Golden Meadow fault zone suggest that subsidence measured at station M may be influenced by the movement along the Golden Meadow fault as a result of fault slip at depth. Chan and Zoback (2006) utilized Poly3D to numerically estimate the impact of reservoir compaction in Lapeyrouse on the Golden Meadow fault. Utilizing a seismic study across the Lapeyrouse field, the shape of all the reservoirs and the Golden Meadow faults were digitized for the numerical models (shown schematically in Figure 12.20). The contours on the fault plane in Figure 12.20 indicate the amount of slip on the fault resulting from compaction of the various reservoirs. Note that as much as 20 cm of slip occurs on the fault immediately above the reservoirs, more than enough to shear well casings or cause other operational problems if wells penetrate the fault in that area.

Figure 12.21 (Chan and Zoback 2006) compares the effect of slip along the Golden Meadow fault on surface subsidence to subsidence without any fault slip. When slip on the fault does not occur (Figure 12.21a), the total surface subsidence caused by reservoir compaction is very similar to that calculated using the Geertsma solution (Figure 12.17). The slight difference between the subsidence bowls in Figure 12.21 is due to the shape of the reservoirs: all the reservoirs are disk-shaped in the Geertsma solution while the reservoirs are irregular shaped in the numerical model. The occurrence of fault slip along the Golden Meadow fault (Figure 12.21b) significantly alters the shape of the subsidence bowl in the vicinity of the fault, but there is less than 5 cm of slip at the surface predicted near station M, far less than that measured (Figure 12.19d). Regardless of whether production of the Lapeyrouse reservoirs induced slip on the Golden Meadow fault, it seems clear that neither reservoir compaction nor fault slip is adequate to explain the apparent 25 cm of subsidence at station M with respect to benchmarks to the south of the field. The reasons for this are not understood. As discussed by Chan and Zoback (2006), possibilities include the effects of production from reservoirs to the north of the Golden Meadow field that were not incorporated in the analysis or benchmark instability, as the benchmarks immediately to the south and north of benchmark M show appreciably less subsidence.

References

Aadnoy, B. S. (1990a). "Inversion technique to determine the In situ stress field from fracturing data." *Journal of Petroleum Science and Engineering*, **4**, 127–141.

Aadnoy, B. S. (1990b). "In situ stress direction from borehole fracture traces." *Journal of Petroleum Science and Engineering*, **4**, 143–153.

Abé, H., Mura, T. *et al.* (1976). "Growth rate of a penny-shaped crack in hydraulic fracture of rocks." *J. Geophys. Res*, **81**, 5335.

Adachi, T. and Oka, F. (1982). "Constitutive equations for normally consolidated clays based on elasto-viscoplasticity." *Soils and Foundations*, **22**(4), 57–70.

Addis, M. A., Cauley, M. B. *et al.* (2001). *Brent in-fill drilling programme: Lost circulation associated with drilling depleted reservoirs, Paper number SPE/IADC 67741*. SPE/IADC Drilling Conference, Amsterdam, Netherlands, Society of Petroleum Engineers.

Alexander, L. L. and Flemings, P. B. (1995). "Geologic evolution of Plio-Pleistocene salt withdrawl minibasin: Eugene Island block 330, offshore Louisiana." *American Association of Petroleum Geologists Bulletin*, **79**, 1737–1756.

Alnes, J. R. and Lilburn, R. A. (1998). "Mechanisms for generating overpressure in sedimentary basins: A reevaluation. Discussion." *American Association of Petroleum Geologists Bulletin*, **82**, 2266–2269.

Amadei, B. and Stephansson, O. (1997). *Rock Stress and its Measurement*. London, Chapman & Hall.

Anderson, E. M. (1951). *The Dynamics of Faulting and Dyke Formation with Applications to Britain*. Edinburgh, Oliver and Boyd.

Anderson, R. N., Flemings, P. *et al.* (1994). "In situ properties of a major Gulf of Mexico growth fault: Implications for behavior as a hydrocarbon migration pathway." *Oil and Gas Journal*, **92**(23), 97–104.

Angelier, J. (1979). "Determination of the mean principal directions of stresses for a given fault population." *Tectonophysics*, **56**, T17–T26.

Angelier, J. (1984). "Tectonic analysis of fault slip data sets." *Journal of Geophysical Research*, **89**, 5835–5848.

Antonellini, M. and Aydin, A. (1994). "Effect of faulting and fluid flow in porous sandstones: Geometry and spatial distribution." *American Association of Petroleum Geologists Bulletin*, **79**(5), 642–671.

Artyushkov, E. V. (1973). "Stresses in the lithosphere caused by crustal thickness inhomogeneities." *Journal of Geophysical Research*, **78**, 7675–7708.

Athy, L. F. (1930). "Density, porosity and compaction of sedimentary rocks." *American Association of Petroleum Geologists Bulletin*, **14**, 1–24.

Atkinson, B. K., Ed. (1987). *Fracture Mechanics of Rock*. Academic Press Geology Series. London, Academic Press.

Baria, R., Baumgaerdner, J. *et al.* (1999). "European HDR research programme at Soultz-sous-Forets (France) 1987–1996." *Geothermics*, **28**, 655–669.

Barton, C. A., Tessler, L. *et al.* (1991). Interactive analysis of borehole televiewer data. In *Automated Pattern Analysis in Petroleum Exploration*, I. Palaz and S. K. Sengupta (eds). New York, Springer Verlag.

Barton, C. A. and Zoback, M. D. (1992). "Self-similar distribution and properties of macroscopic fractures at depth in crystalline rock in the Cajon Pass scientific drill hole." *Journal of Geophysical Research*, **97**, 5181–5200.

Barton, C. A. and Zoback, M. D. (1994). "Stress perturbations associated with active faults penetrated by boreholes: Possible evidence for near-complete stress drop and a new technique for stress magnitude measurements." *J. Geophys. Res.*, **99**, 9373–9390.

Barton, C. A. and Zoback, M. D. (2002). Wellbore Imaging Technologies Applied to Reservoir Geomechanics and Environmental Engineering. *Geological Applications of Well Logs*. M. Lovell and N. Parkinson (eds). AAPG Methods in Exploration, No. 13, 229–239.

Barton, C. A., Zoback, M. D. *et al.* (1988). "In situ stress orientation and magnitude at the Fenton Geothermal site, New Mexico, determined from wellbore breakouts." *Geophysical Research Letters*, **15**(5), 467–470.

Barton, C. A., Zoback, M. D. *et al.* (1995). "Fluid flow along potentially active faults in crystalline rock." *Geology*, **23**, 683–686.

Baumgärtner, J., Carvalho, J. *et al.* (1989). *Fracturing deviated boreholes: An experimental approach.* Rock at Great Depth, Proceedings ISRM-SPE International Symposium, Elf Aquitaine, Pau, A.A.Balkema.

Baumgärtner, J., Rummel, F. *et al.* (1990). Hydraulic fracturing in situ stress measurements to 3 km depth in the KTB pilot hole VB. A summary of a preliminary data evaluation, in *KTB Report 90–6a*, 353–400.

Baumgärtner, J. and Zoback, M. D. (1989). "Interpretation of hydraulic fracturing pressure – time records using interactive analysis methods." *International Journal of Rock Mechanics and Mining Sciences & Geomechanical Abstracts*, **26**, 461–469.

Bell, J. S. (1989). "Investigating stress regimes in sedimentary basins using information from oil industry wireline logs and drilling records." *Geological Applications of Wireline Logs*. Special Publication 48(Geological Society of London), 305–325.

Bell, J. S. and Babcock, E. A. (1986). "The stress regime of the Western Canadian Basin and implications for hydrocarbon production." *Bulletin of Canadian Petroleum Geology*, **34**, 364–378.

Bell, J. S. and Gough, D. I. (1979). "Northeast-southwest compressive stress in Alberta: Evidence from oil wells." *Earth Planet. Sci. Lett.*, **45**, 475–482.

Bell, J. S. and Gough, D. I. (1983). *The use of borehole breakouts in the study of crustal stress, in Hydraulic fracturing stress measurements*. D.C, National Academy Press, Washington.

Berry, F. A. F. (1973). "High fluid potentials in California Coast Ranges and their tectonic significance." *American Association of Petroleum Geologists Bulletin*, **57**, 1219–1245.

Biot, M. A. (1962). "Mechanics of deformation and acoustic propagation in porous media." *Journal of Acoustic Society of America*, **28**, 168–191.

Birch, F. (1961). "Velocity of compressional waves in rocks to 10 kilobars, Part 2." *J. Geophys. Res.*, **66**, 2199–2224.

Boness, N. and Zoback, M. D. (2004). "Stress-induced seismic velocity anisotropy and physical properties in the SAFOD Pilot hole in Parkfield, CA." *Geophysical Research Letters*, **31**, L15S17.

Boness, N. and Zoback, M. D. (2006). "A multi-scale study of the mechanisms controlling shear velocity anisotropy in the San Andreas Fault Observatory at Depth." *Geophysics*, **71**, F131–F136.

Bourbie, T., Coussy, O. and Zinszner, B. (1987). *Acoustics of porous media*. Paris, France, Editions Technip.

Bourgoyne Jr., A. T., Millheim, K. K. *et al.* (2003). *Applied Drilling Engineering*. Richardson, Texas, Society of Petroleum Engineers.

Bowers, G. L. (1994). Pore pressure estimation from velocity data: accounting for overpressure mechanisms besides undercompaction. SPE 27488 Dallas, Texas, Society of Petroleum Engineers, 515–589.

Brace, W. F. (1980). "Permeability of crystalline and argillaceous rocks." *Int'l. J. Rock Mech. Min. Sci. and Geomech. Abstr.*, **17**, 241–251.

Brace, W. F. and Kohlstedt, D. L. (1980). "Limits on lithospheric stress imposed by laboratory experiments." *J. Geophys. Res*, **85**, 6248–6252.

Brace, W. F., Paulding, B. W. *et al.* (1966). "Dilatancy in the fracture of crystalline rocks." *Journal of Geophysical Research*, **71**(16), 3939–3953.

Bradley, W. B. (1979). "Failure of Inclined Boreholes." *J. Energy Res. Tech., Trans. ASME*, **102**, 232.

Bratli, R. K. and Risnes, R. (1981). "Stability and failure of sand arches." *Soc. of Petroleum Engineers Journal*, (April), 236–248.

Bratton, T., Bornemann, T. *et al.* (1999). *Logging-while-drilling images for geomechanical, geological and petrophysical interpretations*. SPWLA 40th Annual Logging Symposium, Oslo, Norway, Society of Professional Well Log Analysts.

Breckels, I. M. and Van Eekelen, H. A. M. (1981). "Relationship between horizontal stress and depth in sedimentary basins: Paper SPE10336, 56th Annual Fall Technical Conference." *Society of Petroleum Engineers of AIME, San Antonio, Texas, October 5–7, 1981*.

Bredehoeft, J. D., Wolf, R. G. *et al.* (1976). "Hydraulic fracturing to determine the regional in situ stress field Piceance Basin colorado." *Geol. Soc. Am. Bull.*, **87**, 250–258.

Brown, D. (1987). The flow of water and displacement of hydrocarbons in fractured chalk reservoirs. *Fluid flow in sedimentary basins and aquifers, Geological Society London Special Publication 34*. J. C. Goff and B. P. Williams. London, The Geological Society. **34**, 201–218.

Brown, K. M., Bekins, B. *et al.* (1994). "Heterogeneous hydrofracture development and accretionary fault dynamics." *Geology*, **22**, 259–262.

Brown, S. R. and Scholz, C. H. (1985). "Closure of random elastic surfaces in contact." *J. Geophys. Res.*, **90**, 5531–5545.

Brown, S. R. and Scholz, C. H. (1986). "Closure of rock joints." *J. Geophy. Res.*, **91**, 4939–4948.

Bruce, C. H. (1984). "Smectite dehydration: Its relation to structural development and hydrocarbon accumulation in northern Gulf of Mexico basin." *Am. Assoc. Petr. Geol. Bull.*, **68**, 673–683.

Brudy, M. and Zoback, M. D. (1993). "Compressive and tensile failure of boreholes arbitrarily-inclined to principal stress axes: Application to the KTB boreholes, Germany." *International Journal Rock Mechanics Mining Sciences*, **30**, 1035–1038.

Brudy, M. and Zoback, M. D. (1999). "Drilling-induced tensile wall-fractures: implications for the determination of in situ stress orientation and magnitude." *International Journal of Rock Mechanics and Mining Sciences*, **136**, 191–215.

Brudy, M., Zoback, M. D. *et al.* (1997). "Estimation of the complete stress tensor to 8 km depth in the KTB scientific drill holes: Implications for crustal strength." *J. Geophys. Res.*, **102**, 18,453–18,475.

Burrus, J. (1998). Overpressure models for clastic rocks, their relation to hydrocarbon expulsion: A critical reevaluation – AAPG Memoir 70. In *Abnormal pressures in hydrocarbon environments*. B. E. Law, G. F. Ulmishek and V. I. Slavin (eds). Tulsa, OK, American Association of Petroleum Geologists, 35–63.

Byerlee, J. D. (1978). "Friction of rock." *Pure & Applied Geophysics*, **116**, 615–626.

Carmichael, R. S. (1982). *Handbook of Physical Properties of Rocks*. Boca Raton, FL, CRC Press.

Carman, P. C. (1961). *L'écoulement des Gaz á Travers le Milieux Poreux*, Bibliothéque des Sciences et Techniques Nucléaires, Presses Universitaires de France, Paris, 198pp.

Castillo, D., Bishop, D. J. *et al.* (2000). "Trap integrity in the Laminaria high-Nancar trough region, Timor Sea: Prediction of fault seal failure using well-constrained stress tensors and fault surfaces interpreted from 3D seismics." *Appea Journal*, **40**, 151–173.

Castillo, D. and Zoback, M. D. (1995). "Systematic stress variations in the southern San Joaquin valley and along the White Wolf fault: Implications for the rupture mechanics of the 1952 Ms 7.8 Kern County earthquake and contemporary seismicity." *Journal of Geophysical Research*, **100**(B4), 6249–6264.

Castillo, D. A. and Zoback, M. D. (1994). "Systematic variations in stress state in the Southern San Joaquin Valley: Inferences based on well-bore data and contemporary seismicity." *American Association Petroleum Geologists Bulletin*, **78**(8), 1257–1275.

Cayley, G. T. (1987). Hydrocarbon migration in the central North Sea. In *Petroleum Geology of North West Europe*. J. Brooks and K. Glennie (eds). London, Graham and Trotman, 549–555.

Chan, A., Hagin, P. *et al.* (2004). *Viscoplastic deformation, stress and strain paths in unconsolidated reservoir sands (Part 2): Field applications using dynamic DARS analysis: ARMA/NARMS 04–568*. Gulf Rocks 2004, The Sixth North American Rock Mechanics Symposium, Houston, TX, American Rock Mechanics Association.

Chan, A. and Zoback, M. D. (2002). *Deformation analysis in reservoir space (DARS): A simple formalism for prediction of reservoir deformation with depletion – SPE 78174*. SPE/ISRM Rock Mechanics Conference, Irving, TX, Society of Petroleum Engineers.

Chan, A. W. and Zoback, M. D. (2006). "The role of hydrocarbon production on land subsidence and fault reactivation in the Louisiana coastal zone." *Journal of Coastal Research*, submitted.

Chang, C. and Haimson, B. (2000). "True triaxial strength and deformability of the German Continental Deep Drilling Program (KTB) deep hole amphibolite." *Journal of Geophysical Research*, **105**, 18999–19013.

Chang, C., Moos, D. *et al.* (1997). "Anelasticity and dispersion in dry unconsolidated sand." *International Journal of Rock Mechanics and Mining Sciences*, **34**(3–4), Paper No. 048.

Chang, C., Zoback, M. D. *et al.* (2006). "Empirical relations between rock strength and physical properties in sedimentary rocks." *Journal of Petroleum Science and Engineering*, **51**, 223–237.

Chapple, W. and Forsythe, D. (1979). "Earthquakes and bending of plates at trenches." *Journal of Geophysical Research*, **84**, 6729–6749.

Charlez, P. A. (1991). *Rock Mechanics: Theoetical Fundamentals*. Paris, Editions Technip.

Chen, S. T. (1988). "Shear-wave logging with dipole sources." *Geophysics*, **53**, 659–667.

Cheng, C. H., Jinzhong, Z. *et al.* (1987). "Effects of in-situ permeabilty on the propagation of Stonely (tube) waves in a borehole." *Geophysics*, **52**, 1279–1289.

Chester, F. M. and Logan, J. M. (1986). "Implications for mechanical properties of brittle faults from observations of the Punchbowl fault zone, California." *Pure Appl. Geophys*, **124**, 79–106.

Chester, J., Chester, F. M. *et al.* (2005). "Fracture energy of the Punchbowl fault, San Andreas system." *Nature*, **437**, 133–136.

Cloetingh, S. and Wortel, R. (1986). "Stress in the Indo-Australian plate." *Tectonophysics*, **132**, 49–67.

Colmenares, L. B. and Zoback, M. D. (2002). "A statistical evaluation of rock failure criteria constrained by polyaxial test data for five different rocks." *International Journal of Rock Mechanics and Mining Sciences*, **39**, 695–729.

Colmenares, L. B. and Zoback, M. D. (2003). "Stress field and seismotectonics of northern South America." *Geology*, **31**, 721–724.

Coulomb, C. A. (1773). "Sur une application des regles de maximums et minimums a quelques problemes de statistique relatifs a larchitesture, Acad. Roy." *Sci. Mem. Mech. Min Sci.,* **7**, 343–382.

Crampin, S. (1985). "Evaluation of anisotropy by shear wave splitting." *Geophysics,* **50**, 142–152.

Crawford, B. R. and Yale, D. P. (2002). *Constitutive modeling of deformation and permeability: relationships between critical state and micromechanics: SPE 78189.* Society of Petroleum Engineers.

Daines, S. R. (1992). "Aquathermal pressuring and geopressure evaluation." *American Association of Petroleum Geologists Bulletin,* **66**, 931–939.

Daneshy, A. A. (1973). "A Study of Inclined Hydraulic Fractures." *Soc. Pet. Eng. J.,* **13**, 61.

Davatzes, N. C. and Aydin, A. (2003). "The formation of conjugate normal fault systems in folded sandstone by sequential jointing and shearing, Waterpocket monocline, Utah." *J. Geophys. Res,* **108**(B10, 2478), ETG 7–1–7–15.

Davies, R. and Handschy, J. Eds. (2003). *Fault Seals.* Tulsa, OK, American Association of Petroleum Geologists.

Davis, S. D., Nyffenegger, P. A. *et al.* (1995). "The April 1993 Earthquake in South Central Texas: Was it Induced by Oil and Gas Production?" *Bull. Seismol. Soc. Am.,* **85**, 1888–1895.

de Waal, J. A. and Smits, R. M. M. (1988). "Prediction of reservoir compaction and surface subsidence: Field application of a new model." *SPE Formation Evaluation* (June), 347–356.

Desai, C. S. and Siriwardane, H. J. (1984). *Constitutive laws for engineering materials with emphasis on geologic materials.* Englewood Cliffs, New Jersey, Prentice-Hall.

Dholakia, S. K., Aydin, A. *et al.* (1998). "Fault-controlled hydrocarbon pathways in the Monterey Formation, California." *Amer. Assoc. Pet. Geol. Bull.,* **82**, 1551–1574.

Dickinson, G. (1953). "Geological aspects of abnormal reservoir pressures in Gulf Coast Lousiana." *American Association of Petroleum Geologists Bulletin,* **37**, 410–432.

Donath, F. A. (1966). "Experimental study of shear failure in anisotropic rock." *Bulletin of Geological Soc. America,* **72**, 985–990.

Dore, A. G. and Jensen, L. N. (1996). "The impact of late Cenozoic uplift and erosion on hydrocarbon exploration: offshore Norway and some other uplifted basins." *Global and Planetary Change,* **12**, 415–436.

Doser, D. I., Baker, M. R. *et al.* (1991). "Seismicity in the War-Wink Gas Field, West Texas, and its Relationship to Petroleum Production." *Bull. Seismol. Soc. Am.,* 971.

Drucker, D. and Prager, W. (1952). "Soil mechanics and plastic analysis or limit design." *Quantitative and Applied Mathematics,* **10**, 157–165.

du Rouchet, J. (1981). "Stress fields, a key to oil migration." *American Association of Petroleum Geologists Bulletin,* 74–85.

Dudley, J. W. I., Meyers, M. T. *et al.* (1994). *Measuring compaction and compressibilities in unconsolidated reservoir materials via time-scaling creep.* Eurock '94, Delft, Netherlands, Balkema.

Dugan, B. and Flemings, P. B. (1998). Pore pressure prediction from stacking velocities in the Eugene Island 330 Field (Offshore Lousiana). Chicago, Ill., Gas Research Institute, 23.

Dullien, F. A. L. (1992). *Porous Media: Fluid Transport and Pore Structure.* San Diego, Academic Press.

Dvorkin, J., Mavko, G. *et al.* (1995). "Squrt flow in fully saturated rocks." *Geophysics,* **60**, 97–107.

Eaton, B. A. (1969). "Fracture gradient prediction and its application in oilfield operations." *Journal of Petroleum Technology,* **246**, 1353–1360.

Eberhart-Phillips, D., Han, D.-H. *et al.* (1989). "Empirical relationships among seismic velocity, effective presure, porosity and clay content in sandstone." *Geophysics,* **54**, 82–89.

Economides, M. J. and Nolte, K. G. Eds. (2000). *Reservoir Simulation.* West Sussex, England, John Wiley & Sons, Ltd.

Ekstrom, M. P., Dahan, C. A. *et al.* (1987). "Formation imaging with microelectrical scanning arrays." *The Log Analyst.*, **28**, 294–306.

Engelder, T. (1987). Joints and shear fractures in rock. In *Fracture Mechanics of Rock*. B. K. Atkinson. London, Academic Press, 534.

Engelder, T. (1993). *Stress regimes in the lithosphere*. Princeton, New Jersey, Princeton.

Engelder, T. and Leftwich, J. T. (1997). A pore-pressure limit in overpressured south Texas oil and gas fields. *Seals, traps and the petroleum system: AAPG Memoir 67*. R. C. Surdam. Tulsa, OK, AAPG, 255–267.

Engelder, T. and Sbar, M. L. (1984). "Near-surface in situ stress: Introduction." *Journal of Geophysical Research*, **89**, 9321–9322.

England, W. A., MacKenzie, A. S. *et al.* (1987). "The movement and entrapment of petroleum fluids in the subsurface." *Journal of the Geological Society*, **144**, 327–347.

Eoff, L., Funkhauser, G. P. *et al.* (1999). *High-density monomer system for formation consolidation/water shutoff applications: SPE 50760*. International symposium on oilfield chemistry, Houston, TX, Society of Petroleum Engineers.

Ewy, R. (1999). "Wellbore-stability predictions by use of a modified Lade criterion." *SPE Drilling and Completion*, **14**(2), 85–91.

Ewy, R., Stankowich, R. J. *et al.* (2003). *Mechanical behavior of some clays and shales from 200 m to 3800 m depth, Paper 570*. 39th U.S. Rock Mechanics Symposium/12th Panamerican Conference on Soil Mechanics and Geotechnical Engineering, Cambridge, MA.

Færseth, R. B., Sjøblom, R. J. *et al.* (1995). "Sequence Stratigraphy on the Northwest European Margin." *Elsevier, Amsterdam*.

Faybishenko, B., Witherspoon, P. A. *et al.*, Eds. (2000). *Dynamics of fluids in fractured rock*. Geophysical Monograph Series. Washington, D.C., American Geophysical Union.

Fehler, M., Jupe, A. *et al.* (2001). "More than a cloud: new techniques for characterizing reservoir structures using induced seismicity." *Leading Edge*, **20**, 324–328.

Feignier, B. and Grasso, J.-R. (1990). "Seismicty Induced by Gas ProductionI: Correlation of Focal Mechanism & Dome Structure." *134 Pure & Applied Geophys.*, 405.

Finkbeiner, T. (1998). In situ stress, pore pressure and hydrocarbon migration and accumulation in sedimentary basins. *Geophysics*. Stanford, CA, Stanford University, 193.

Finkbeiner, T., Barton, C. B. *et al.* (1997). "Relationship between in-situ stress, fractures and faults, and fluid flow in the Monterey formation, Santa Maria basin, California." *Amer. Assoc. Petrol. Geol. Bull.*, **81**(12), 1975–1999.

Finkbeiner, T., Zoback, M. D. *et al.* (2001). "Stress, pore pressure and dynamically-constrained hydrocarbon column heights in the south Eugene Island 330 field, Gulf of Mexico." *Amer. Assoc. Petrol. Geol. Bull.*, **85**(June), 1007–1031.

Fisher, N. I., Lewis, T. *et al.* (1987). *Statistical analysis of spherical data*. Cambridge, Cambridge University Press.

Fisher, Q. J., Casey, M. *et al.* (2003). "Fluid-flow properties of faults in sandstone: The importance of temperature history." *Geology*, **31**(11), 965–968.

Fisher, Q. J. and Knipe, R. J. (1998). Fault sealing processes in siliciclastic sediments. In *Faulting and fault sealing in hydrocarbon reservoirs*, G. Jones *et al.* (eds). London, Geological Society (London), **147**, 117–134.

Fjaer, E., Holt, R. M. *et al.* (1992). *Petroleum Related Rock Mechanics*. Amsterdam, Elsevier.

Fleitout, L. and Froidevaux, C. (1983). "Tectonics and topography for a lithosphere containing density heterogenieties." *Tectonics*, **2**, 315–324.

Flemings, P. B., Stump, B. B. *et al.* (2002). "Flow focusing in overpressured sandstones: theory, observations and applications." *American Journal of Science*, **302**, 827–855.

Forsyth, D. and Uyeda, S. (1975). "On the relative importance of the driving forces of plate motion." *Geophys. J. R. Astr. Soc.*, **43**, 163–200.

Fowler, C. M. R. (1990). *The solid earth.* Cambridge, U.K., Cambridge University Press.

Fredrich, J. T., Coblentz, D. D. *et al.* (2003). *Stress perturbations adjacent to salt bodies in the deepwater Gulf of Mexico: SPE 84554.* SPE Annual Technical Conference and Exhibition, Denver, CO, Society of Petroleum Engineers.

Freyburg, D. (1972). "Der Untere und mittlere Buntsandstein SW-Thuringen in seinen gesteinstechnicschen Eigenschaften." *Ber. Dte. Ges. Geol. Wiss. A; Berlin*, **17**(6), 911–919.

Fuh, G.-A., Morita, N. *et al.* (1992). *A new approach to preventing loss circulation while drilling.* SPE 24599, Soc. Petr. Eng. 67th Annual Tech. Conf. and Exhib, Washington, D.C.

Gaarenstroom, L., Tromp, R. A. J. *et al.* (1993). *Overpressures in the Central North Sea: implications for trap integrity and drilling safety.* Petroleum Geology of Northwest Europe: Proceedings of the 4th Conference, London.

Geertsma, J. (1973). "A basic theory of subsidence due to reservoir compaction: the homogeneous case." *Trans. Royal Dutch Soc. of Geologists and Mining Eng.*, **28**, 43–62.

Gephart, J. W. (1990). "Stress and the direction of slip on fault planes." *Tectonics*, **9**, 845–858.

Gephart, J. W. and Forsyth, D. W. (1984). "An improved method for determining the regional stress tensor using earthquake focal mechanism data: application to the San Fernando earthquake sequence." *Journal of Geophysical Research*, **89**, 9305–9320.

Germanovich, L. N. and Dyskin, A. V. (2000). "Fracture mechanisms and instability of openings in compression." *International Journal of Rock Mechanics and Mining Sciences*, **37**, 263–284.

Germanovich, L. N., Galybin, A. N. *et al.* (1996). *Borehole stability in laminated rock.* Prediction and performance in rock mechanics and rock engineering, Torino, Italy, A. A. Balkema.

Golubev, A. A. and Rabinovich, G. Y. (1976). "Resultay primeneia apparatury akusticeskogo karotasa dija predeleina procontych svoistv gornych porod na mestorosdeniaach tverdych isjopaemych." *Prikladnaja GeofizikaMoskva*, **73**, 109–116.

Gordon, D. S. and Flemings, P. B. (1998). "Generation of overpressure and compaction-driven fluid flow in a Plio-Pleistocene grwoth-faulted basin, Eugene Island 330, offshore Louisiana." *Basin Research*, **10**, 177–196.

Grasso, J. R. (1992). "Mechanics of seismic instabilities induced the recovery of hydrocarbons." *Pure & Applied Geophysics*, **139**, 507–534.

Grasso, J. R. and Wittlinger, G. (1990). "10 Years of Seismic Monitoring over a Gas Field." *80 Bull. Seismo. Soc. Am.*, 450.

Griffith, J. (1936). Thermal Expansion of Typical American Rocks. *Iowa State College of Agriculture and Mechanic Arts, Iowa Engineering Experiment*, **35**(19), 24.

Grollimund, B., Zoback, M. D. *et al.* (2001). "Regional synthesis of stress orientation, pore pressure and least principal stress data in the Norwegian sector of the North Sea." *Petroleum Geoscience*, **7**, 173–180.

Grollimund, B. R. and Zoback, M. D. (2000). "Post glacial lithospheric flexure and induced stresses and pore pressure changes in the northern North Sea." *Tectonophysics*, **327**, 61–81.

Grollimund, B. R. and Zoback, M. D. (2001). "Impact of glacially-induced stress changes on hydrocarbon exploration offshore Norway." *American Association of Petroleum Geologists Bulletin*, **87**(3), 493–506.

Grollimund, B. R. and Zoback, M. D. (2003). "Impact of glacially induced stress changes on fault-seal integrity offshore Norway." *American Association of Petroleum Geologists Bulletin*, **87**, 493–506.

Gudmundsson, A. (2000). "Fracture dimensions, displacements and fluid transport." *Journal of Structural Geology*, **22**, 1221–1231.

Guenot, A. (1989). "Borehole breakouts and stress fields." *Int. J. Rock Mech. Min. Sci. & Geomech. Abstr.*, **26**, 185–195.

Guo, G., Morgenstern, N. R. *et al.* (1993). "Interpretation of hydraulic fracturing pressure: A comparison of eight methods used to identify shut-in pressures." *International Journal of Rock Mechanics and Mining Sciences*, **30**, 627–631.

Hagin, P. and Zoback, M. D. (2004a). *Viscoplastic deformation in unconsolidated reservoir sands (Part 1): Laboratory observations and time-dependent end cap models ARMA/NARMS 04–567.* Gulf Rocks, Houston, TX, American Rock Mechanics Association.

Hagin, P. and Zoback, M. D. (2004b). "Viscous deformation of unconsolidated sands-Part 1: Time-dependent deformation, frequency dispersion and attenuation." *Geophysics*, **69**, 731–741.

Hagin, P. and Zoback, M. D. (2004c). "Viscous deformation of unconsolidated reservoir sands-Part 2: Linear viscoelastic models." *Geophysics*, **69**, 742–751.

Hagin, P. and Zoback, M. D. (2007). "Characterization of time-dependent deformation of unconsolidated reservoir sands." *Geophysics*, in press.

Haimson, B. and Fairhurst, C. (1967). "Initiation and Extension of Hydraulic Fractures in Rocks." *Soc. Petr. Eng. Jour.*, Sept.: 310–318.

Haimson, B. and Fairhurst, C. (1970). In situ stress determination at great depth by means of hydraulic fracturing. In *11th Symposium on Rock Mechanics*. W. Somerton, Society of Mining Engineers of AIME, 559–584.

Haimson, B. C. (1989). "Hydraulic fracturing stress measurements." *Rock Mech. and Min. Sci. and Geomech. Abstr* Special Issue: *Inter. Jour.*, 26.

Haimson, B. C. and Herrick, C. G. (1989). *Borehole breakouts and in situ stress*. 12th Annual Energy-Sources Technology Conference and Exhibition, Houston, Texas.

Hall, P. L. (1993). Mechanisms of overpressuring-an overview. *Geochemistry of clay-pore fluid interactions*. D. A. C. Manning, P. L. Hall and C. R. Hughes. London, Chapman and Hall, 265–315.

Han, D., Nur, A. *et al.* (1986). "Effects of porosity and clay content on wave velocities in sandstones." *Geophysics*, **51**, 2093–2107.

Handin, J., Hager, R. V. *et al.* (1963). "Experimental deformation of sedimentary rocks under confining pressure: pore pressure effects." *Bulletin American Assoc. Petrol. Geology*, 717–755.

Haneberg, W. C., Mozley, P. S. *et al.*, Eds. (1999). *Faults and subsurface fluid flow in the shallow crust*. Geophysical Monograph. Washington, D.C., American Geophysical Union.

Haney, M. M., Snieder, R. *et al.* (2005). "A fault caught in the act of burping." *Nature*, **437**, 46.

Harrison, A. R., Randall, C. J. *et al.* (1990). *Acquisition and analysis of sonic waveforms from a borehole monopole and dipole source for the determination of compressional shear speeds and their relation to rock mechanic propoerties and surface seismic data – SPE 20557*. SPE Annual Technical Conference and Exhibition, New Orleans.

Harrold, T. W., Swarbrick, R. E. *et al.* (1999). "Pore pressure estimation from mudrock porosities in Tertiary basins, Southeast Asia." *American Association of Petroleum Geologists Bulletin*, **83**, 1057–1067.

Hart, B. S., Flemings, P. B. *et al.* (1995). "Porosity and pressure: Role of compaction disequilibrium in the development of geopressures in a Gulf Coast Pleistocene basin." *Geology*, **23**, 45–48.

Hayashi, K. and Haimson, B. C. (1991). "Characteristics of shut-in curves in hydraulic fracturing stress measurements and determination of in situ minimum compressive stress." *Journal of Geophysical Research*, **96**, 18311–18321.

Healy, J. H., Rubey, W. W. *et al.* (1968). "The Denver earthquakes." *Science*, **161**, 1301–1310.

Heppard, P. D., Cander, H. S. *et al.* (1998). Abnormal pressure and the occurrence of hydrocarbons in offshore eastern Trinidad, West Indies. In *Abnormal pressures in hydrocarbon environments – AAPG Memoir 70*, B. E. Law, G. F. Ulmishek and V. I. Slavin (eds). Tulsa, OK, American Association of Petroleum Geologists. Memoir, 70, 215–246.

Hickman, S. (1991). "Stress in the lithosphere and the strength of active faults, U.S. National Report International Union Geodesy and Geophys. 1987–1990." *Geophys.*, **29**, 759–775.

Hickman, S., Sibson, R. *et al.* (1995a). "Introduction to special section: Mechanical involvement of fluids in faulting." *J. Geophys. Res.*, **100**, 12831–12840.

Hickman, S. and Zoback, M. D. (2004). "Stress measurements in the SAFOD pilot hole: Implications for the frictional strength of the San Andreas fault." *Geophysical Research Letters*, **31**, L15S12.

Hickman, S. H., Barton, C. A. *et al.* (1997). "In situ stress and fracture permeability along the Stillwater fault zone, Dixie Valley, Nevada." *Int. J. Rock Mech. and Min. Sci.*, **34**, 3–4, Paper No. 126.

Hickman, S. H. and Zoback, M. D. (1983). *The interpretation of hydraulic fracturing pressure-time data for in situ stress determination. Hydraulic Fracturing Measurements*. Washington, D.C, National Academy Press.

Hoak, T. E., Klawitter, A. L. *et al.*, Eds. (1997). *Fractured Reservoirs: Characterization and Modeling*. Denver, The Rocky Mountain Association of Geologists.

Hoek, E. and Brown, E. (1997). "Practical estimates of rock strength." *International Journal of Rock Mechanics and Mining Sciences*, **34**(8), 1165–1186.

Hoek, E. and Brown, E. T. (1980). "Empirical strength criterion for rock masses." *J. Geotechnical Engineering Div.*, **106**, 1013–1035.

Hofmann, R. (2006). Frequency dependent elastic and anelastic properties of clastic rocks. *Geophysics*. Golden, CO., Colorado School of Mines. Ph.D., 166.

Holbrook, P. W., Maggiori, D. A. *et al.* (1993). *Real-time pore pressure and fracture gradient evaluation in all sedimentary lithologies, SPE 26791*. Offshore European Conference, Aberdeen, Scotland, Society of Petroleum Engineers.

Holland, D. S., Leedy, J. B. *et al.* (1990). "Eugene Island Block 330 Field – U.S.A. offshore Louisiana, Structural Traps III: Tectonic Fold and Fault Traps, Atlas of Oil and Gas Fields, E. Beaumont and N. Foster (eds.). American Assoc. of Petroleum Geologists, Tulsa." 103–143.

Holt, R. M., Flornes, O. *et al.* (2004). Consequences of depletion-induced stress changes on reservoir compaction and recovery. *Gulf Rocks 2004, the 6th North America Rock Mechanics Symposium (NARMS): Rock Mechanics Across Borders and Disciplines – ARMA/NARMS 04–589*. Houston.

Horsrud, P. (2001). "Estimating Mechanical Properties of Shale from Empirical Correlations." *SPE Drilling and Completion*, **16**(2), 68–73.

Hottman, C. E., Smith, J. E. *et al.* (1979). "Relationship among earth stresses, pore pressure, and drilling problems offshore Gulf of Alaska." *Journal of Petroleum Technology*, November, 1477–1484.

Hubbert, M. D. and Rubey, W. W. (1959). "Role of fluid pressure in mechanics of overthrust faulting." *Geol. Soc. Am. Bull.*, **70**, 115–205.

Hubbert, M. K. and Willis, D. G. (1957). "Mechanics of hydraulic fracturing." *Petr. Trans. AIME*, **210**, 153–163.

Hudson, J. A. (1981). "Wave propagation and attenuation of elastic waves in material containing cracks." *Geophys. J. Roy. Astr. Soc.*, **64**, 122–150.

Hudson, J. A. and Priest, S. D. (1983). "Discontinuity frequency in rock masses." *Int. J. Rock Mech. Min. Sci. & Geomech. Abstr.*, **20**(2), 73–89.

Huffman, A. and Bowers, G. L., Eds. (2002). *Pressure regimes in sedimentary basins and their prediction: AAPG memoir 76*. Tulsa, OK, American Association of Petroleum Geologists.

Ingebritsen, S. E., Sanford, W. *et al.* (2006). *Groundwater in Geologic Processes*. Cambridge, United Kingdom, Cambridge Press.

Ito, T., Zoback, M. D. *et al.* (2001). "Utilization of mud weights in excess of the least principal stress to stabilize wellbores: Theory and practical examples." *Soc. of Petroleum Engineers Drilling and Completions*, **16**, 221–229.

Jaeger, J. C. and Cook, N. G. W. (1971). *Fundamentals of Rock Mechanics*. London, Chapman and Hall.

Jaeger, J. C. and Cook, N. G. W. (1979). *Fundamentals of Rock mechanics*, 2nd edn. New York, Chapman and Hall.

Jarosinski, M. (1998). "Contemporary stress field distortion in the Polish part of the western outer Carpathians and their basement." *Tectonophysics*, **297**, 91–119.

Jizba, D. (1991). Mechanical and acoustical properties of Sandstones and shales. PhD dissertation, Stanford University.

Jones, G., Fisher, Q. J. *et al.*, Eds. (1998). *Faulting, fault sealing and fluid flow in hydrocarbon resrvoirs*. London, Geological Society.

Kamb, W. B. (1959). "Ice petrofabric observations from Blue Glacier, Washington, in relation to theory and experiment." *J. Geophys. Res.*, **64**, 1891–1910.

Kimball, C. V. and Marzetta, T. M. (1984). "Semblance processing of borehole acoustic array data." *Geophysics*, **49**, 264–281.

Kirsch, G. (1898). "Die Theorie der Elastizitat und die Bedurfnisse der Festigkeitslehre, Zeitschrift des Verlines Deutscher Ingenieure." **42**, 707.

Klein, R. J. and Barr, M. V. (1986). Regional state of stress in western Europe. In *Proc. of International Symposium on Rock Stress and Rock Stress Measurements*. Lulea, Sweden, Stockholm, Centek Publ, 694 pp.

Kosloff, D. and Scott, R. F. (1980). "Finite element simulation of Wilmington oilfield subsidence: I, Linear modeling." *Tectonophysics*, **65**, 339–368.

Kranz, R. L., Frankel, A. D. *et al.* (1979). "The permeability of whole and jointed Barre granite." *Int. J. Rock Mech. Min. Sci. Geomech. Abst.*, **16**, 225–234.

Kristiansen, G. (1998). Geomechanical characterization of the overburden above the compacting chalk reservoir at Valhall. In *Eurock '98, SPE/ISRM Rock Mechanics in Petroleum Engineering*. Trondheim, Norway, The Norwegian University of Science and Technology, 193–202.

Kuempel, H. J. (1991). "Poroelasticity: parameters reviewed." *Geophysical Journal International*, **105**, 783–799.

Kwasniewski, M. (1989). *Laws of brittle failure and of B-D transition in sandstones*. Rock at Great Depth, Proceedings ISRM-SPE International Symposium, Elf Aquitaine, Pau, France, A. A. Balkema.

Kwon, O., Kronenberg, A. K. *et al.* (2001). "Permeability of Wilcox shale and its effective pressure law." *Journal of Geophysical Research*, **106**, 19339–19353.

Labenski, F., Reid, P. *et al.* (2003). *Drilling fluids approaches for control of wellbore instability in fractured formations SPE/IADC 85304*. SPE/IADC Middle East Drilling Technology Conference and Exhibition, Abu Dhabi, UAE, Society of Petroleum Engineers.

Lachenbruch, A. H. and Sass, J. H. (1992). "Heat flow from Cajon Pass, fault strength and tectonic implications." *J. Geophys. Res.*, **97**, 4995–5015.

Ladc, P. (1977). "Elasto-plasto stress-strain theory for cohesionless soil with curved yield surfaces." *International Journal of Solids and Structures*, **13**, 1019–1035.

Lal, M. (1999). *Shale stability: drilling fluid interaction and shale strength, SPE 54356*. SPE Latin American and Carribean Petroleum Engineering Conference, Caracas, Venezuela, Society of Petroleum Engineering.

Lama, R. and Vutukuri, V. (1978). *Handbook on Mechanical Properties of Rock*. Clausthal, Germany, Trans Tech Publications.

Lashkaripour, G. R. and Dusseault, M. B. (1993). *A statistical study of shale properties; relationship amnog principal shale properties*. Proceedings of the Conference on Probabilistic Methods in Geotechnical Engineering, Canberra, Australia.

Laubach, S. E. (1997). "A method to detect natural fracture strike in sandstones." *Amer. Assoc. Petrol. Geol. Bull.*, **81**(4), 604–623.

Law, B. E., Ulmishek, G. F. *et al.*, Eds. (1998). *Abnormal pressures in hydrocarbon environments*. AAPG Memoir 70, American Association of Petroleum Geologists.

Lekhnitskii, S. G. (1981). *Theory of elasticity of an anisotropic body*. Moscow, Mir.

Leslie, H. D. and Randall, C. J. (1990). "Eccentric dipole sources in fluid-filled boreholes: Experimental and numerical results." *Journal of Acoustic Society of America*, **87**, 2405–2421.

Li, X., Cui, L. *et al.* (1998). *Thermoporoelastic modelling of wellbore stability in non-hydrostatic stress field*. 3rd North American Rock Mechanics Symposium.

Ligtenberg, J. H. (2005). "Detection of fluid migration pathways in seismic data: implications for fault seal analysis." *Basin Research*, **17**, 141–153.

Lindholm, C. D., Bungum, H. *et al.* (1995). *Crustal stress and tectoncis in Norwegian regions determined from earthquake focal mechanisms*. Proceedings of the Workshop on Rock Stresses in the North Sea, Trondheim, Norway.

Lockner, D. A. (1995). Rock Failure. *Rock physics and phase relations*. Washington, D.C., American Geophysical Union, 127–147.

Lockner, D. A., Byerlee, J. D. *et al.* (1991). "Quasi-static fault growth and shear fracture energy in granite." *Nature*, **350**, 39–42.

Long, C. S. and a. others (1996). *Rock fractures and fluid flow*. Washington, D.C., National Academy Press.

Lorenz, J. C., Teufel, L. W. *et al.* (1991). "Regional fractures I: A mechanism for the formation of regional fractures at depth in flat-lying reservoirs." *Amer. Assoc. Petrol. Geol. Bull.*, **75**(11), 1714–1737.

Losh, S., Eglinton, L. *et al.* (1999). "Vertical and lateral fluid flow related to a large growth fault, South Eugene Island Block 330 field, Offshore Louisiana." *American Association of Petroleum Geologists Bulletin*, **83**(2), 244–276.

Lucier, A., Zoback, M. D. *et al.* (2006). "Geomechanical aspects of CO_2 sequestration in a deep saline reservoir in the Ohio River Valley region." *Environmental Geosciences*, **13**(2), 85–103.

Lund, B. and Zoback, M. D. (1999). "Orientation and magnitude of in situ stress to 6.5 km depth in the Baltic Shield." *International Journal of Rock Mechanics and Mining Sciences*, **36**, 169–190.

Luo, M. and Vasseur, G. (1992). "Contributions of compaction and aquathermal pressuring to geopressure and the influence of environmental conditions." *American Association of Petroleum Geologists Bulletin*, **76**, 1550–1559.

MacKenzie, D. P. (1969). "The relationship between fault plane solutions for earthquakes and the directions of the principal stresses." *Seismological Society of Amererica Bulletin*, **59**, 591–601.

Mallman, E. P. and Zoback, M. D. (2007). "Subsidence in the Louisiana Coastal Zone due to hydrocarbon production." *Journal of Coastal Research*, in press.

Mastin, L. (1988). "Effect of borehole deviation on breakout orientations." *J. Geophys. Res.*, **93**(B8), 9187–9195.

Matthews, W. R. and Kelly, J. (1967). "How to predict formation pressure and fracture gradient." *Oil and Gas Journal*, February, 92–106.

Maury, V. and Zurdo, C. (1996). "Drilling-induced lateral shifts along pre-existing fractures: A common cause of drilling problems." *SPE Drilling and Completion* (March), 17–23.

Mavko, G., Mukerjii, T. *et al.* (1998). *Rock Physics Handbook*. Cambridge, United Kingdom (GBR), Cambridge University Press.

Mavko, G. and Nur, A. (1997). "The effect of percolation threshold in the Kozeny-Carman relation." *Geophysics*, **622**, 1480–1482.

Maxwell, S. C. (2000). *Comparison of production-induced microseismicity from Valhall and Ekofisk*. Passive Seismic Method in E&P of Oil and Gas Workshop, 62nd EAGE Conference.

Maxwell, S. C., Urbancic, T. I. *et al.* (2002). *Microseismic imaging of hydraulkic fracture complexity in the Barnett shale, Paper 77440*. Society Petroleum Engineering Annual Technical Conference, San Antonio, TX, Society of Petroleum Engineers.

McGarr, A. (1991). "On a Possible Connection Between Three Major Earthquakes in California and Oil Production." *Bull. of Seismol. Soc. of Am.*, **948**, 81.

McGarr, A. and Gay, N. C. (1978). "State of stress in the earth's crust." *Ann. Rev. Earth Planet. Sci.*, **6**, 405–436.

McLean, M. and Addis, M. A. (1990). *Wellbore stability: the effect of strength criteria on mud weight recommendations*. 65th Annual Technical Conference and Exhibition of the Society of Petroleum Engineers, New Orleans, Society of Petroleum Engineers.

McNally, G. H. N. (1987). "Estimation of coal measures rock strength using sonic and neutron logs." *Geoexploration*, **24**, 381–395.

McNutt, M. K. and Menard, H. W. (1982). "Constraints on yield strength in the oceanic lithosphere derived from observations of flexure." *Geophys. J. R. Astron. Soc.*, **71**, 363–394.

Mereu, R. F., Brunet, J. *et al.* (1986). "A study of the Microearthquakes of the Gobbles Oil Field Ara of Southwestern Ontario." *Bulle. Seismol. Soc. Am.*, **1215**, 76.

Michael, A. (1987). "Use of focal mechanisms to determine stress: A control study." *Journal of Geophysical Research*, **92**, 357–368.

Militzer, M. a. S., P. (1973). "Einige Beitrageder geophysics zur primadatenerfassung im Bergbau." *Neue Bergbautechnik, Lipzig*, **3**(1), 21–25.

Mitchell, A. and Grauls, D. Eds. (1998). *Overpressure in petroleum exploration*. Pau, France, Elf-Editions.

Mody, F. K. and Hale, A. H. (1993). *A borehole stability model to couple the mechanics and chemistry of drilling-fluid/shale interactions*. SPE/IADC Drilling Conference, Amsterdam.

Mogi, K. (1971). "Effect of the triaxial stress system on the failure of dolomite and limestone." *Tectonophysics*, **11**, 111–127.

Mollema, P. N. and Antonellini, M. A. (1996). "Compaction bands; a structural analog for anti-mode I cracks in aeolian sandstone." *Tectonophysics*, **267**, 209–228.

Moore, D. E. and Lockner, D. A. (2006). Friction of the smectite clay montmorillonite: A review and interpretation of data. *The Seismogenic Zone of Subduction Thrust Faults, MARGINS Theoretical and Experimental Science Series*. C. M. and T. Dixon. New York, Columbia University Press. 2.

Moos, D., Peska, P. *et al.* (2003). Comprehensive wellbore stability analysis using quantitative risk assessment. *Jour. Petrol. Sci. and Eng., Spec. Issue on Wellbore Stability, 38*. B. S. Aadnoy and S. Ong, 97–109.

Moos, D. and Zoback, M. D. (1990). "Utilization of Observations of Well Bore Failure to Constrain the Orientation and Magnitude of Crustal Stresses: Application to Continental Deep Sea Drilling Project and Ocean Drilling Program Boreholes." *J. Geophys. Res.*, **95**, 9305–9325.

Moos, D. and Zoback, M. D. (1993). "State of stress in the Long Valley caldera, California." *Geology*, **21**, 837–840.

Moos, D., Zoback, M. D. *et al.* (1999). *Feasibility study of the stability of openhole multilaterals, Cook Inlet, Alaska.* 1999 SPE Mid-continent Operations Symposium, Oklahoma City, OK, Society of Petroleum Engineers.

Morita, N., Black, A. D. *et al.* (1996). "Borehole Breakdown Pressure with Drilling Fluids – I. Empirical Results." *Int. J. Rock Mech. and Min. Sci.*, **33**, 39.

Morita, N. and McLeod, H. (1995). "Oriented perforation to prevent casing collapse for highly inclined wells." *SPE Drilling and Completion*, (September), 139–145.

Morrow, C., Radney, B. *et al.* (1992). *Frictional strength and the effective pressure law of montmorillonite and illite clays. Fault Mechanics and Transport Properties of Rocks*, Academic, San Diego, Calif.

Mouchet, J. P. and Mitchell, A. (1989). *Abnormal pressures while drilling. Manuels techniques Elf Aquitaine, 2*. Boussens, France, Elf Aquitaine.

Mount, V. S. and Suppe, J. (1987). "State of stress near the San Andreas fault: Implications for wrench tectonics." *Geology*, **15**, 1143–1146.

Mueller, M. C. (1991). "Prediction of lateral variability in fracture intensity using multicomponent shear-wave seismic as a precursor to horizontal drilling." *Geophysical Journal International*, **107**, 409–415.

Munns, J. W. (1985). "The Valhall field: a geological overview." *Marine and Petroleum Geology*, **2**, 23–43.

Murrell, S. A. F. (1965). "Effect of triaxial systems on the strength of rocks at atmospheric temperatures." *Geophysical Journal Royal Astron. Soc.*, **106**, 231–281.

Nakamura, K., Jacob, K. H. *et al.* (1977). "Volcanoes as possible indicators of tectonic stress orientation – Aleutians and Alaska." *Pure and Applied Geophysics*, **115**, 87–112.

Nashaat, M. (1998). Abnormally high formation pressure and seal impacts on hydrocarbon accumulations in the Nile Delta and North Sinai basins, Egypt. *Abnormal pressure in hydrocarbon environments: AAPG Memoir 70*. B. E. Law, G. F. Ulmishek and V. I. Slavin. Tulsa, OK, AAPG, 161–180.

Nolte, K. G. and M. J. Economides (1989). Fracturing diagnosis using pressure analysis. *Reservoir Simulation*. M. J. Economides and K. G. Nolte. Englewood Cliffs, N.J., Prentice Hall.

Nur, A. and Byerlee, J. D. (1971). "An exact effective sress law for elastic deformation of rock with fluids." *J. Geophys. Res.*, 6414–6419.

Nur, A. and Simmons, G. (1969). "Stress induced velocity anisotropy in rock: An experimental study." *J. Geophys. Res.*, **74**, 6667–6674.

Nur, A. and Walder, J. (1990). Time-Dependent Hydraulics of the Earth's Crust. *The role of fluids in crustal processes*. Washington D.C., National Research Council, 113–127.

Okada, Y. (1992). "Internal deformation due to shear and tensile faults in a half space." *Bulletin of Seismological Society of America*, **82**, 1018–1040.

Ortoleva, P., Ed. (1994). *Basin Compartments and Seals*. Tulsa, American Association of Petroleum Geologists.

Ostermeier, R. M. (1995). "Deepwater Gulf of Mexico turbidites – compaction effects on porosity and permeabilty." *Soc. of Petroleum Engineers Formation Evaluation*, 79–85.

Ostermeier, R. M. (2001). "Compaction effects on porosity and permeability: deepwater Gulf of Mexico turbidites." *Journal of Petroleum Technology*, **53**(Feb. 2001), 68–74.

Ott, W. K. and Woods, J. D. (2003). *Modern Sandface Completion Practises*. Houston, Texas, World Oil.

Ottesen, S., Zheng, R. H. *et al.* (1999). *Wellbore Stability Assessment Using Quantitative Risk Analysis, SPE/IADC 52864*. SPE/IADC Drilling Conference, Amsterdam, The Netherlands, Society of Petroleum Engineers.

Paterson, M. S. and Wong, T.-f. (2005). *Experimental rock deformation – The brittle field*. Berlin, Springer.

Paul, P. and Zoback, M. D. (2006). *Wellbore Stability Study for the SAFOD Borehole through the San Andreas Fault: SPE 102781*. SPE Annual Technical Conference, San Antonio, TX.

Pennington, W. D., Davis, S. D. *et al.* (1986). "The Evolution of Seismic Barriers and Asperities Caused by the Depressuring of Fault Planes in Oil & Gas Fields of South Texas." *Bull. Seism. Soc. Am.*, **188**, 78.

Pepin, G., Gonzalez, M. *et al.* (2004). "Effect of drilling fluid temperature on fracture gradient." *World Oil*, October, 39–48.

Perzyna, P. (1967). Fundamental Problems in Viscoplasticity. *Advances in Applied Mechanics*, **9**, 244–368.

Peska, P. and Zoback, M. D. (1995). "Compressive and tensile failure of inclined wellbores and determination of in situ stress and rock strength." *Journal of Geophysical Research*, **100**(B7), 12791–12811.

Peska, P. and Zoback, M. D. (1996). "Stress and failure of inclined boreholes SFIB v2.0: Stanford Rock Physics and Borehole Geophysics Annual Report, 57, Paper H3." *Stanford University Department of Geophysics*.

Pine, R. J. and Batchelor, A. S. (1984). "Downward migration of shearing in jointed rock during hydraulic injections." *Int. Journ. Rock Mech. Min. Sci. & Geomech. Abst.*, **21**(5), 249–263.

Pine, R. J., Jupe, A. *et al.* (1990). An evaluation of in situ stress measurements affecting different volumes of rock in the Carnmenellis granite. *Scale Effects in Rock Masses*. P. d. Cunha. Rotterdam, Balkema, 269–277.

Plumb, R. A. and Cox, J. W. (1987). "Stress directions in eastern North America determined to 4.5 km from borehole elongation measurements." *Journal of Geophysical Research*, **92**, 4805–4816.

Plumb, R. A. and Hickman, S. H. (1985). "Stress-induced borehole elongation: A comparison between the four-arm dipmeter and the borehole televiewer in the Auburn geothermal well." *Journal of Geophysical Research*, **90**, 5513–5521.

Pollard, D. and Aydin, A. (1988). "Progress in understanding jointing over the past century." *Geological Society of America Bulletin*, **100**, 1181–1204.

Pollard, D. and Fletcher, R. C. (2005). *Fundamentals of Structural Geology*. Cambridge, United Kingdom, Cambridge University Press.

Pollard, D. and Segall, P. (1987). Theoretical displacements and stresses near fractures in rock: With applications to faults, joints, veins, dikes and solution surfaces. *Fracture Mechanics of Rock*. B. K. Atkinson, Academic Press.

Powley, D. E. (1990). "Pressures and hydrogeology in petroleum basins." *Earth Sci. Rev.*, **29**, 215–226.

Qian, W., Crossing, K. S. *et al.* (1994). "Corrections to "Inversion of borehole breakout orientation data by Wei Qian and L.B. Pedersen." *Journal of Geophysical Research*, **99**, 707–710.

Qian, W. and Pedersen, L. B. (1991). "Inversion of borehole breakout orientation data." *Journal of Geophysical Research*, **96**, 20093–20107.

Raaen, A. M. and Brudy, M. (2001). *Pump in/Flowback tests reduce the estimate of horizontal in situ stress significantly, SPE 71367*. SPE Annual Technical Conference, New Orleans, LA, Society of Petroleum Engineers.

Raleigh, C. B., Healy, J. H. *et al.* (1972). Faulting and crustal stress at Rangely, Colorado., *Flow and Fracture of Rocks*. J. C. Heard. Washington, D.C., American Geophysical Union, 275–284.

Raleigh, C. B., Healy, J. H. *et al.* (1976). "An experiment in earthquake control at Rangely, Colorado." *Science*, **191**, 1230–1237.

Reid, P. and Santos, H. (2003). *Novel drilling, completion and workover fluids for depleted zones: Avoiding losses, formation damage and stuck pipe: SPE/IADC 85326*. SPE/IADC Middle East Drilling Technology Conference and Exhibition, Abu Dhabi, UAE, Society of Petroleum Engineers.

Richardson, R. (1992). "Ridge forces, absolute plate motions and the intraplate stress field." *Journal of Geophysical Research*, **97**(B8), 11739–11748.

Richardson, R. M. (1981). *Hydraulic fracture in arbitrarily oriented boreholes: An analytic approach*. Workshop on Hydraulic Fracturing Stress Measurements, Monterey, California, National Academy Press.

Riis, F. (1992). "Dating and measuring of erosion, uplift and subsidence in Norway and the Norwegian shelf in glacial periods." *Norsk Geologisk Tidsskrift*, **72**, 325–331.

Ritchie, R. H. and Sakakura, A. Y. (1956). "Asymptotic expansions of solutions of the heat conduction equation in internally bounded cylindrical geometry." *J. Appl. Phys.*, **27**, 1453–1459.

Roegiers, J. C. and Detournay, E. (1988). *Considerations on failure initiation in inclined boreholes*. Key Questions in Rock Mechanics, Balkeema, Brookfield, Vermont.

Rogers, S. (2002). Critical stress-related permeability in fractured rocks. *Fracture and in situ stress characterization of hydrocarbon reservoirs*. M. Ameen. London, The Geological Society, **209**, 7–16.

Rojas, J. C., Clark, D. E. *et al.* (2006). *Optimized salinity delivers improved drilling performance: AADE-06-DF-HO-11*. AADE 2006 Fluids Conference, Houston, Texas, American Association of Drilling Engineers.

Rudnicki, J. W. (1999). Alteration of regional stress by reservoirs and other inhomogeneities: Stabilizing or destabilizing? *Proceedings International Congress on Rock Mechanics*. G. Vouille and P. Berest, International Society of Rock Mechanics, 3, 1629–1637.

Rummel, F. and Hansen, J. (1989). "Interpretation of hydrofrac pressure recordings using a simple fracture mechanics simulation model, Inter. Jour. Rock Mech. and Min." *Sci. and Geomech. Abstr.*, **26**, 483–488.

Rummel, F. and Winter, R. B. (1983). "Fracture mechanics as applied to hydraulic fracturing stress measurements." *Earthq. Predict. Res.*, **2**, 33–45.

Rutledge, J. T., Phillips, W. S. *et al.* (2004). "Faulting induced by forced fluid injection and fluid flow forced by faulting: An interpretation of hydraulic-fracture microseismicity, Carthage Cotten Valley Gas Field, Texas." *Bulletin of the Seismological Society of America*, **94**(5), 1817–1830.

Rzhevsky, V. and Novick, G. (1971). *The physics of rocks*. Moscow, Russia, Mir.

Sayers, C. M. (1994). "The elastic anisotropy of shales." *J. Geophys. Res.*, **99**, 767–774.

Schmitt, D. R. and Zoback, M. D. (1992). "Diminished pore pressure in low-porosity crystalline rock under tensional failure: Apparent strengthening by dilatancy." *J. Geophys. Res.*, **97**, 273–288.

Schowalter, T. T. (1979). "Mechanics of Secondary Hydrocarbon Migration and Entrapment." *American Association of Petroleum Geologists Bulletin*, **63**(5), 723–760.

Schutjens, P. M. T. M., Hanssen, T. H. *et al.* (2001). *Compaction-induced porosity/permeability reduction in sandstone reservoirs: Data and model for elasticity-dominated deformation, SPE*

71337. SPE Annual Technical Conference and Exhibition, New Orleans, LA, Society of Petroleum Engineers.

Secor, D. T. (1965). "Role of fluid pressure in jointing." *American Journal of Science*, **263**, 633–646.

Segall, P. (1985). "Stress and Subsidence from Subsurface Fluid Withdrawal in the Epicentral Region of the 1983 Coalinga Earthquake." *J. Geophys. Res.,* **6801**, 90.

Segall, P. (1989). "Earthquakes Triggered by Fluid Extraction." *Geology*, **17**, 942–946.

Segall, P. and Fitzgerald, S. D. (1996). "A note on induced stress changes in hydrocarbon and geothermal reservoirs." *Tectonophysics*, **289**, 117–128.

Seldon, B. and Flemings, P. B. (2005). "Reservoir pressure and seafloor venting: Predicting trap integrity in a Gulf of Mexico deepwater turbidite minibasin." *American Association of Petroleum Geologists Bulletin*, **89**(2), 193–209.

Shamir, G. and Zoback, M. D. (1992). "Stress orientation profile to 3.5 km depth near the San Andreas Fault at Cajon Pass California." *Jour. Geophys Res.*, **97**, 5059–5080.

Sibson, R. H. (1992). "Implications of fault valve behavior for rupture nucleation and recurrence." *Tectonophysics*, **211**, 283–293.

Sinha, B. K. and Kostek, S. (1996). "Stress-induced azimuthal anisotropy in borehole flexural waves." *Geophysics*, **61**, 1899–1907.

Sinha, B. K., Norris, A. N. *et al.* (1994). "Borehole flexural modes in anisotropic formations." *Geophysics*, **59**, 1037–1052.

Sonder, L. (1990). "Effects of density contrasts on the orientation of stresses in the lithosphere: Relation to principal stress directions in the Transverse Ranges, California." *Tectonics*, **9**, 761–771.

Stein, R. S., King, G. C. *et al.* (1992). "Change in failure stress on the southern San Andreas fault system caused by the 1992 magnitude 7.4 Landers earthquake." *Science*, **258**, 1328–1332.

Stein, S. and Klosko, E. (2002). Earthquake mechanisms and plate tectonics. *International Handbook of Earthquake and Engineering Seismology Part A*. W. H. K. Lee, H. Kanamori, P. C. Jennings and K. Kisslinger. Amsterdam, Academic Press, 933.

Stephens, G. and Voight, B. (1982). "Hydraulic fracturing theory for conditions of thermal stress." *International Journal of Rock Mechanics and Mining Sciences*, **19**, 279–284.

Sternlof, K. R., Karimi-Fard, M., Pollard, D. D. and Durlofsky, L. J. (2006). "Flow and transport effects of compaction bands in sandstone at scales relevant to aquifer and reservoir management." *Water Resources Research*, **42**, Wo7425.

Sternlof, K. R., Rudnicki, J. W. *et al.* (2005). "Anticrack inclusion model for compaction bands in sandstone." *J. Geophys. Res*, **110**(B11403), 1–16.

Stock, J. M., Healy, J. H. *et al.* (1985). "Hydraulic fracturing stress measurements at Yucca Mountain, Nevada, and relationship to the regional stress field." *Journal of Geophysical Research*, **90**(B10), 8691–8706.

Stump, B. B. (1998). Illuminating basinal fluid flow in Eugene Island 330 (Gulf of Mexico) through in situ observations, deformation experiments, and hydrodynamic modeling. *Geosciences*, Pennsylvania State, 121.

Sulak, R. M. (1991). "Ekofisk field: The first 20 years." *Journal of Petroleum Technology*, **33**, 1265–1271.

Swarbrick, R. E. and Osborne, M. J. (1998). Mechanisms that generate abnormal pressures: An overview. *Abnormal pressures in hydrocarbon environments, AAPG Memoir 70*. B. E. Law, G. F. Ulmishek and V. I. Slavin. Tulsa, OK, American Association of Petroleum Geologists, 13–34.

Takahashi, M. and Koide, H. (1989). *Effect of the intermediate principal stress on strength and deformation behavior of sedimentary rock at the depth shallower than 2000 m*. Rock at Great Depth, Pau, France, Balkema, Rotterdam.

Tang, X. M. and Cheng, C. H. (1996). "Fast inversion of formation permeability from borehole Stonely wave logs." *Geophysics*, **61**, 639–645.

Terzaghi, K. (1923). *Theoretical Soil Mechanics*. John Wiley, New York.

Teufel, L. W. (1992). *Production-induced changes in reservoir stress state: Applications to reservoir management*. Society of Exploration Geophysicists 62nd Annual International Meeting, New Orleans, SEG, Tulsa.

Teufel, L. W., Rhett, D. W. *et al.* (1991). Effect of reservoir depletion and pore pressure drawdown on In situ stress and deformation in the Ekofisk field, North Sea. *Rock Mechanics as a Multidisciplinary Science*. J. C. Roegiers. Rotterdam, Balkema.

Tezuka, K. (2006). *Hydraulic injection and microseismic monitoring in the basement gas reservoir in Japan*. 2006 SPE Forum Series in Asia Pacific – Hydraulic Fracturing Beyond 2010, Macau, China, Society of Petroleum Engineers.

Thompson, A. L. (1993). Poly3D: A three-dimensional, polygonal element, displacement discontinuity boundary element computer program iwth applications to fractures, faults and cavities in the earth's crust. *Geology*. Stanford, CA, Stanford University.

Thomsen, L. (1986). "Weak elastic anisotropy." *Geophysics*, **51**, 1954–1966.

Toublanc, A., Renaud, S. *et al.* (2005). "Ekofisk field: fracture permeability evaluation and implementation in the flow model." *Petroleum Geoscience*, **11**, 321–330.

Townend, J. (2003). Mechanical constraints on the strength of the lithosphere and plate-bounding faults. *Geophysics*. Stanford, CA, Stanford University. PhD, 135.

Townend, J. and Zoback, M. D. (2000). "How faulting keeps the crust strong." *Geology*, **28**(5), 399–402.

Townend, J. and Zoback, M. D. (2001). Implications of earthquake focal mechanisms for the frictional strength of the San Andreas fault system. *The Nature and Tectonic Significance of Fault Zone Weakening*. R. E. Holdsworth, R. A. Strachan, J. J. Macloughlin and R. J. Knipe. London, Special Publication of the Geological Society of London, **186**, 13–21.

Townend, J. and Zoback, M. D. (2004). "Regional tectonic stress near the San Andreas fault in Central and Northern California." *Geophysical Research Letters*, **31**, L15–18.

Traugott, M. O. and Heppard, P. D. (1994). *Prediction of pore pressure before and after drilling- taking the risk out of drilling overpressured prospects*. AAPG Hedberg Research Conference, American Association of Petroleum Geologists.

Tsvankin, I. (2001). *Seismic Signatures and Analysis of Reflection Data in Anisotropic Media*. Cambridge, MA, Elsevier Science.

Turcotte, D. L. and Schubert, G. (2002). *Geodynamics*. Cambridge, Cambridge.

Tutuncu, A. N., Podio, A. L. *et al.* (1998). "Nonlinear viscoelastic behavior of sedimentary rocks: Part I, Effect of frequency and strain amplitude." *Geophysics*, **63**(1), 184–194.

Tutuncu, A. N., Podio, A. L. *et al.* (1998). "Nonlinear viscoelastic behavior of sedimentary rocks: Part II, Hysteresis effects and influence of type of fluid on elastic moduli." *Geophysics*, **63**(1), 195–203.

Twiss, R. J. and Moores, E. M. (1992). *Structural Geology*. New York, W. H. Freeman and Company.

Van Balen, R. T. and Cloetingh, S. A. (1993). Stress-induced fluid flow in rifted basins. *Diagenesis and basin development*. A. D. Horbury and A. G. Robinson. Tulsa, American Association of Petroleum Geologists, **36**, 87–98.

van Oort, E., Gradisher, J. *et al.* (2003). *Accessing deep reservoirs by drilling severly depleted formations: SPE 79861*. SPE/IADC Drilling Conference, Amsterdam, Society of Petroleum Engineers.

Van Oort, E., Hale, A. H. *et al.* (1995). *Manipulation of coupled osmotic flows for stabilization of shales exposed to water-based drilling fluids: SPE 30499*. SPE Annual Technical Conf. and Exhibition, Dallas, Texas, Society of Petroleum Engineers.

Vardulakis, I., S. J. *et al.* (1988). "Borehole instabilities as bifurcation phenomena." *Intl. J. Rock Mech. Min. & Geomech. Abstr.*, **25**, 159–170.

Veeken, C., Walters, J. *et al.* (1989). Use of plasticity models for predicting borehole stability. *Rock at Great Depth, Vol. 2*. V. Maury and D. Fourmaintraux. Rotterdam, Balkema, 835–844.

Vernik, L., Bruno, M. *et al.* (1993). "Empirical relations between compressive strength and porosity of siliciclastic rocks. *Int. J. Rock Mech. Min. Sci. & Geomech. Abstr.*, **30**,**7**, 677–680.

Vernik, L., Lockner, D. *et al.* (1992). "Anisotropic strength of some typical metamorphic rocks from the KTB pilot hole, Germany." *Scientific Drilling*, **3**, 153–160.

Vernik, L. and Zoback, M. D. (1990). *Strength anisotropy of crystalline rock: Implications for assessment of In situ stresses from wellbore breakouts. Rock Mechanics Contributions and Challenges.* Balkema, Rotterdam.

Vigneresse, J., Ed. (2001). *Fluids and fractures in the lithosphere*. Tectonophysics. Amsterdam, Elsevier.

Walls, J. and Nur, A. (1979). *Pore pressure and confining pressure dependence of permeability in sandstone*. 7th Formation Evaluation Symposium, Calgary, Canada, Canadian Well Logging Society.

Wang, H. F. (2000). *Theory of linear poroelasticity with applications to geomechanics and hydrogeology*. Princeton, NJ, Princeton University Press.

Ward, C. D. and Beique, M. (1999). "How to identify Lost Circulation Problems with Real-time Pressure Measurement: Downhole Pressure Sensing heads off Deepwater Challenge." *Offshore*, August.

Ward, C. D. and Clark, R. (1998). *Bore hole ballooning diagnosis with PWD*. Workshop on Overpressure, Pau, France., Elf EP-Editions.

Warpinski, N. R. and R. E. Peterson *et al.* (1988). In Situ Stress and Moduli, Comparison of Values Derived from Multiple Techniques, Proc. 1998 SPE Annual Technical Conference and Exhibition, New Orleans, LA, 27–30 September, 1998, SPE 49190, 569–583.

Warren, W. E. and Smith, C. W. (1985). "In situ stress estimates from hydraulic fractruring and direct observation of crack orientation." *Journal of Geophysical Research*, **90**, 6829–6839.

Webb, S., Anderson, T. *et al.* (2001). *New treatments substantially increase LOT/FIT pressures to solve deep HTHP drilling challenges: SPE 71390*. Annual Technical Conference and Exhibition, New Orleans, LA, Society of Petroleum Engineers.

Weng, X. (1993). *Fracture Initiation and Propagation from Deviated Wellbores*. paper SPE 26597 presented at the 1993 Annual Technical Conference and Exhibition, Houston, 3–6 October.

Whitehead, W., Hunt, E. R. *et al.* (1986). *In-Situ Stresses: A comparison between log-derived values and actual field-measured values in the Travis Peak formation of east Texas: SPE 15209*. Unconventional Gas Technology Symposium, Louisville, Kentucky, Society of Petroleum Engineers.

Wiebols, G. A. and Cook, N. G. W. (1968). "An energy criterion for the strength of rock in polyaxial compression." *International Journal of Rock Mechanics and Mining Sciences*, **5**, 529–549.

Willson, S. M., Last, N. C. *et al.* (1999). *Drilling in South America: A wellbore stability approach for complex geologic condtions, SPE 53940*. 6th LACPEC Conference, Caracas, Venezuela, Society of Petroleum Engineers.

Winterstein, D. F. and Meadows, M. A. (1995). "Analysis of shear-wave polarization in VSP data: A tool for reservoir development SPE 234543." *Dec., SPE Formation Evaluation*, **10**, No. **4**, 223–231.

Wiprut, D., Zoback, M. *et al.* (2000). "Constraining the full stress tensor from observations of drilling-induced tensile fractures and leak-off tests: Application to borehole stability and sand production on the Norwegian margin." *Int. J. Rock Mech. & Min. Sci.*, **37**, 317–336.

Wiprut, D. and Zoback, M. D. (2000). "Fault reactivation and fluid flow along a previously dormant normal fault in the northern North Sea." *Geology*, **28**, 595–598.

Wiprut, D. and Zoback, M. D. (2002). *Fault reactivation, leakage potential and hydrocarbon column heights in the northern North Sea*. Hydrocarbon Seal Quantification, Stavanger, Norway, Elsevier.

Wolhart, S. L., Berumen, S. *et al.* (2000). *Use of hydraulic fracture diagnostics to optimize fracturing jobs in the Arcabuz-Calebra field, SPE 60314*. 2000 SPE Rocky Mountain Region/Low Permeability Reservoirs Symposium, Denver, CO, Society of Petroleum Engineers.

Wong, T.-f., David, C. *et al.* (1997). "The transition from brittle faulting to cataclastic flow in porous sandstones: Mechanical deformation." *Journal of Geophysical Research*, **102**(B2), 3009–3025.

Wood, D. M. (1990). *Soil behaviour and critical state soil mechanics*. Cambridge, England, Cambridge University.

Wright, C. A. and Conant, R. A. (1995). *Hydraulic fracture reorientation in primary and secondary recovery from low permeability reservoirs, SPE 30484*. 1995 SPe Technical Conference and Exhibition, Dallas, TX, Society of Petroleum Engineers.

Wright, C. A., Stewart, D. W. *et al.* (1994). *Reorientation of propped refracture treatments in the Lost Hills field, SPE 27896*. 1994 SPE Western Regional Meeting, Long Beach, California.

Yale, D. P. (2002). *Coupled geomechanics-fluid flow modeling: effects of plasticity and permeability alteration: SPE 78202*. Society of Petroleum Engineers.

Yale, D. P. (2003). Fault and stress magnitude controls on variations in the orientation of in situ stress. *Fracture and in-situ stress characterization of hydrocarbon reservoirs*. M. Ameen. London, Geological Society, **209**, 55–64.

Yale, D. P., Nabor, G. W. *et al.* (1993). *Application of variable formation compressibility for improved reservoir analysis: SPE 26647*, Society of Petroleum Engineers.

Yale, D. P., Rodriguez, J. M., *et al.* (1994). In-Situ *stress orientation and the effects of local structure – Scott Field, North Sea*. Eurock '94, Delft, Netherlands, Balkema.

Yassir, N. A. and Bell, J. S. (1994). "Relationships between pore pressure, stresses and present-day geodynamics in the Scotian shelf, offshore eastern Canada." *American Association of Petroleum Geology Bulletin*, **78**(12), 1863–1880.

Yassir, N. A. and Zerwer, A. (1997). "Stress regimes in the Gulf Coast, offshore Lousiana from wellbore breakout analysis." *American Association of Petroleum Geologists Bulletin*, **81**(2), 293–307.

Yew, C. H. and Li, Y. (1988). "Fracturing of a deviated well." *SPE Production Engineering*, **3**, 429–437.

Zajac, B. and Stock, J. M. (1992). "Using borehole breakouts to constrain complete stress tensor." *AGU 1992 Fall Meeting Abstract Supplement to EOS, Dec. 1992*, 559.

Zemanek, J., Glenn, E. E. *et al.* (1970). "Formation evaluation by inspection with the borehole televiewer." *Geophysics*, **35**, 254–269.

Zheng, Z., Kemeny, J. *et al.* (1989). "Analysis of borehole breakouts." *Journal of Geophysical Research*, **94**(B6), 7171–7182.

Zhou, S. (1994). "A program to model initial shape and extent of borehole breakout." *Computers and Geosciences*, **20**, 7/8, 1143–1160.

Zimmer, M. (2004). Controls on the seismic velocities of unconsolidated sands: Measurements of pressure, porosity and compaction effects. *Geophysics*. Stanford, CA., Stanford University.

Zinke, J. C. and Zoback, M. D. (2000). "Structure-related and stress-induced shear-wave velocity anisotropy: Observations from microearthquakes near the Calaveras fault in central California." *Bulletin of Seismological Society of America*, **90**, 1305–1312.

Zoback, M. D., Apel, R. *et al.* (1993). "Upper crustal strength inferred from stress measurements to 6 km depth in the KTB borehole." *Nature*, **365**, 633–635.

Zoback, M. D., Barton, C. B. *et al.* (2003). "Determination of stress orientation and magnitude in deep wells." *International Journal of Rock Mechanics and Mining Sciences*, **40**, 1049–1076.

Zoback, M. D. and Byerlee, J. D. (1975). "Permeability and effective stress." *Am. Assoc.Petr. Geol. Bull.*, **59**, 154–158.

Zoback, M. D. and Byerlee, J. D. (1976). "A note on the deformational behavior and permeability of crushed granite." *Int'l. J. Roch Mech.*, **13**, 291–294.

Zoback, M. D., Day-Lewis, A. and Kim S.-M. (2007). Predicting changes in hydrofrac orientation in depleting oil and gas reservoirs, Patent Application Pending.

Zoback, M. D. and Haimson, B. C. (1982). *Status of the hydraulic fracturing method for in-situ stress measurements*. 23rd Symposium on Rock Mechanics, Soc. Mining Engineers, New York.

Zoback, M. D. and Haimson, B. C. (1983). "Workshop on Hydraulic Fracturing Stress Measurements." *National Academy Press*, 44–54.

Zoback, M. D. and Harjes, H. P. (1997). "Injection induced earthquakes and crustal stress at 9 km depth at the KTB deep drilling site, Germany." *J. Geophys. Res.*, **102**, 18477–18491.

Zoback, M. D. and Healy, J. H. (1984). "Friction, faulting, and "in situ" stresses." *Annales Geophysicae*, **2**, 689–698.

Zoback, M. D. and Healy, J. H. (1992). "In situ stress measurements to 3.5 km depth in the Cajon Pass Scientific Research Borehole: Implications for the mechanics of crustal faulting." *J. Geophys. Res.*, **97**, 5039–5057.

Zoback, M. D., Mastin, L. *et al.* (1987). In situ *stress measurements in deep boreholes using hydraulic fracturing, wellbore breakouts and Stonely wave polarization*. In *Rock Stress and Rock Stress Measurements*. Stockholm, Sweden, Centrek Publ., Lulea.

Zoback, M. D., Moos, D. *et al.* (1985). "Well bore breakouts and In situ stress." *Journal of Geophysical Research*, **90**(B7), 5523–5530.

Zoback, M. D. and Peska, P. (1995). "In situ stress and rock strength in the GBRN/DOE 'Pathfinder' well, South Eugene Island, Gulf of Mexico." *Jour. Petrol Tech.*, **37**, 582–585.

Zoback, M. D. and Pollard, D. D. (1978). *Hydraulic fracture propagation and the interpretation of pressure-time records for in-situ stress determinations*. 19th U.S. Symposium on Rock Mechanics, MacKay School of Mines, Univ. of Nevada, Reno, Nevada.

Zoback, M. D. and Townend, J. (2001). "Implications of hydrostatic pore pressures and high crustal strength for the deformation of intraplate lithosphere." *Tectonophysics*, **336**, 19–30.

Zoback, M. D., Townend, J. *et al.* (2002). "Steady-state failure equilibrium and deformation of intraplate lithosphere." *International Geology Review*, **44**, 383–401.

Zoback, M. D. and Zinke, J. C. (2002). "Production-induced normal faulting in the Valhall and Ekofisk oil fields." *Pure & Applied Geophysics*, **159**, 403–420.

Zoback, M. D. and Zoback, M. L. (1991). Tectonic stress field of North America and relative plate motions. In *The Geology of North America. Neotectonics of North America*. D. B. a. o. Slemmons. Boulder, Colo, Geological Society of America, 339–366.

Zoback, M. D., Zoback, M. L. *et al.* (1987). "New evidence on the state of stress of the San Andreas fault system." *Science*, **238**, 1105–1111.

Zoback, M. L. (1992). "First and second order patterns of tectonic stress: The World Stress Map Project." *Journal of Geophysical Research*, **97**, 11,703–11,728.

Zoback, M. L. and Mooney, W. D. (2003). "Lithospheric buoyancy and continental intraplate stress." *International Geology Review*, **7**, 367–390.

Zoback, M. L. and a. others (1989). "Global patterns of intraplate stresses; a status report on the world stress map project of the International Lithosphere Program." *Nature*, **341**, 291–298.

Zoback, M. L. and Zoback, M. D. (1980). "State of stress in the conterminous United States." *J. Geophys. Res.*, **85**, 6113–6156.

Zoback, M. L. and Zoback, M. D. (1989). "Tectonic stress field of the conterminous United States." *Geol. Soc. Am. Memoir.*, **172**, 523–539.

Zoback, M. L., Zoback, M. D. *et al.* (1989). "Global patterns of tectonic stress." *Nature*, **341**, 291–298. (28 September 1989).

Index

Printed in the United States
By Bookmasters